NUCLEAR MAGNETIC RESONANCE

NUCLEAR MAGNETIC RESONANCE
Concepts and Methods

Daniel Canet
Université H. Poincaré, Nancy I, France

Translated by the Author

JOHN WILEY & SONS
Chichester · New York · Brisbane · Toronto · Singapore

National 01243 779777
International (+44) 1243 779777

Other Wiley Editorial Offices

John Wiley & Sons, Inc., 605 Third Avenue,
New York, NY 10158-0012, USA

Jacaranda Wiley Ltd, 33 Park Road, Milton,
Queensland 4064, Australia

John Wiley & Sons (Canada) Ltd, 22 Worcester Road,
Rexdale, Ontario M9W 1L1, Canada

John Wiley & Sons (Asia) Pte Ltd, 2 Clementi Loop #02-01,
Jin Xing Distripark, Singapore 0512

Library of Congress Cataloging-in-Publication Data

Canet, D.
[La RMN: Concepts et Methodes. French]
Nuclear magnetic resonance : concepts and methods / D. Canet.
p. cm.
Includes bibliographical references and index.
ISBN 0-471-94234-0 (cloth : alk. paper).—ISBN 0-471-96145-0
(pbk. : alk. paper)
1. Nuclear magnetic resonance. I. Title.
QC762.C3614 1996
543′.0887—dc20 95-39900
CIP

British Library Cataloguing in Publication Data

A catalogue record for this book is available from the British Library

ISBN 0 471 94234 0 (cloth)
ISBN 0 471 96145 0 (paper)

Typeset in 10/12pt Times by Keytec Typesetting Ltd, Bridport, Dorset, UK
Printed and bound in Great Britain by Bookcraft (Bath) Limited
This book is printed on acid-free paper responsibly manufactured from sustainable forestation, for which at least two trees are planted for each one used for paper production.

CONTENTS

PREFACE

This was intended to be a translation of a book that I wrote in French in the period 1990–1991. I quickly became aware of many errors and weaknesses which made the French version a sort of draft. In this English version, I have attempted to correct mistakes and misprints, but also to revise completely some parts of the book by including new methods and concepts which came out since the first writing, or with which I have become more familiar. Nevertheless, I have tried to keep the initial spirit, that is, to provide the advanced undergraduate student, the graduate student or the researcher in the field of NMR with a sort of self-sufficient textbook essentially oriented toward 'modern' NMR spectroscopy and 'spin engineering'. Needless to say, owing to the universality and potentiality of NMR, this book cannot be exhaustive. Some aspects are covered only briefly or even just briefly mentioned (solid state NMR, NMR of paramagnetic systems, exchange phenomena NMR imaging . . .). Nevertheless, I have tried to put forward the major features, in terms of mathematical and physical concepts, which should allow the reader to understand the basic experiments of high-resolution NMR at the present time, and not to be discouraged during the examination of a research paper.

Examples and figures furnished throughout the text should be helpful in allowing the reader to digest some rather difficult concepts. On the other hand, it may appear that mathematical or quantum mechanical developments are not strictly necessary for practising advanced experiments. These developments are indicated by a recognizable typography (smaller characters) and can possibly be skipped. Hopefully, the remainder of the text should be self-consistent and should contain, without proof, the main results required for further reading.

Chapter 1 should be readable by an undergraduate student, as it includes the basic approach to the interpretation of common NMR spectra either in the liquid state or in an anisotropic medium. The next two chapters involve a little more mathematics and quantum mechanics, but contain the necessary material to understand methodologies which have appeared over the last 20 years (Fourier transformation, product operators, signal processing etc). In Chapter 4, I have attempted to rationalize some concepts of spin relaxation and, in a more general way, to deal with spin dynamics in relation to molecular motions, including rotational and translational motions. The last chapter can be thought of as a survey of the major multipulse and multidimensional methods of present day NMR, hopefully presented in an unified way. Such methods include selective excitation, correlation spectroscopies and NMR imaging. Their understanding rests, of course, on the concepts developed in Chapters 2 and 3.

The bibliography and references given in this book could well be incomplete. This is my own responsibility, and I would like to present my sincere apologies to those I may have forgotten.

I am indebted to the following colleagues and coworkers, who helped me in preparing the manuscript either by discussions or by figures they kindly provided: B. Ancian, J.-C. Boubel, J. Brondeau, B. Diter, D. Grandclaude, M. Decorps, F. Humbert, P. Mutzenhardt, R. Raulet, C. Roumestand and P. Tekely.

Many thanks to my son-in-law, Adrian Sargeant, who had the difficult task of understanding my 'French-English' and removing many linguistic errors.

Finally, I am especially grateful to my secretary, Mrs Jocelyne Devienne, who, for almost two years, had to bear the burden of processing the text, preparing most of the illustrations and . . . putting up with the author's bad temper.

Nancy, June 1995.

INTRODUCTION

During the four decades which have followed its discovery (in 1945), nuclear magnetic resonance (NMR) has become a science *per se*, covering a broad field whose applications range from chemical synthesis to medicine with the advent, in the late seventies, of MRI (magnetic resonance imaging).

The initial experiment, whose potential applications appeared rather modest, was designed essentially for satisfying the curiosity of scientists: was it possible to induce transitions between the quantum states of nuclear spins embedded in a magnetic field? Physicists gave a positive answer to that question; thereafter, physicochemists, chemists, biochemists, biologists and physicians also recognized the wealth of information contained in a nuclear magnetic resonance spectrum, or rather, in the different spectra that can be obtained by a variety of techniques which permit the determination or the correlation of a number of structural or dynamic parameters.

The evolution of NMR involves important milestones, each representing a crucial innovation, and delaying, sometimes unexpectedly, the moment where the technique can be considered as having reached its steady state and full achievement.

Until the end of the 1960s, numerous problems of stereochemistry had already been essentially solved by proton (the nucleus of hydrogen) NMR, via the measurement of chemical shifts and spin–spin couplings. The method is nowadays so widespread that one can hardly imagine an organic chemist without an NMR spectrometer to hand.

The structure of NMR spectra and their analysis are considered in the first chapter of this book. If one is dealing with an isotropic medium, one generally obtains high-resolution spectra which are characterized by (i) the location of resonance lines, rationalized in terms of chemical shift, which reflects the electronic environment of the nucleus considered, and (ii) the multiplet structure arising from the so-called spin–spin couplings (or indirect couplings, or J couplings), which occur via the spins of bonding electrons. Conversely, direct interactions (much larger than indirect couplings) dominate the NMR spectra of anisotropic media and lead generally to broad absorption bands. Clever techniques, combining high power decoupling and sample rotation at the magic angle, had by the end of the 1970s enabled the elimination of the broadening associated with the anisotropic character of such interactions. These techniques entirely revitalized NMR in the solid state and generated new interest not extinguished since then.

Until a relatively recent period (again around 1970), NMR was rightly accused of poor sensitivity by comparison with optical spectroscopies also devoted to structural studies, such as electronic spectroscopy (UV–visible) or vibrational spectroscopy.

The renaissance of pulse techniques (Chapter 2 of this book) and their corollary, data treatment by Fourier transformation, which became possible with the advent of low-cost mini-computers (Chapter 3), partially overcame the sensitivity problem. In parallel, technological advances in the design of superconductor magnets capable of delivering ever more intense magnetic fields lead to appreciable gains in sensitivity and resolution (chemical shift dispersion). The magnetic field of an NMR spectrometer must be stable and very homogeneous (its variation from one point to the other in the sample must not exceed 10^{-9} in order to satisfy the resolution required for most spectra in solution). The highest field of 'NMR quality' lies at the present time around 17–18 tesla. With all these improvements, the use of NMR spectroscopy of isotopes other than proton has become routine; this is especially the case with carbon-13 or nitrogen-15, whose observation was earlier hampered by their low natural abundance and their low resonance frequency.

Pulse techniques also enabled the routine measurement of spin relaxation times, which constitute the other class of parameters accessible by NMR (Chapter 4). These quantities are directly related to molecular mobility (rotational or translational) and thus represent a unique source of characterization for the understanding of the liquid (or even the solid) state and for the existence of internal or local motions in molecules which can be as complex as proteins.

An explosion of new techniques, based on the concept of two-dimensional spectroscopy (later extended to more than two dimensions), had occurred by the end of the 1970s (although initially quoted in the first chapter, these 2D techniques are considered in detail in Chapter 5). The idea was to devise experiments involving two time variables and, after a double Fourier transform, to extract correlations between quantities pertaining to the frequency domain. This novel concept led at first to the systematic assignment of resonances in complex spectra and later to the determination of interatomic distances via the nuclear Overhauser effect (NOE). Through various improvements which have benefited these techniques, it is nowadays possible to attack structural problems of very complicated molecules in solution, such as proteins or nucleic acids. Hence NMR has become one of the essential tools of biochemists.

The idea of correlation between two spectral quantities, inherent in two-dimensional spectroscopy, triggered the development of another field of research and applications: magnetic resonance imaging (MRI), where correlated spectral variables are substituted by correlated spatial variables. The method involves the application of magnetic field gradients which act as space encoders. The two-dimensional map so obtained is indicative of the distribution of nuclear spins (often water protons) in space. Images can be further contrasted by the effect of relaxation times. MRI has been mostly applied in medicine and biology but is now increasingly used in material science and, whenever small objects are investigated, it is known as NMR microscopy.

This brief introduction suggests that NMR encompasses a considerable variety of methods. The field of applications grows relentlessly. This book aims at underlining the essential methodological aspects without claiming any sort of exhaustivity (several volumes would not be sufficient); it is, however, intended to provide the reader with the basic tools adapted to NMR at the present time.

1 STRUCTURE OF NUCLEAR MAGNETIC RESONANCE SPECTRA

Very soon after the discovery of nuclear magnetic resonance, it appeared that nuclei of the same isotopic nature (e.g. protons) and belonging to a given molecule did not lead to a single absorption band but rather to a series of multiplets. It was, for instance, observed that the proton spectrum of ethanol includes a triplet, a quartet and a singlet respectively assigned to the CH_3, CH_2 and OH groupings of this molecule (Figure 1.1).

It was therefore inferred that this technique, initially part of nuclear physics, could become an extremely powerful means of structural investigation, as long as the position (or shift) of the various multiplets, and their fine structure, could be correlated with the chemical nature of the constituents of the molecule under consideration. This optimistic view proved perfectly correct since, nowadays, thanks to the methodological developments of nuclear magnetic resonance (widely known by the acronym NMR), it is quite possible to determine the structure of extremely complicated molecules such as proteins in solution.

The objective of this chapter is to present intuitively the essential characteristics

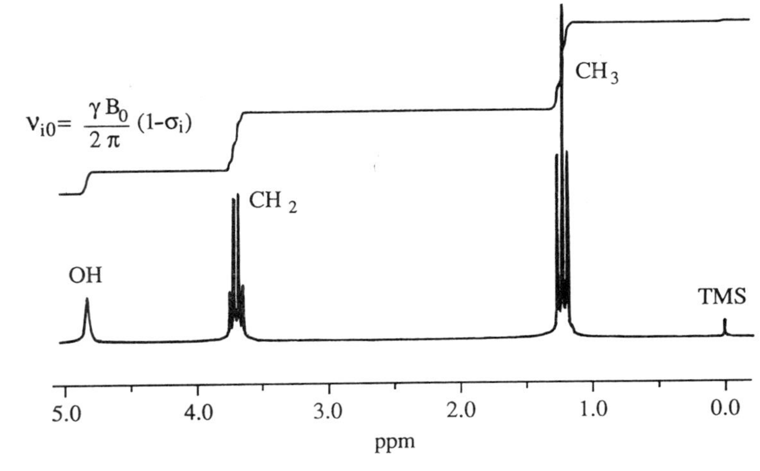

Figure 1.1 The proton NMR spectrum of ethanol measured at 200 MHz. The rightmost signal arises from a widely used reference substance (TMS: Tetramethylsilane). The inset formula relates the resonance frequency of each type of proton (ν_{0i}) with its shielding factor σ_i. The upper trace represents the integration (area) of the successive multiplets in the spectrum

of NMR spectra in solution and in the solid state. An optional quantum mechanics approach is also proposed. This approach is useful for a thorough understanding of the structure of NMR spectra but may be skipped for an interpretation at a 'first sight' level.

1.1 Nuclear Spin and its Associated Magnetic Moment

The appearance of NMR spectra, and consequently the molecular structure they are able to provide, arises from the discrete nature (quantification) of the energy levels pertaining to a nuclear spin system. The concept of electron spin is familiar because it is involved, through the Pauli principle, in the numbering and characterization of electronic states in atoms and molecules. It turns out that the proton (^1H) is also a particle possessing a spin and so is the neutron. Annihilation of spin between protons and neutrons (the particles making up atomic nuclei) may occur. This explains the absence of spin for certain isotopes like carbon-12; fortunately this is not the case for carbon-13, whose non-zero natural abundance, although weak (of the order of 1%), led to the development of a spectroscopy complementary to proton spectroscopy. Each isotope is characterized by an integer or half-integer I, called the spin number, which defines the number of different states in which the corresponding spin can be found (see below). Some examples can illustrate this feature: the spin number (or more simply the spin) of the electron, of the proton and of the neutron is equal to 1/2, corresponding to two distinct states; the spin of deuterium is 1, corresponding to three distinct states, and so on. The following properties can be verified for all existing isotopes:

- if the mass number A is odd, the nuclear spin is a half-integer;
- if the mass number A and the atomic number Z are even, the nuclear spin is zero (indicating the absence of spin: ^{12}C, ^{16}O, etc.);
- if the mass number A is even and the atomic number Z is odd, the nuclear spin is an integer.

What is important for the understanding of the phenomenon of magnetic resonance is less the exact meaning of the spin momentum $\boldsymbol{I}$ (it is merely the intrinsic angular momentum of the particle, reflecting the rotation of the particle with respect to its center of mass) but rather the simultaneous existence of a magnetic moment $\boldsymbol{\mu}$ collinear with $\boldsymbol{I}$:

$$\boldsymbol{\mu} = \gamma \hbar \boldsymbol{I} \tag{1.1}$$

where $\hbar$ is the Planck constant divided by 2π ($\hbar = 1.055 \times 10^{-34}$ J s); γ, which is a constant specific for each isotope, is termed the gyromagnetic ratio. The spin number of some of the usual isotopes and their resonance frequency (proportional to the gyromagnetic ratio) are given in Table 1.1.

The proton gyromagnetic ratio is equal to 26.75×10^7 rad T^{-1} s^{-1}. It can be recalled that, as far as the electron is concerned, γ is usually expressed in the form of a product involving the two quantities g_e (the Landé factor) and β_e (the Bohr

Table 1.1 Some usual isotopes: spin number, mean resonance frequency for an induction of 2.35 T, proton resonance at 100 MHz, and natural abundance. (T stands for Telsa, the MKSA magnetic field unit with 1 Telsa = 10 000 Gauss; the earth's magnetic field value is around 0.5 G)

Nucleus	I	ν (MHz)	Natural abundance (%)
^{1}H	1/2	100	99.98
^{2}H	1	15.351	1.5×10^{-2}
^{13}C	1/2	25.144	1.108
^{17}O	5/2	13.557	3.7×10^{-2}
^{14}N	1	7.224	99.63
^{15}N	1/2	10.133	0.37
^{31}P	1/2	40.481	100
^{19}F	1/2	94.077	100
^{29}Si	1/2	19.865	4.7
^{27}Al	5/2	26.057	100
^{23}Na	3/2	26.451	100

magneton): $\gamma_e = g_e\beta_e/\hbar$ with $g_e = 2.00232$—for the free electron—and $\beta_e = 9.2740 \times 10^{-24}\,\mathrm{J\,T^{-1}}$. The electron gyromagnetic ratio is seen to be nearly 700 times as large as the proton gyromagnetic ratio.

1.2 The Zeeman Effect. The Resonance Phenomenon

1.2.1 THE INTUITIVE APPROACH

Because of its associated magnetic moment, an $I = 1/2$ spin can be viewed as a microscopic magnetized needle (like the one of a compass) whose orientation is arbitrary until it is placed in a magnetic field B_0; it can then be oriented either parallel to B_0 (the most stable configuration) or antiparallel (the less stable configuration). Thus two distinct energy levels are created, whose splitting $\Delta E = \gamma\hbar B_0$ can be calculated from the energy interaction $-\boldsymbol{\mu}.\boldsymbol{B}$ between a magnetic field and a magnetic moment (Figure 1.2).

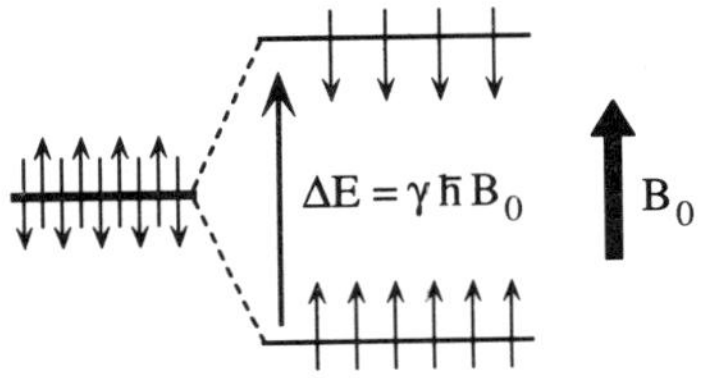

Figure 1.2 Splitting of the energy levels of a spin 1/2 which occurs due to the application of a static magnetic field B_0. The intensity of the transition which can be induced between these two levels is proportional to the population excess of the lowest level with respect to the highest level

As for any spectroscopy, a transition may be induced between these two levels if an electromagnetic wave is applied to the system, provided that its frequency obeys the Bohr relationship $\Delta E = h\nu$ and that a proper experimental arrangement is used (Figure 1.3).

One of the unique features of magnetic resonance is that the transition frequency $\nu_0 = \gamma B_0/2\pi$, called the resonance frequency or Larmor frequency, is proportional to the applied field. In a more phenomenological way, we can consider the excess of spins oriented along the magnetic field B_0 (with respect to those oriented in the opposite direction, or antiparallel); this results in a macroscopic nuclear magnetization $\boldsymbol{M}$ aligned with $\boldsymbol{B}_0$, which is called equilibrium magnetization. It can be displaced from this equilibrium position by an appropriate perturbation (see below); it is then subjected to a precessional motion around $\boldsymbol{B}_0$. This motion, also called Larmor precession, can be explained on the basis of classical mechanics (angular momentum theorem) and occurs at a frequency precisely equal to ν_0 (Figure 1.4). A perturbation which brings $\boldsymbol{M}$ into a plane perpendicular to $\boldsymbol{B}_0$ allows the observation of the Larmor precession through the electromotive force (emf) which occurs in a coil whose axis is contained in that plane (Figure 1.3). The amplitude of the induced voltage is proportional to M while its frequency yields ν_0.

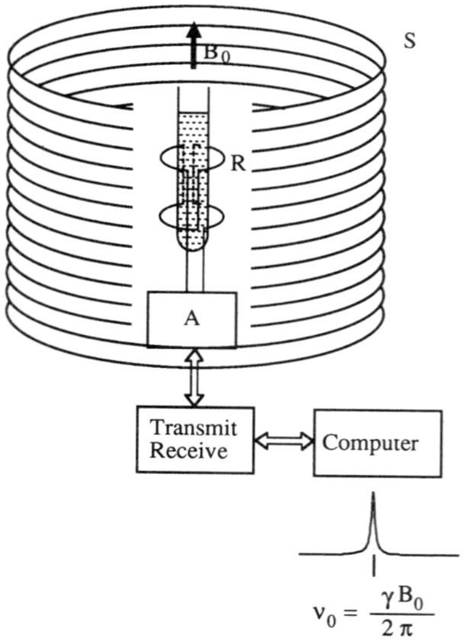

Figure 1.3 A static magnetic field B_0 as high as possible—generally delivered by a superconducting coil S—leads to the splitting of the energy levels (Figure 1.2) of the spins inside the sample (at the center of the magnet). The coil R, tuned at the Larmor frequency ν_0 (the relevant electrical circuitry is symbolized by A), has a double purpose:

- to create an alternative magnetic field perpendicular to B_0 (transmit function) which will induce the transition between the two levels of Figure 1.2;
- to detect the resonance signal (receive function). This signal is processed by a computer which also controls the whole spectrometer

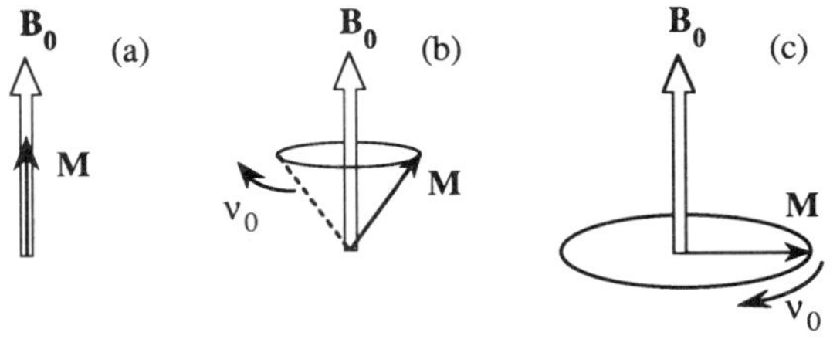

Figure 1.4 (a) At thermal equilibrium, the nuclear magnetization $\boldsymbol{M}$ is collinear with $\boldsymbol{B}_0$. (b) If a perturbation displaces $\boldsymbol{M}$ from its equilibrium position, it then undergoes a precessional motion around $\boldsymbol{B}_0$ at the Larmor frequency ν_0. (c) When $\boldsymbol{M}$ is brought in a plane perpendicular to $\boldsymbol{B}_0$ the induced voltage in the coil R (Figure 1.3) is of maximum amplitude

This detection procedure is equivalent, at a microscopic level, to the mode of action of the magnetic bar in a dynamo.

For sensitivity reasons and to take advantage of the largest possible chemical shift dispersion (see below), high values of the magnetic field B_0 are generally favored. As a consequence, most NMR spectrometer magnets are of the superconducting type. The highest value so far obtained, with the stability and homogeneity required by a NMR experiment, is 18.8 T. Rather than providing the actual B_0 value, one usually indicates the proton resonance frequency: 100 MHz corresponds to a B_0 value of 2.35 T, 400 MHz to 9.4 T (considered as the 'standard spectrometer' at the present time) and 800 MHz to 18.8 T.

1.2.2 QUANTUM MECHANICAL APPROACH

The quantum mechanical approach for the magnetic moment μ associated with the nuclear spin calls for the consideration of the spin operators and, at the outset, for their fundamental eigenvalue and eigenvector relationships given in equations (1.2) and (1.3). They concern the $\hat{I}_z$ operator corresponding to the z component of the spin operator $\hat{\boldsymbol{I}}$ and to the $\hat{I}^2$ operator corresponding to the square of its length. The z axis is arbitrary but, as will be seen shortly, is conveniently chosen as the direction of the static field B_0. In Dirac notation, $|m\rangle$ represents an eigenvector of both $\hat{I}_z$ and $\hat{I}^2$, with different eigenvalues, however, for these two operators: m for $\hat{I}_z$ and $I\,(I+1)$ for $\hat{I}^2$.

$$\hat{I}_z|m\rangle = m|m\rangle \tag{1.2}$$

$$\hat{I}^2|m\rangle = I(I+1)|m\rangle \tag{1.3}$$

I, which is an integer or a half-integer, is in fact the spin number defined before (Table 1.1) m goes from $-I$ to $+I$ by steps of 1. The vector $|m\rangle$ represents *an eigenstate* of the system for which the values of $\hat{I}_z$ and $\hat{I}^2$ are perfectly determined. In the case of a spin 1/2, $|1/2\rangle$ is often replaced by α or $|\alpha\rangle$, and $|-1/2\rangle$ by β or $|\beta\rangle$, thus we can write:

$$\begin{aligned} \hat{I}_z|\alpha\rangle &= (1/2)|\alpha\rangle \qquad & \hat{I}_z|\beta\rangle &= (-1/2)|\beta\rangle \\ \hat{I}^2|\alpha\rangle &= (3/4)|\alpha\rangle \qquad & \hat{I}^2|\beta\rangle &= (3/4)|\beta\rangle \end{aligned} \tag{1.4}$$

From the above relationships, it can be seen that the view of parallel and antiparallel orientations of the magnetic moment with respect to B_0 was oversimplified. Because of the length of the $\boldsymbol{I}$ vector and the value of its component along the z axis, the magnetic moment actually makes an angle of 54.74° with the z direction, pointing toward $+z$ or $-z$ for the parallel and antiparallel configurations, respectively. Since, at equilibrium, the orientation with respect to the axes x and y is not specified, individual magnetic moments lie on a cone, as shown in Figure 1.5.

The set of vectors $|m\rangle$ is orthonormalized in the sense of the scalar product defined by $\langle m|m'\rangle = \delta_{mm'}$. Here, $\delta_{mm'}$ is the Kronecker symbol and equal to 1 if $m' = m$ and 0 if $m' \neq m$. It can also be recalled that the scalar product is linear: if λ, λ', μ and μ' are complex numbers and ψ, ψ', ϕ and ϕ' are arbitrary vectors, then:

$$\langle \lambda\psi + \mu\phi|\lambda'\psi' + \mu'\phi'\rangle = \lambda^*\lambda'\langle\psi|\psi'\rangle + \lambda^*\mu'\langle\psi|\phi'\rangle + \mu^*\lambda'\langle\phi|\psi'\rangle + \mu^*\mu'\langle\phi|\phi'\rangle$$

where λ^* is the complex conjugate of λ.

In the context of this brief introduction to the properties of spin operators, the concept of *hermiticity* must be stated. An operator $\hat{G}$ is *hermitian* whenever the two scalar products $\langle\psi|\hat{G}|\phi\rangle$ and $\langle\hat{G}\psi|\phi\rangle$ are identical. The former is the scalar product of ψ with the vector obtained by applying $\hat{G}$ to ϕ, whereas the latter is the scalar product, with ϕ, of the vector obtained by applying $\hat{G}$ to ψ. It can be shown that all operators associated with an observable, thus possessing real eigenvalues, are hermitian. This will therefore be true for the hamiltonian which is the operator associated with energy and for the spin operators $\hat{I}_x$, $\hat{I}_y$, $\hat{I}_z$ and $\hat{I}^2$.

Other useful relationships concern the behavior of the operators $\hat{I}_x$ and $\hat{I}_y$ associated with the x and y components of the spin momentum. It is convenient to define the so-called raising and lowering operators

$$\begin{aligned}\hat{I}_+ &= \hat{I}_x + i\hat{I}_y \\ \hat{I}_- &= \hat{I}_x - i\hat{I}_y\end{aligned} \tag{1.5}$$

which are non-hermitian, and which have the property of respectively raising and lowering the eigenvalue m according to the following equation (whose proof can be found in quantum mechanics textbooks)

$$\hat{I}_\pm|m\rangle = \sqrt{I(I+1) - m(m \pm 1)}\,|m \pm 1\rangle \tag{1.6}$$

with necessarily $\hat{I}_+|I\rangle = 0$ and $\hat{I}_-|-I\rangle = 0$.

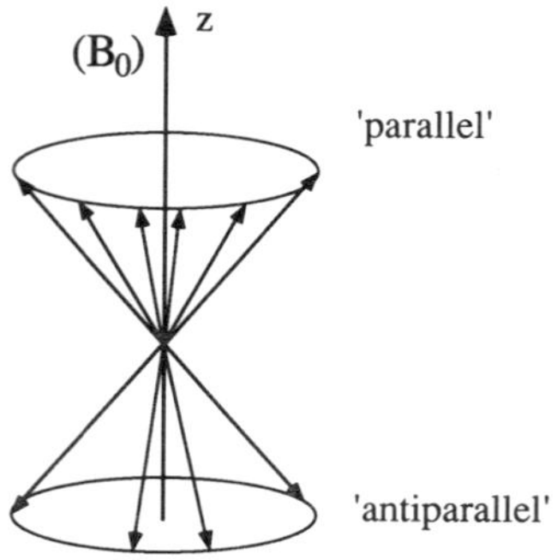

Figure 1.5 Orientations of individual nuclear magnetic moments (spin 1/2) at thermal equilibrium

In the case of spin 1/2, equations derived from (1.6) are especially simple:

$$\hat{I}_+|\alpha\rangle = 0 \qquad \hat{I}_-|\alpha\rangle = |\beta\rangle$$
$$\hat{I}_+|\beta\rangle = |\alpha\rangle \qquad \hat{I}_-|\beta\rangle = 0 \tag{1.7}$$

Very often, we shall need the commutation rules of spin operators. These rules tell us that we must take care of the order in which the spin operators act. Hence, applying first $\hat{I}_x$ and then $\hat{I}_y$ to an arbitrary vector φ can be expressed as $\hat{I}_y\hat{I}_x\varphi$, whereas applying first $\hat{I}_y$ and then $\hat{I}_x$ is written $\hat{I}_x\hat{I}_y\varphi$; the results of these two operations are not identical and one actually gets: $\hat{I}_x\hat{I}_y\varphi - \hat{I}_y\hat{I}_x\varphi = i\hat{I}_z\varphi$. One can say that the two operators $\hat{I}_x$ and $\hat{I}_y$ do not commute, the preceding equation being symbolized by:

$$[\hat{I}_x, \hat{I}_y] = i\hat{I}_z \tag{1.8a}$$

which is complemented by two further relationships deduced from (1.8a) by rotating the x, y, z subscripts:

$$[\hat{I}_y, \hat{I}_z] = i\hat{I}_x \tag{1.8b}$$
$$[\hat{I}_z, \hat{I}_x] = i\hat{I}_y \tag{1.8c}$$

The operator associated with the square of the spin momentum commutes with operators associated with each of its components:

$$[\hat{I}_x, \hat{I}^2] = [\hat{I}_y, \hat{I}^2] = [\hat{I}_z, \hat{I}^2] = 0 \tag{1.9}$$

Finally, it can be recalled that it is always possible to devise a common set of eigenvectors shared by two commuting operators.

We turn now to the hamiltonian operator $\mathscr{H}$ which governs the behavior of a spin system in the presence of a static magnetic field. The form of the hamiltonian is deduced from the expression of energy in classical mechanics, $-\mu B_0$, by substituting the classical quantities by their associated operators. Defining the z axis as the direction of B_0, we obtain

$$\mathscr{H} = -\gamma\hbar B_0\hat{I}_z \tag{1.10}$$

$\mathscr{H}$ possesses the same eigenvectors as $\hat{I}_z$, and this property makes it possible to establish easily the energy diagram for a spin $I = 1/2$ and a spin $I = 1$ (Figure 1.6).

Selection rules are better handled by the raising and lowering operators (equations (1.5)). Consider again the example of a spin $I = 1/2$ and let us suppose that, in addition to the static field $\boldsymbol{B}_0$, an alternative magnetic field $\boldsymbol{B}_1\cos(2\pi\nu_0 t)$, whose frequency ν_0 obeys the Bohr equation, is applied to the spin system. The hamiltonian

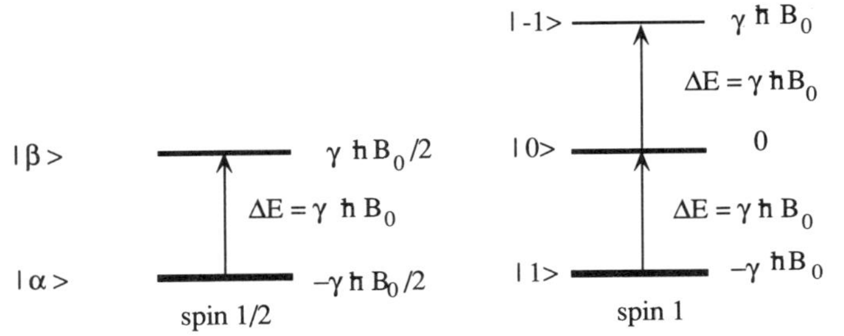

Figure 1.6 Energy diagram and allowed transitions for spin numbers $I = 1/2$ and $I = 1$, respectively

$\hat{\mathcal{H}}'$, describing the interaction between a spin and the relevant alternative field, is obtained by replacing the classical quantities in $\boldsymbol{\mu}\boldsymbol{B}_1 \cos(2\pi\nu_0 t)$ by their associated operators:

$$\hat{\mathcal{H}}' = -\gamma\hbar\cos(2\pi\nu_0 t)[B_{1x}\hat{I}_x + B_{1y}\hat{I}_y + B_{1z}\hat{I}_z]$$

or with raising and lowering operators

$$\hat{\mathcal{H}}' = -\gamma\hbar\cos(2\pi\nu_0 t)[\tfrac{1}{2}(B_{1x} - iB_{1y})\hat{I}_+ + \tfrac{1}{2}(B_{1x} + iB_{1y})\hat{I}_- + B_{1z}\hat{I}_z]$$

From time-dependent perturbation theory, it can be stated that a transition between two states ψ and φ is allowed provided that $\langle\psi|\hat{\mathcal{H}}'|\varphi\rangle \neq 0$. Moreover, the line intensity is proportional to $|\langle\psi|\hat{\mathcal{H}}'|\varphi\rangle|^2$. Now, let us determine whether a transition can occur between the two states α and β:

$$\langle\alpha|\hat{\mathcal{H}}'|\beta\rangle = -\gamma\hbar\cos(2\pi\nu_0 t)[(1/2)\langle\alpha|(B_{1x} - iB_{1y})\hat{I}_+|\beta\rangle + (1/2)\langle\alpha|(B_{1x} + iB_{1y})\hat{I}_-|\beta\rangle + \langle\alpha|B_{1z}\hat{I}_z|\beta\rangle]$$

It is clear that, among the three terms in the above equation, only $\langle\alpha|\hat{I}_+|\beta\rangle = \langle\alpha|\alpha\rangle = 1$ leads to a non-zero result [see (1.7)]. This can be expressed by the necessary condition for the occurrence of a transition: *the alternative magnetic field* B_1 *must be polarized perpendicularly to the static field* B_0. Therefore, provided that this condition is fulfilled, the transition $|\alpha\rangle \rightarrow |\beta\rangle$ is indeed possible at a frequency satisfying the Bohr equation $h\nu_0 = \Delta E = \gamma\hbar B_0$, in agreement with the Larmor frequency:

$$\nu_0 = \gamma B_0/2\pi \tag{1.11}$$

Concerning a spin $I = 1$, similar calculations show that only the transitions $|1\rangle \rightarrow |0\rangle$ and $|0\rangle \rightarrow |-1\rangle$ are allowed and occur at the same frequency, still given by (1.11).

1.3 Isotropic Phase High-Resolution Spectra

The above considerations do not, however, explain the spectrum of ethanol (Figure 1.1): neither the existence of separate patterns corresponding to the different groupings of this molecule (which are relevant to the so-called chemical shift phenomenon), nor the splittings within these patterns and which arise from the so-called spin–spin couplings. These two features will be now considered in detail after we have examined some factors capable of modifying the resonance frequency.

1.3.1 RESONANCE FREQUENCY AND THE PRINCIPLE OF MAGNETIC RESONANCE IMAGING

The value of the externally applied magnetic field B_0 may be altered at the level of the various nuclei (or nuclear spins) within the molecule under investigation; a perturbation may originate from the local 'electronic cloud' (or electronic distribution) and leads to a slight modification of B_0 which becomes $B_0\,(1 - \sigma)$. σ is called the *shielding constant* and is positive when the modification of B_0 has for its main origin the precession around B_0 of the electronic angular momentum (which describes the rotation of the electronic cloud); in that case, the magnetic field induced by this precessional motion is opposed to B_0. Generally speaking, the resonance frequency is no longer equal to $\gamma B_0/2\pi$ but must be expressed as

$$\nu_0 = \gamma(1 - \sigma)B_0/2\pi \tag{1.12}$$

which has therefore to be substituted for $\gamma B_0/2\pi$ in (1.11); this amounts to correcting the gyromagnetic ratio by $(1 - \sigma)$ and leads to a differentiation of resonance frequencies as a function of the electronic environment, thus as a function of the nature of the chemical grouping to which the spin considered belongs.

This latter effect, known as '*chemical shift*', is rather small (σ, which is dimensionless, is of the order of 10^{-6}) and requires, to be observed, a field B_0 that is very homogeneous across the sample. It proves essential to correct the unavoidable inhomogeneity of B_0 by means of additional coils called '*shims*' which generate small corrective fields for compensating the deviations of B_0 from its central value; this can be viewed by an expansion with respect to the spatial coordinates X, Y, Z:

$$B_0(X, Y, Z) = B_0(0, 0, 0) + X\left(\frac{\partial B_0}{\partial X}\right)_0 + Y\left(\frac{\partial B_0}{\partial Y}\right)_0 + Z\left(\frac{\partial B_0}{\partial Z}\right)_0 + X^2\left(\frac{\partial^2 B_0}{\partial X^2}\right)_0$$
$$+ Y^2\left(\frac{\partial^2 B_0}{\partial Y^2}\right)_0 + Z^2\left(\frac{\partial^2 B_0}{\partial Z^2}\right)_0 + XY\left(\frac{\partial^2 B_0}{\partial X \partial Y}\right)_0 + \ldots$$

where $B_0(0, 0, 0)$ is the field at the center of the sample. Shims are designed for creating *gradients* (for example $-\partial B_0/\partial X$) which are expected to compensate, as exactly as possible, the genuine gradients as they appear in the above expansion. 'Shimming' (adjusting the current in each coil) is tedious, especially when dealing with heterogeneous samples, as in the case of *in vivo* spectroscopy. It is unfortunately the prerequisite of any high-resolution NMR measurement and no satisfactory automated procedure has yet been found.

One of the most spectacular applications of NMR, namely MRI (for magnetic resonance imaging), relies on the utilization of gradients similar to those created by shims. The basic idea is to produce a spectrum in which the conventional frequency scale is substituted by a scale representing one of the spatial directions, say X, the amplitude (line intensity) indicating the spin density at abscissa X (Figure 1.7).

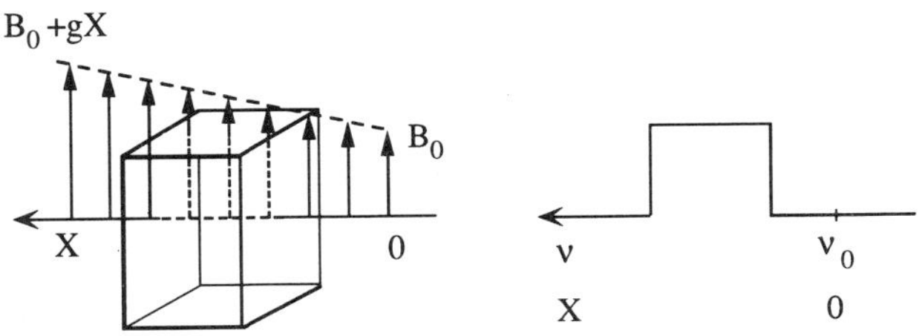

Figure 1.7 A box supposed to be filled with a liquid containing a single chemical shift species and subjected to a static field B_0 homogeneous in all directions but the X direction, for which its amplitude increases linearly with X. The spectrum obtained in these conditions reflects exactly the spin distribution along X

The considerations of the previous sections suggest that, in a general way, the amplitude of the NMR signal occurring at a frequency ν is proportional to the number of spins (or to the density of these spins) resonating at this frequency. Let us assume that we are dealing with a single chemical species whose resonance frequency is ν_0 in a homogeneous magnetic field B_0 (for instance, water protons in biological tissues) and let us try to make this resonance frequency spatially dependent or, better, to make it proportional to X, that is, to devise a space–frequency conversion (frequency encoding). In order to achieve this goal, we can superimpose onto B_0 (supposed to be perfectly homogeneous) a *constant gradient* in the X direction, $g = \partial B_0/\partial X$, so that the magnetic field varies along X according to $B_0 + gX$. As a consequence, the resonance frequency for a spin located at abscissa X becomes

$$\nu = \gamma(1 - \sigma)(B_0 + gX)/2\pi = \nu_0 + kX \tag{1.13}$$

with $k = \gamma\,(1 - \sigma)g/2\pi$.

The resulting spectrum therefore shows up a profile which reflects the spin distribution along X. Repeating this process for the two other spatial directions Y and Z makes it possible to reconstruct a three-dimensional image of the object under investigation. In fact, two-dimensional images (spin distribution in the X, Y plane) are generally produced. To do this, one has to perform a 'slice selection' in the Z direction by means of selective excitation applied simultaneously to a gradient along Z (experimental procedures will be detailed in Chapter 5). A typical image obtained in that way is shown in Figure 1.8.

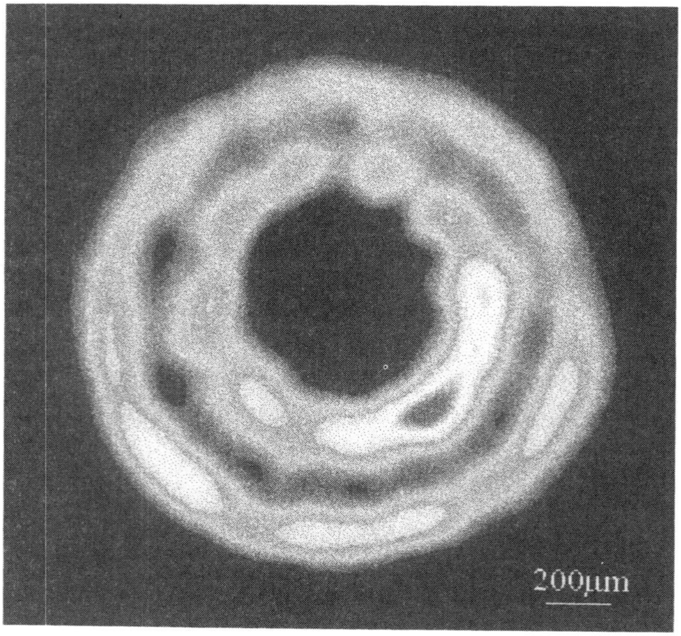

Figure 1.8 A two-dimensional image from the stem of a small plant (≈ 2 mm o.d.)

1.3.2 THE CHEMICAL SHIFT

Provided that field inhomogeneities have been corrected, chemical shift effects become visible, and a procedure aiming at chemical shift characterization is required. Because the absolute measurement of B_0 with an accuracy of 10^{-6} or better is impossible, ν_0 can only be measured with respect to the resonance frequency ν_{ref} of a reference substance. Moreover, for comparing spectra obtained with spectrometers operating at different B_0 values, one generally uses a reduced variable denoted δ and defined as:

$$\delta_{ppm} = \frac{\nu_0 - \nu_{ref}}{\nu_{ref}} \times 10^6 \tag{1.14}$$

This quantity is expressed in ppm because of the 10^6 factor, which simply avoids handling very small numbers. Neglecting σ_{ref} (of the order of 10^{-6}) with respect to 1, we obtain from (1.12):

$$\delta_{ppm} = (\sigma_{ref} - \sigma_0) \times 10^6 \tag{1.15}$$

Thus the δ variable actually represents a measurement of the shielding constant relative to that of a reference, independently of the spectrometer in use. *In practice, δ is calculated by determining experimentally the frequency difference ($\nu_0 - \nu_{ref}$) and by dividing this latter quantity expressed in Hz by the measurement frequency ν_{ref} expressed in MHz*. For historical reasons (spectrometers in the 1950s and even in the 1960s were operating in the field sweep mode) spectra are presented in such a way that *frequency increases from right to left* (Figure 1.9). This is the same for the δ scale, but the opposite for the σ scale; shielding is more important for resonances located in the right part of the spectrum. Those signals are also said to resonate at 'high field' by reference to the field sweep procedure, in the opposite direction of the frequency sweep. The substance chosen as reference should ideally exhibit a single resonance, be easily soluble, be insensitive to solvent effects and appear at an extremity of the spectrum. For proton and carbon-13 spectroscopies, TMS (tetramethylsilane, $Si(CH_3)_4$) is commonly employed.

Good correlations (although not totally unambiguous) exist between the chemical shift and the nature of the molecular grouping to which the considered spin belongs. These correlations are shown schematically in Figures 1.10 to 1.12 for the most important isotopes. The first of these diagrams (Figure 1.10) demonstrates that carbon-13 spectroscopy is very often a mandatory complement to proton spectroscopy because of a much more important chemical shift range (200 ppm as compared with the 10 ppm proton chemical shift range). In these diagrams, overlap

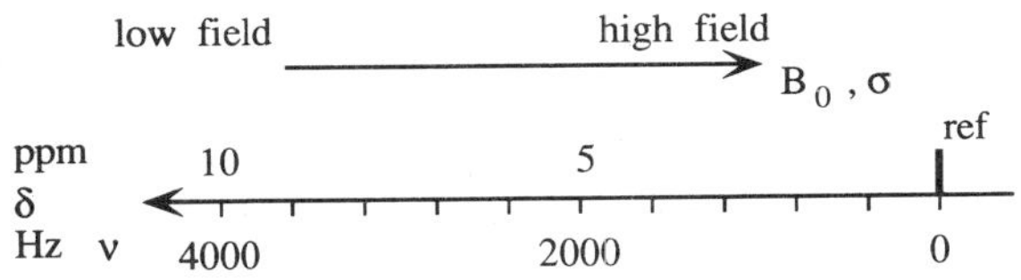

Figure 1.9 Chemical shift scale for $B_0 = 9.4$ T (proton resonance frequency = 400 MHz), expressed in Hz and in ppm

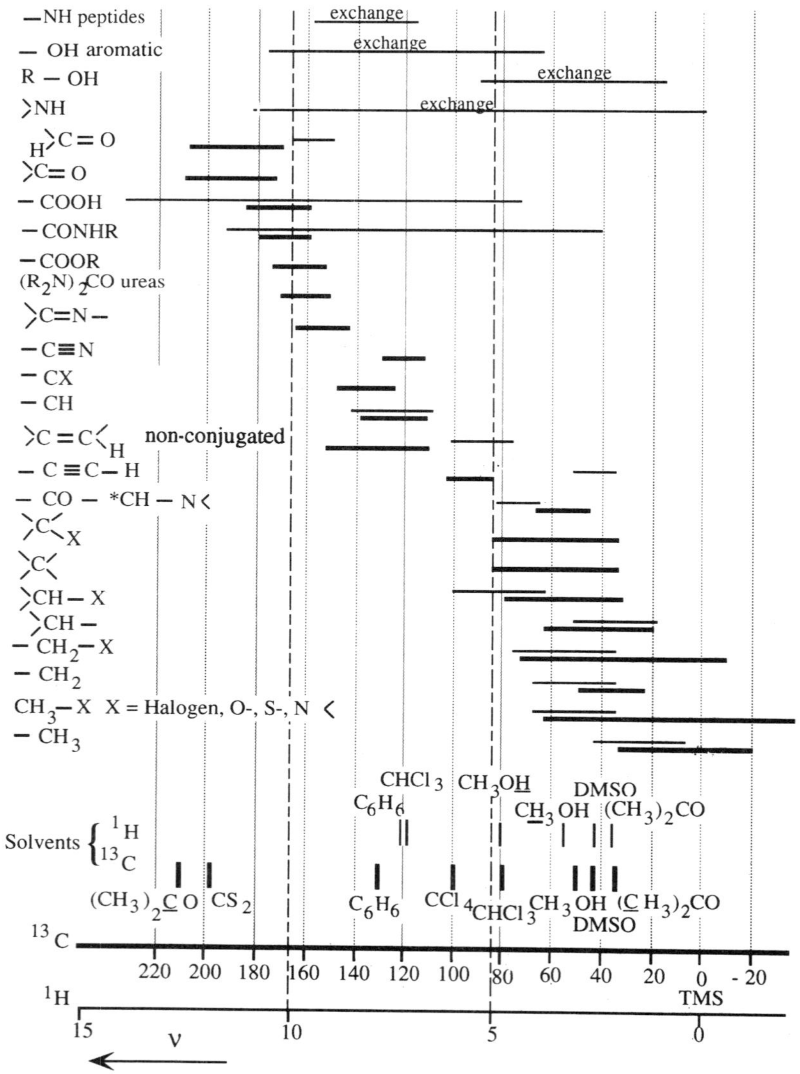

Figure 1.10 Comparison of proton and carbon-13 chemical shifts. Data concerning carbon are in bold face

frequently occurs which, of course, precludes absolute assignments. However, the simultaneous consideration of proton and carbon-13 chemical shifts may lift some ambiguities. Proton, carbon-13 and phosphorus-31 chemical shifts of the usual solvents or typical molecules are given in Tables 1.2 and 1.3.

The shielding constant σ (we shall see later that it arises from a tensorial quantity) can be evaluated by quantum chemistry calculations. Here, we shall limit ourselves to a qualitative discussion about the secondary magnetic field at the level of the considered nucleus arising from the electronic distribution. In a first stage,

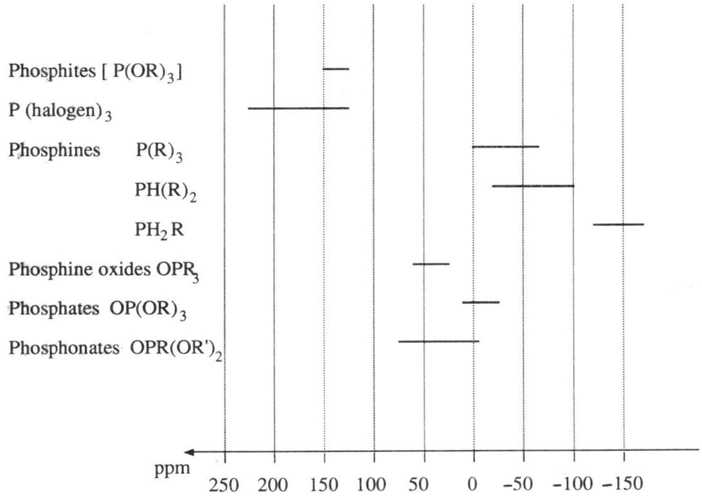

Figure 1.11 Phosphorus-31 chemical shifts, with respect to phosphoric acid (85% in H_2O)

we can take into account the precession of the 'electronic cloud' around the magnetic field $\boldsymbol{B}_0$. As already mentioned, we are concerned with the precession of the vector $\boldsymbol{L}$, which represents the angular momentum describing the motion of electrons. As for an electric current in a loop, the electronic cloud induces a secondary magnetic field $\boldsymbol{B}'$ which is opposed to $\boldsymbol{B}_0$ (Figure 1.13). This is therefore a *diamagnetic* effect and the relevant part of the shielding constant will be denoted σ_d ($\sigma_d > 0$). Proceeding further in the analysis, we can realize that the electron velocity depends on the value of B_0. It can be shown that this latter dependence entails an additional field in the same direction as $\boldsymbol{B}_0$. This yields a *paramagnetic* contribution ($\sigma_p < 0$) and the total screening constant can be written $\sigma = \sigma_d + \sigma_p$.

It turns out that σ_d depends only on the fundamental state of the electronic system and can be directly related to the electric charge of the atom being considered. Shielding increases with negative charge and produces a shift toward high fields; conversely, 'acidic' protons will appear at low fields. It can be shown that σ_p depends on electronic excited states and strongly on the symmetry of valence orbitals. It is zero for s orbitals (which are of spherical symmetry) and is therefore of little relevance for proton shielding constants. On the contrary, it overhelms σ_d if valence orbitals are not of spherical symmetry. This is the case for carbon-13, fluorine-19 and nitrogen-15 (and nitrogen-14), to cite a few, whose valence orbitals are p orbitals; the chemical shift range of these isotopes is very large.

Whenever the chemical shift range is weak (as is the case for protons), other small effects have to be taken into account. At first, an *electric field of intra-molecular or intermolecular origin* (arising for instance from hydrogen bonds) produces a distortion of the electronic cloud and generally a deshielding effect (shift toward low fields). *Anisotropy of molecular magnetic susceptibility* may also play an important role. Let us suppose for simplicity that the system under

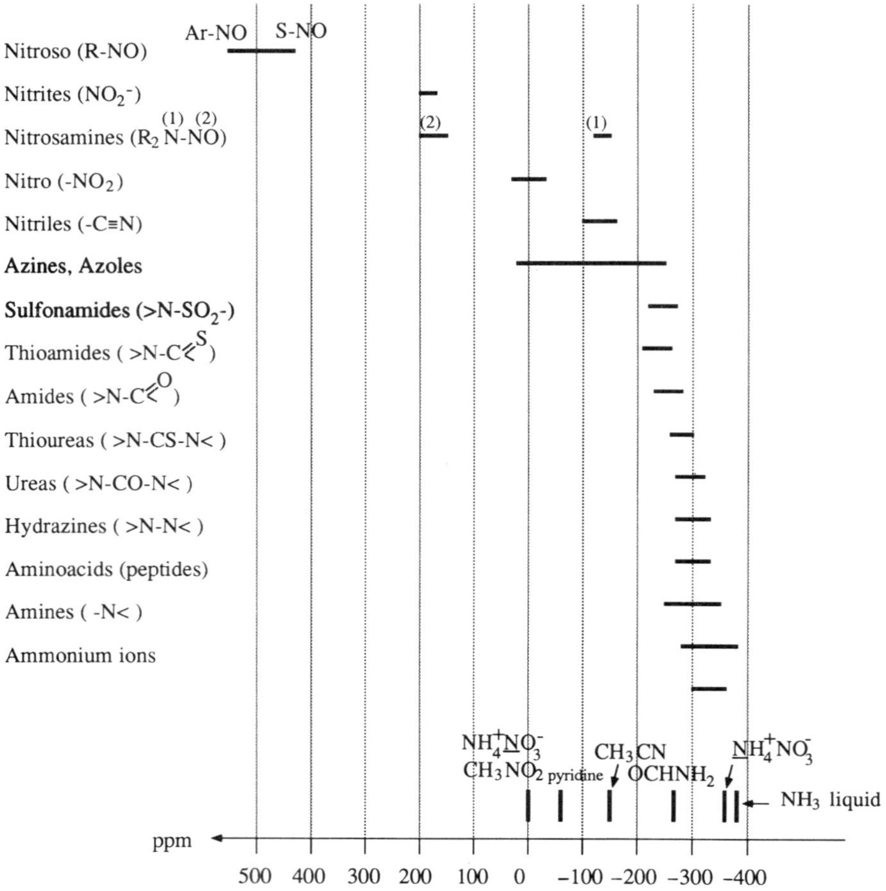

Figure 1.12 Nitrogen-15 (or nitrogen-14) chemical shifts

investigation is of axial symmetry, and let us define two directions $\parallel$ and $\perp$ corresponding to the magnetic susceptibilities $\chi_{\parallel}$ and $\chi_{\perp}$.

Let R be the distance from the considered nucleus (a proton for instance) to the point in the molecule where the anisotropy takes place and θ the angle between the $\parallel$ and R directions (Figure 1.14). It can be shown that the shielding constant is modified according to:

$$\Delta\sigma = \frac{\chi_{\parallel} - \chi_{\perp}}{3R^3}(1 - 3\cos^2\theta)/4\pi \tag{1.16}$$

When applied to acetylene (Figure 1.15), this equation yields $\Delta\sigma = +10$ ppm ($\chi_{\parallel} \approx 0$, $\theta = 0$).

In the case of ethylene the angle θ is different from 0 and $\Delta\sigma$ is modified through the factor $(1 - 3\cos^2\theta)$, whose absolute value is smaller; this explains why ethylene resonates at higher frequency (Figure 1.15). In practice, one often defines a 'deshielding cone' which encompasses the region where $(1 - 3\cos^2\theta)$ is positive.

Table 1.2 ^1H and ^{13}C chemical shifts of some common solvents, given with respect to the TMS resonance

Solvent	Formula	δ^1H (ppm)	$\delta^{13}C$ (ppm)
Methanol	CH_3OH	3.35	49.8
	CH_3OH (italic H)	4.89	
Acetone	$(CH_3)_2CO$	2.00	205.3
	$(CH_3)_2CO$		30.4
Benzene	C_6H_6	7.15	128.5
DMSO	C_2H_6SO	2.5	40.7
Chloroform	$CHCl_3$	7.02	77.3
Carbon tetrachloride	CCl_4		96.2
Carbon disulfide	CS_2		192.4
Ethanol	CH_3CH_2OH (italic H_3)	1.18	18.2
	CH_3CH_2OH (italic H_2)	3.63	57.7
Cyclohexane	C_6H_{12}	1.43	27.5
Dioxane	$C_4H_8O_2$	3.53	67.4
Methylene dichloride	CH_2Cl_2	5.03	53.6
Diethyl ether	$(CH_3CH_2)_2O$ (italic H_3)	1.12	15.6
	$(CH_3CH_2)_2O$ (italic H_2)	3.36	66.1

Table 1.3 ^{31}P chemical shifts of some common molecules given with respect to 85% H_3PO_4 (phosphoric acid) in H_2O

Solvent	Formula	$\delta^{31}P$ (ppm)
Hexamethylphosporamide	$((CH_3)_2N)_3PO$	23
Trimethyl phosphite	$(CH_3O)_3P$	141
Trimethyl phosphate	$(CH_3O)_3PO$	−2.4
Phosphorous anhydride	P_4O_6	113
Trimethyl phosphine oxide	$(CH_3)_3PO$	36

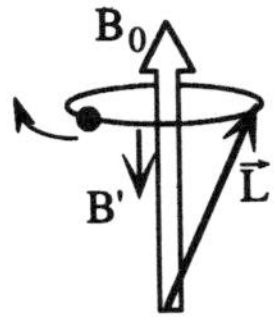

Figure 1.13 The rotation of the electronic cloud induces a field $\boldsymbol{B'}$ opposed to $\boldsymbol{B}_0$

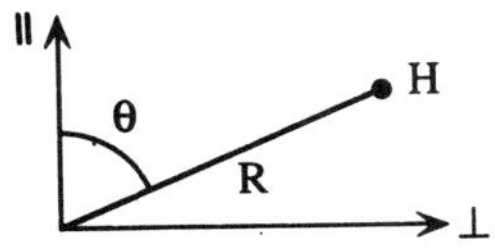

Figure 1.14 The two directions $\parallel$ and $\perp$. H stands for the proton subjected to magnetic susceptibility anisotropy

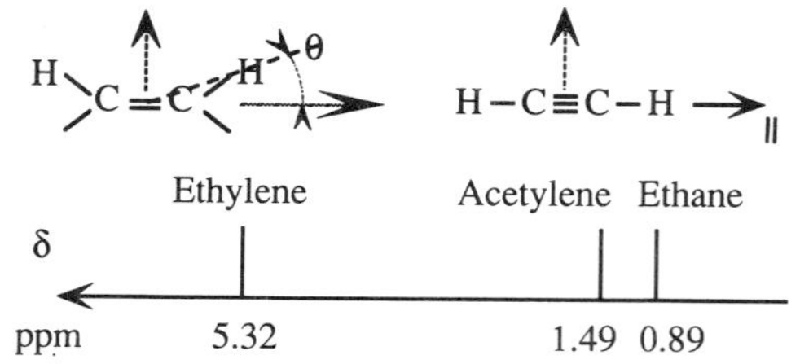

Figure 1.15 Chemical shift hierarchy in ethane, acetylene (right) and ethylene (left) (gaseous state)

The amazingly large chemical shift of benzene protons (around 7 ppm) can be explained again by considering the precession of the electronic cloud with respect to $\boldsymbol{B}_0$. Here, we have to account for the specificity and the preponderence of π electrons. As shown in Figure 1.16, this effect is maximized when the molecule is oriented with its plane perpendicular to $\boldsymbol{B}_0$ and disappears completely when $\boldsymbol{B}_0$ is in the molecular plane. As we are dealing with the liquid state, we must calculate an average over all molecular orientations. The result is non-zero and is just scaled down with respect to the maximum; the important feature is that the field $\boldsymbol{B}'$ induced by the rotation of π electrons adds to $\boldsymbol{B}_0$ at the level of a benzenic proton (Figure 1. 16).

In addition to the intermolecular information that they provide, chemical shifts are sensitive to intermolecular effects. Some examples are given below:

- Exchange, as exemplified by the broadening of the OH resonance in the proton spectrum of ethanol (Figure 1.1). This proton is labile and may exchange with water protons or between two different ethanol molecules.
- The vicinity of an aromatic ring which leads to effects similar to the ones explained above in the intermolecular case. This property has lead to a method known as '*aromatic solvent induced shift*' (ASIS).
- The vicinity of a paramagnetic species. The interaction with the unpaired electron causes a shift which may be considerable.

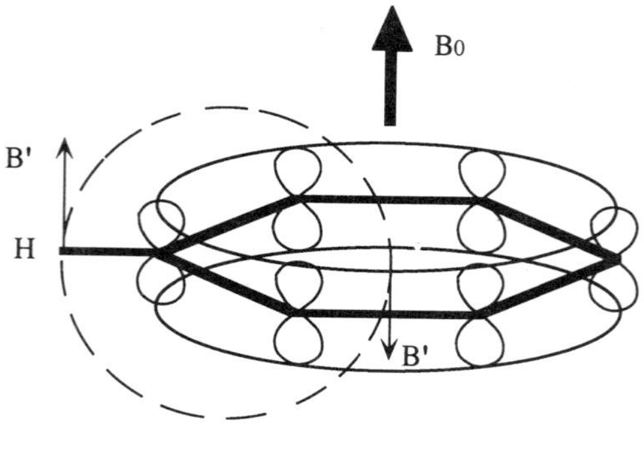

Figure 1.16 Ring current in the benzene molecule

- pH. For instance, the chemical shift of inorganic phosphate in a ^{31}P spectrum yields the pH of the relevant solution. A popular application is the measurement of intracellular pH (Figure 1.17); the method relies upon a difference of 2.4 ppm between the chemical shifts of HPO_4^{2-} and $H_2PO_4^-$. Under conditions of rapid exchange between these two forms, an averaged chemical shift is observed which reflects their relative concentrations.

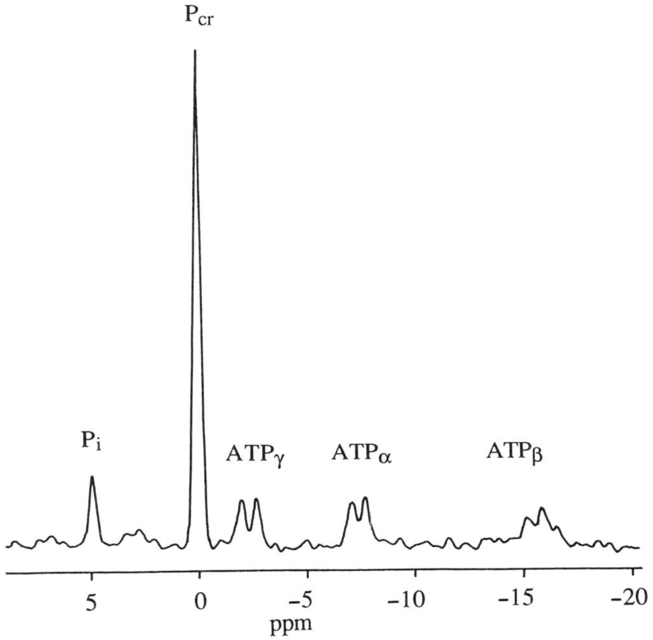

Figure 1.17 ^{31}P spectrum of a biological tissue. P_i: inorganic phosphate; P_{cr}: phosphocreatine; ATP: adenosine triphosphate

1.3.3 THE SPIN–SPIN INDIRECT COUPLING OR *J* COUPLING. QUALITATIVE APPROACH

The fine structure exhibited by some patterns (see the spectra shown up to now) arise necessarily from an additional interaction between nuclear spins. It *cannot* however be assigned to the *direct* interaction between the magnetic moments associated with the spins because, in non-viscous liquids, the latter can be shown to average to zero owing to molecular reorientation. In fact, this interaction occurs via bonding electrons (paired with opposite spins in diamagnetic systems) and does not therefore undergo the averaging process mentioned above. It will therefore be termed *indirect* coupling or, because of a widely used notation, *J* coupling. Figure 1.18 shows schematically this process for two nuclei of spin 1/2. We consider one orientation for spin A and the two possible orientations for spin X. In the situation of Figure 1.18(a), the nuclear spin A and the electronic spin e_1 are in an antiparallel configuration, and e_2 and X are in a parallel configuration. This

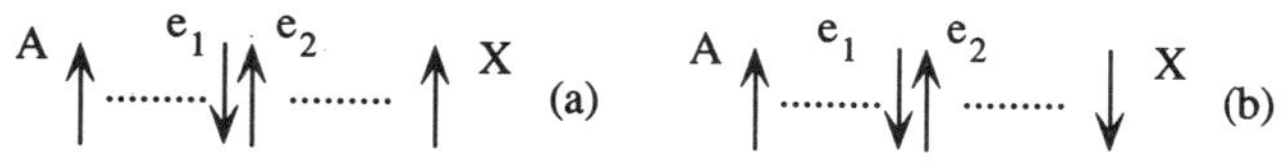

Figure 1.18 Schematic (and simplified) presentation of the interaction between two nuclear spins (A and X) relayed by the two electron spins e_1 and e_2. (a) and (b) correspond to two possible configurations

corresponds to a higher energy than for the situation of Figure 1.18(b), where both (A, e_1) and (e_2, X) are antiparallel. It can be further noticed that A and X are parallel and antiparallel for 1.18(a) and 1.18(b) respectively.

Because of the energy shift associated with these two situations, a relevant frequency shift will appear with respect to ν_A (which is the position of the signal in the absence of coupling between A and X). If we denote by $J/2$ and $-J/2$ the frequency shifts corresponding to the parallel and the antiparallel configurations, this will result in a *doublet* of splitting J, centered on ν_A. Since this interaction is mutual, there exists a similar doublet centered on ν_X (Figure 1.19).

The J coupling is obviously independent of B_0 and is therefore expressed in Hz (its measurement leads to the same value regardless of the working frequency). If one is dealing with more than two spins, the multiplet structure can be accounted for by successive splittings, as indicated in Figure 1.20 for a three-spin system

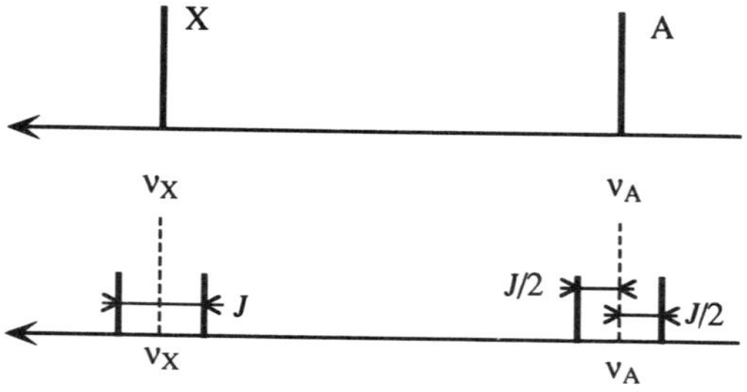

Figure 1.19 Schematic spectrum of two coupled spin 1/2 nuclei

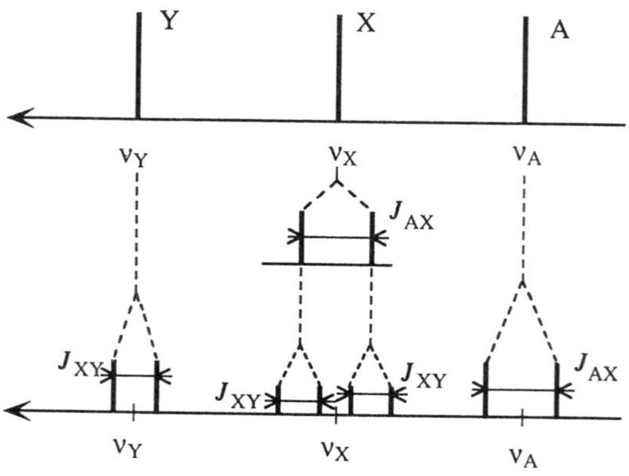

Figure 1.20 Schematic spectrum of three coupled spin 1/2 nuclei. It has been assumed that the coupling between A and Y is negligible and that $J_{AX} > J_{XY}$

(denoted by AXY) where, for simplicity, it has been assumed that $J_{AY} \approx 0$; as a consequence, the X resonance is the only one to be doubly split, firstly according to J_{AX}, secondly according to J_{XY}. Consequently, the part of the spectrum relative to X is a 'doublet of doublets', from which it is easy to retrieve the values of the couplings J_{AX} and J_{XY}. It can be noticed that the same result would have been obtained by considering first the splitting due to J_{XY} and then by further splitting each branch of this doublet according to J_{AX}.

Let us now consider the case where J_{AX} becomes equal to J_{XY}. It is clear that the two inner lines of the 'doublet of doublets' coalesce, yielding a triplet-centered on ν_X with a common splitting $J = J_{AX} = J_{XY}$ and relative intensities in the ratio 1:2:1 (Figure 1.21).

These considerations can be extended to the case of n nuclei of spin 1/2, A_i ($i = 1$ to n), interacting in the same way with X, that is, with the same coupling J ($J_{A_iX} = J$ for any i). When accounting for an additional A_i, this amounts to transforming each line in the current spectrum into a doublet of splitting J. It can be observed, by going from $n = 2$ to $n = 3$, $n = 4$, and so on, that *the resulting multiplet involves* $(n + 1)$ *lines, with an identical separation between two consecutive lines, equal to J, and intensities distributed according to the binomial expansion coefficients* C_n^p (Pascal triangle, Table 1.4).

A more formal derivation of this property is based on the consideration of a subset of nuclei A_i, denoted $(p; n - p)$, for which p spins have a given orientation

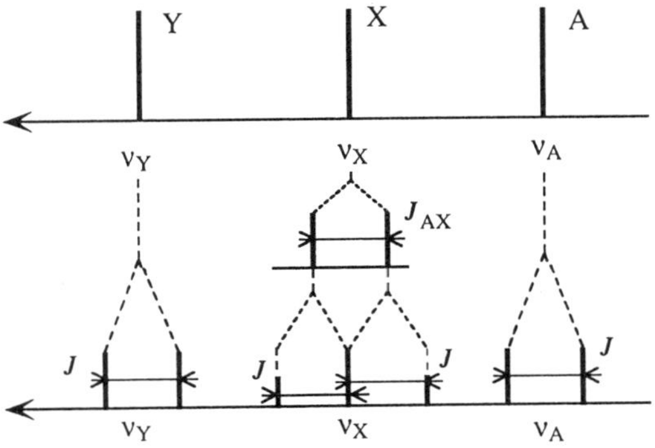

Figure 1.21 System of three nuclei of spin 1/2 for which $J_{AY} \approx 0$ and $J_{AX} = J_{XY} = J$

Table 1.4 Relative intensities within a multiplet, calculated with the Pascal triangle

n	Relative intensities							Multiplet
1	1	1						doublet
2	1	2	1					triplet
3	1	3	3	1				quartet
4	1	4	6	4	1			quintet
5	1	5	10	10	5	1		sextet
6	1	6	15	20	15	6	1	septet

and $(n - p)$ spins the opposite orientation. Such a subset leads to an X transition whose location with respect to ν_X is given by

$$p(J/2) + (n - p)(-J/2) = -n(J/2) + pJ$$

The next line of the multiplet arises from the subset $(p - 1;\ n - p + 1)$ and is located at $-n(J/2) + (p - 1)J$ from ν_X. Hence the separation between two consecutive lines in a multiplet is always equal to J. The number of lines within the X multiplet is equal to the number of distinct configurations $(n - p;\ p)$. This number is necessarily $(n + 1)$ since the possible values of p are n, $n - 1$, $n - 2, \ldots, -n + 2, -n + 1, -n$. The weight of the subset $(n - p;\ p)$, that is, the intensity of the corresponding line, is obtained by calculating the number of combinations of p elements among n; this is none other than $C_n^p = n!/[p!(n - p)!]$.

A situation frequently encountered is that of *magnetically equivalent nuclei*. A set of nuclei A_i is said to include magnetically equivalent nuclei if the two following conditions are met:

- The resonance frequencies of all nuclei A_i are identical;
- Each A_i is coupled in an *identical* manner to any nucleus X, Y . . ., outside the set $\{A_i\}$.

The two examples given in Figure 1.22 should help to understand the concept of magnetic equivalence. For both molecules (1,2,3-trichlorobenzene and *ortho*-dichlorobenzene), chlorines can be discarded (see below) so that one is left with the proton system. Concerning 1,2,3-trichlorobenzene, the only spin outside the set (A, A′) is X and, for symmetry reasons, $J_{AX} = J_{A'X}$ with, of course, $\nu_A = \nu_{A'}$. The two above conditions are therefore satisfied and A and A′ are indeed magnetically equivalent. By contrast, in *ortho*-dichlorobenzene, although the condition $\nu_A = \nu_{A'}$ is satisfied, it is clear that $J_{AX} \neq J_{A'X}$ and similarly $J_{AX'} \neq J_{A'X'}$. Thus, for that molecule, A and A′ are not magnetically equivalent. An essential consequence of the property of magnetic equivalence is the absence, *in the NMR spectrum, of splittings due to J coupling within the set of A_i nuclei*. Hence, $J_{AA'}$ cannot be determined from the spectrum of 1,2,3-trichlorobenzene, whereas it can actually be measured from the spectrum of *ortho*-dichlorobenzene.

Although the latter property can only be derived from a quantum mechanical treatment (see Section 1.3.6), it must always be borne in mind when interpreting an NMR spectrum. Let us go back to the spectrum of ethanol shown at the beginning of this chapter (Figure 1.1). Because of fast rotation around the C–C

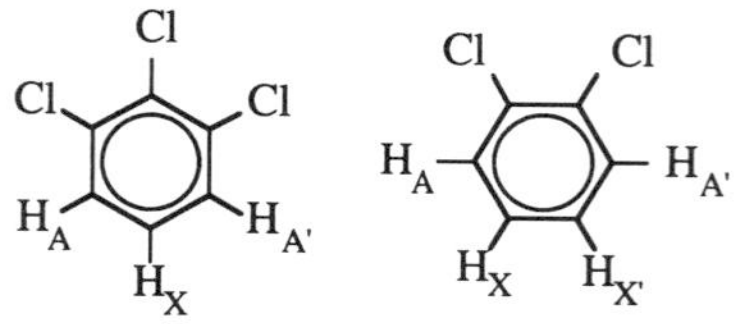

Figure 1.22 Left: 1,2,3-trichlorobenzene; right: *ortho*-dichlorobenzene

bond and the relevant averaging over the different rotamers, it is obvious that methyl protons on the one hand, and methylene protons on the other hand, are magnetically equivalent. As a consequence, the NMR spectrum is totally insensitive to the *J* couplings within each of these two groupings. Nevertheless, each methyl proton 'sees' (or interacts with) the two methylene protons and appears in the form of a triplet. The superposition of three identical triplets is actually observed (a triplet for each proton of the methyl grouping). Likewise, each of the two methylene protons interacts with the three protons of the methyl grouping and appears in the form of a quartet. As already mentioned, any coupling with the hydroxyl proton disappears because this proton is labile and that *J* coupling is necessarily of *intramolecular* nature.

1.3.4 HOMONUCLEAR *J* COUPLINGS

Indirect couplings are related to the electronic density existing between the two interacting nuclei. As far as proton–proton couplings are concerned, it is generally recognized that beyond three covalent bonds, *J* couplings get very weak and most of the time unobservable, unless a particular geometrical arrangement or the occurrence of multiple bonds produces an enhancement of these couplings.

The term *geminal* coupling is generally used for two protons on the same carbon: they lie in the range (−12 Hz)–(−20 Hz) for sp_3 carbon and in the range 0–3.5 Hz for sp_2 carbon (Figure 1.23). *Vicinal* couplings involve two adjacent carbons. Some typical values are given in Figure 1.24.

A double or a triple bond favors the indirect coupling, which thus may become detectable beyond three bonds (Figure 1.25). Couplings in saturated rings have typical values (Figure 1.26).

Concerning homonuclear couplings between isotopes other than the proton, the

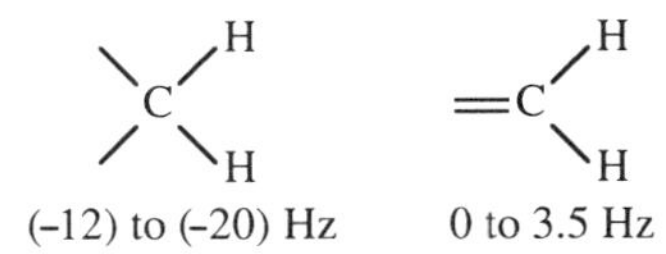

Figure 1.23 Geminal couplings

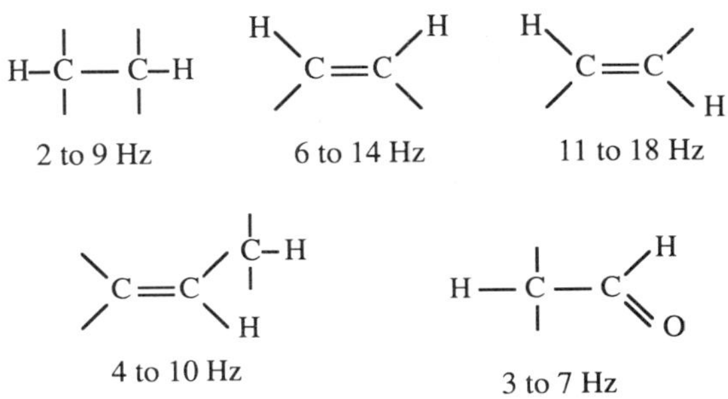

Figure 1.24 Vicinal couplings

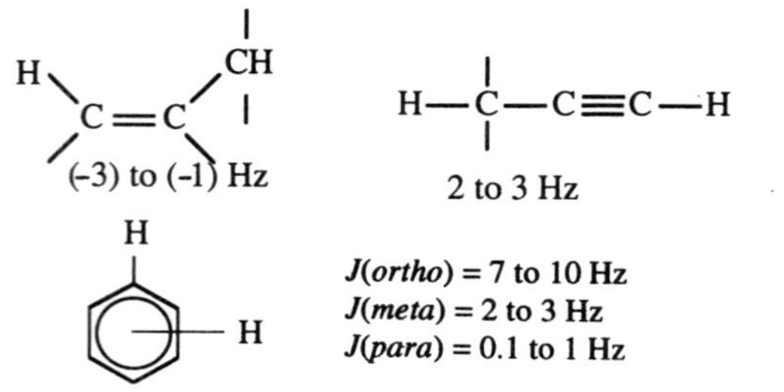

Figure 1.25 Indirect couplings involving multiple bonds

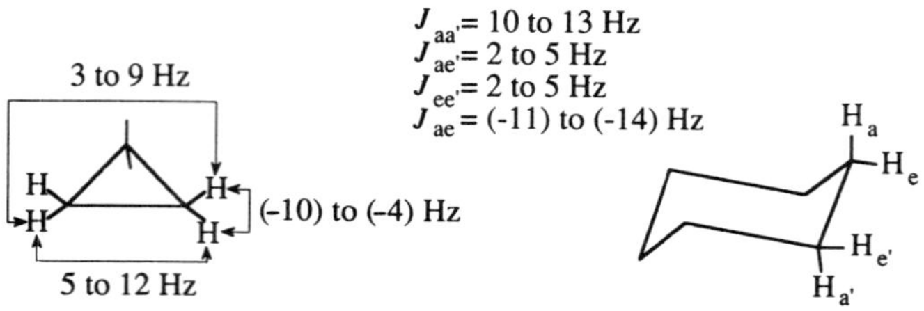

Figure 1.26 Couplings in saturated rings

reader is referred to specialized textbooks or compilations. Here, we shall consider only the coupling between two adjacent carbons, which is a one-bond coupling denoted by $^1J_{CC}$ and whose value lies around 35 Hz. This coupling is of particular importance because it allows one to establish connectivities in a carbon backbone or, in other words, to assign each carbon signal with respect to its nearest neighbors (Chapter 5, Section 5.3.1).

It may have been noticed that a negative sign for some couplings has been mentioned. The sign of J coupling can indeed be determined in some instances (as in strongly coupled spin systems; see Section 1.3.6). The values of J coupling (such as those of shielding constants) can be estimated by quantum chemical calculations, which indicate among other things that a zig-zag path, for passing from one nucleus to its partner, tends to enhance the relevant indirect coupling. This latter property can be related to orbital overlaps. Regarding ethanoic fragments, one can rely on the famous Karplus law which applies to vicinal proton–proton couplings (denoted by ^3J so as to indicate that the two protons considered are three bonds apart). This relationship, given in an explicit form in Figure 1.27, has been widely used for elucidating stereochemical problems; a similar relationship, applied to CH–NH couplings, can contribute to the determination of the tertiary structure of proteins.

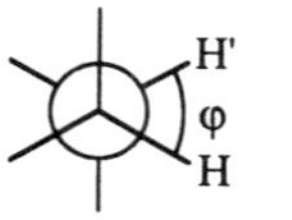

$$J_{HH'} = 9\cos^2\varphi - 0.5\cos\varphi - 0.28 \quad \text{(Hz)}$$

Figure 1.27 Three bond proton–proton coupling and the Karplus law

The interpretation of a proton spectrum may be relatively critical if the molecule under investigation involves many protons. Firstly, the multiplets so far considered may be distorted by the so-called second order effects which arise from the fact that the difference in resonance frequencies $|\nu_A - \nu_X|$ becomes of the same order of magnitude as the J coupling between the two relevant nuclei A and X. These distortions manifest themselves by alterations of line intensities, sometimes called the 'rooftop effect' (as illustrated in Figure 1.28 for a two-spin system) and, for systems of more than two spins, by the occurrence of new transitions.

Second order effects and overlaps of multiplets tend to disappear when measurements are performed at higher B_0 values. In addition to an appreciable sensitivity gain, this constitutes the impetus for going to higher and higher magnetic fields. The improved resolution of the 600 MHz spectrum (top of Figure 1.29) due to an increased chemical shift dispersion illustrates this feature.

The determination of the number of protons corresponding to each multiplet, by means of the integral curves (see Figure 1.1), represents a first piece of information. However, the most acute problem concerns the search for partners of a given nucleus: where is the X nucleus (nuclei) responsible for the multiplet structure of spin A? Whenever a simple examination of multiplets and their structure is not sufficient for assigning all nuclei in a spectrum, physical methods can be considered for solving this problem. The first method relies on a technique dubbed 'spin decoupling', or double resonance, since it consists in inducing *simultaneously* the transitions of two spins A and X interacting by J coupling. In a more explicit way, let us say that A transitions are induced for being observed, whereas X transitions are induced for annihilating the effects of J_{AX} on A resonances. It has been recognized above that a doublet of splitting J_{AX} centered on ν_A was due to the existence of two distinct spin states of nucleus X, denoted respectively as parallel or antiparallel, associated with the two spin functions $|\alpha_X\rangle$ and $|\beta_X\rangle$. If X is continuously irradiated, transitions between $|\alpha_X\rangle$ and $|\beta_X\rangle$ occur relentlessly so that the two states $|\alpha_X\rangle$ and $|\beta_X\rangle$ lose their identity and reduce to a mean state. As the A doublet originated from the existence of these two states, it disappears and thus transforms into a singlet (Figure 1.30). This method requires a secondary radio-frequency field and, in principle, as many experiments as mutually coupled nuclei. For instance, if one is dealing with three coupled spins denoted A, X and

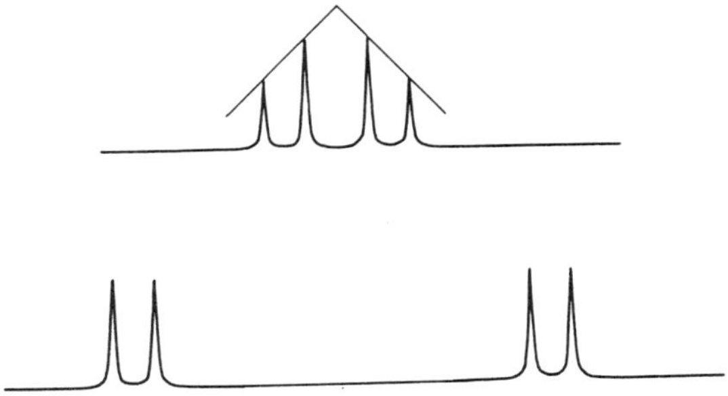

Figure 1.28 Evolution of the spectrum of two nuclei of spin 1/2 as a function of the difference of their resonance frequencies. The intensity of the inner lines increases at the expense of the intensity of outer lines; the straight lines joining the peak maxima schematize the so-called 'rooftop effect'

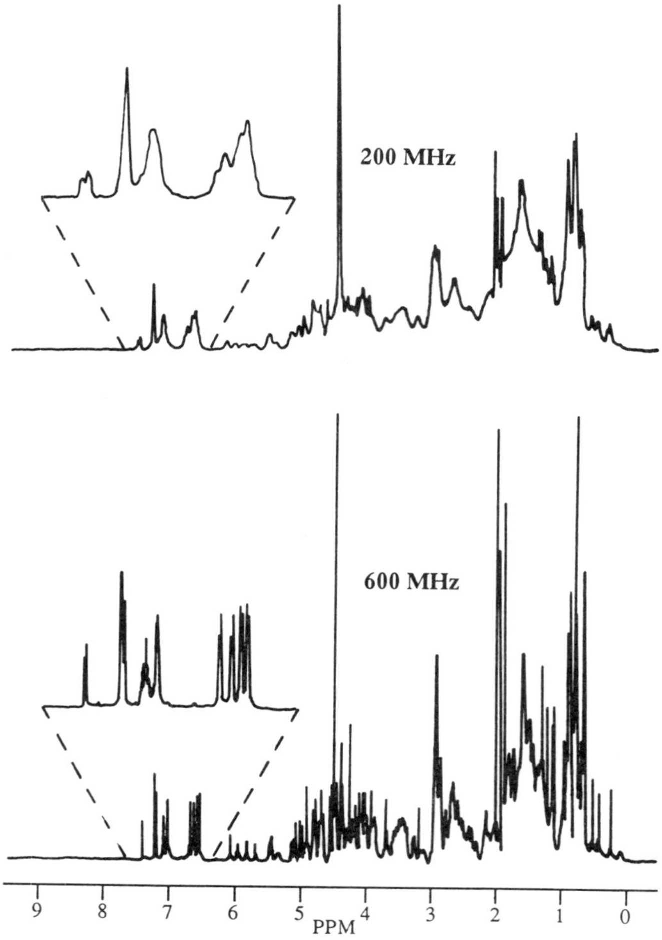

Figure 1.29 Proton spectra of a small protein (60 residues) obtained at 200 MHz (top) and 600 MHz (bottom), respectively. The resolution enhancement is due to a three-fold increase in chemical shift dispersion

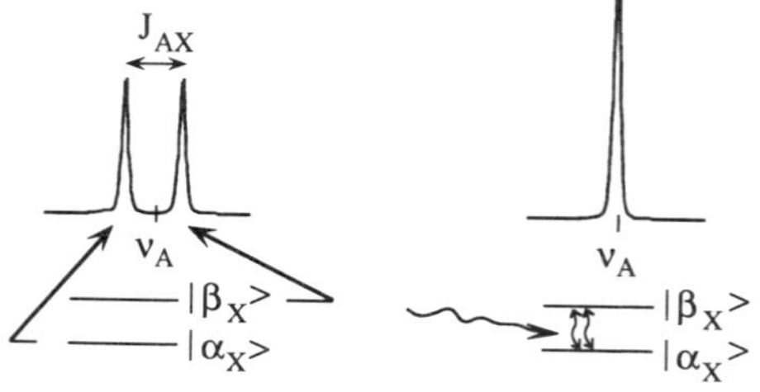

Figure 1.30 A doublet centered on ν_A arises from the two distinct states of spin X (left); upon continuous irradiation of X transitions, the existence of these two states becomes meaningless and the A doublet coalesces into a singlet

Y, irradiating A suppresses the effects of J_{AX} and J_{AY}, and X and Y will become two doublets of splitting J_{XY} centered respectively on ν_X and ν_Y; for confirming this result, it would be recommended to irradiate X and thereafter Y.

In practice, if one has to deal with a complex spectrum, one looks for those patterns which simplify upon irradiation of a given nucleus; this leads to the identification of the spins coupled to the irradiated nucleus. Step by step, and in the manner of a puzzle, the coupling network is reconstituted, yielding the assignment of the various patterns of the spectrum and possibly the molecular structure. This procedure, applied to the proton spectrum of *trans*-crotonaldehyde (Figure 1.31), is illustrated by the series of experiments shown in Figure 1.32. The

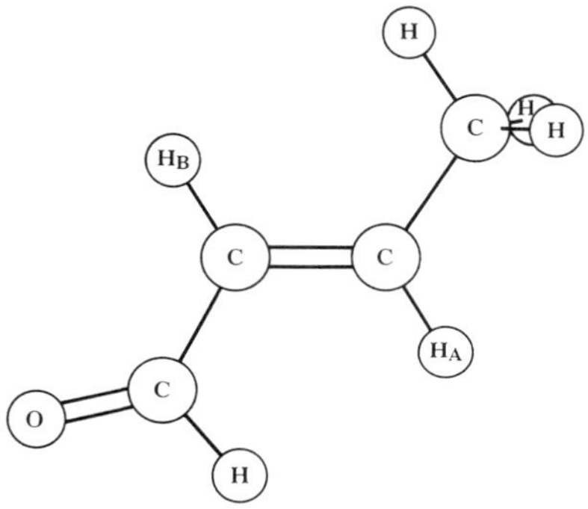

Figure 1.31 The *trans*-crotonaldehyde molecule

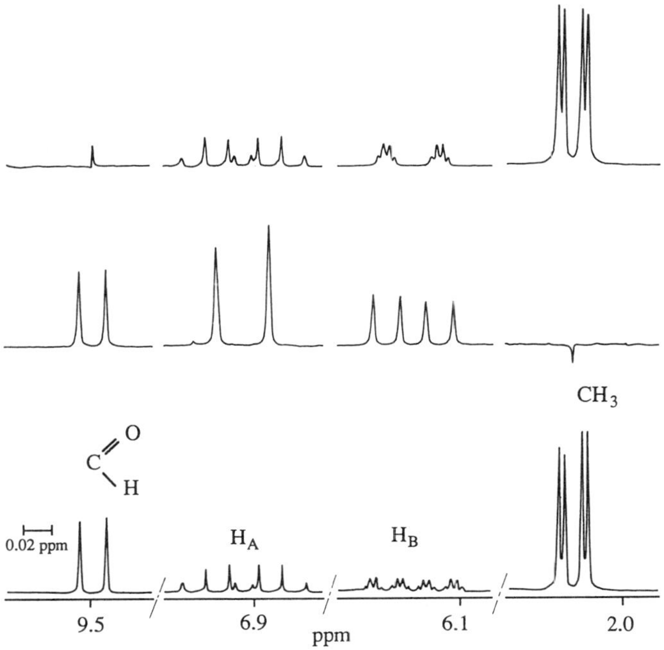

Figure 1.32 Spin decoupling experiments applied to the 400 MHz proton spectrum of *trans*-crotonaldehyde, confirming mutual couplings and the assignment of the different multiplets. Bottom: reference spectrum; middle: spectrum obtained under irradiation of methyl pattern; top: spectrum obtained under irradiation of the aldehydic proton

assignment problem is seen to be solved in a straightforward manner. However, for more complicated molecules, the large number of experiments required would put some limit on the method.

With the advent of *two-dimensional NMR* (2D NMR), it has become possible to substitute, for a whole set of spin decoupling experiments, a single experiment of correlation spectroscopy (COSY), whose principles are detailed in Chapter 5 (Section 5.2.4). For now, only some basic information regarding this method will be given. A one-dimensional NMR experiment consists of measuring the signal following a radio-frequency (r.f.) pulse. The signal is actually sampled as a function of time and can be written as $S(t_2)$ if the time variable is denoted by t_2. It can be shown that a mathematical operation (Fourier transformation) applied to the NMR signal $S(t_2)$ leads to the conventional spectrum (Chapter 3), denoted as $F(\nu_2)$. This latter notation emphasizes the fact that the Fourier transformation acts on a function of time and provides a function of frequency. The 2D COSY experiment involves two r.f. pulses separated by a time interval t_1; the NMR signal is acquired immediately after the second pulse during a time t_2 so that it can be represented by a function $S(t_1, t_2)$. After a double Fourier transformation, the first one with respect to t_2, the second one with respect to t_1, a function $F(\nu_1, \nu_2)$ is obtained. The result is displayed in the form of a map (Figures 1.33 and 1.34) where each of the two dimensions represents the conventional frequency domain. The relevant spectrum is in fact a volume whose peaks are generally figured out by means of a contour plot. The diagonal which runs from the upper rightmost corner to the lower leftmost corner is in principle identical with the conventional one-dimensional spectrum. The interest of such a diagram lies in the cross peaks (peaks appearing outside the diagonal) which indicate mutual couplings. The coupled spins are identified by a horizontal straight line and the vertical straight line intersecting at the cross peak position. This procedure of analysing a 2D map is shown schematically in Figure 1.33. Diagonal peaks and cross peaks involve in fact a replication of the two doublets (in the case of a simple two-spin system); this structure will be analyzed in more detail in Section 5.2.4.

It can be noted that the projection of the entire map along each axis leads again to a 1D spectrum. Figure 1.34 demonstrates the ease with which a spectrum such

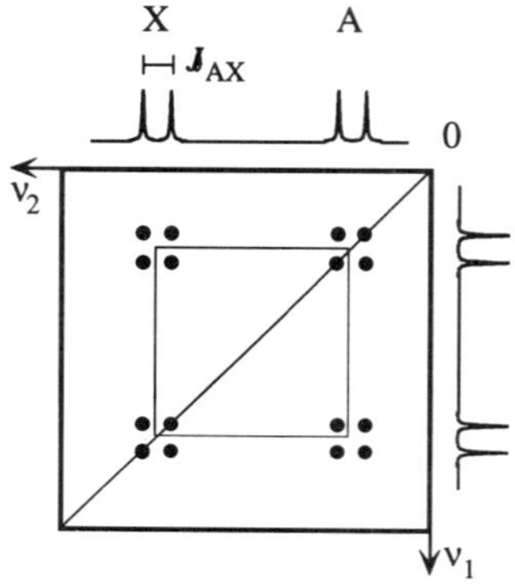

Figure 1.33 Schematized COSY spectrum of a two spin system

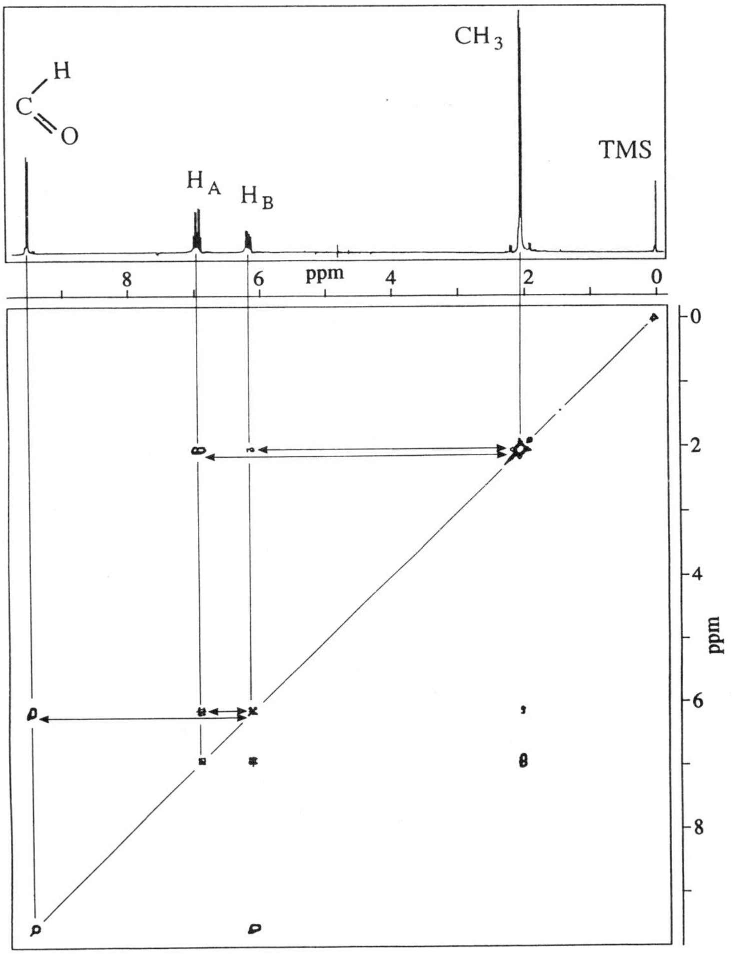

Figure 1.34 COSY spectrum of *trans*-crotonaldehyde. Cross peaks of lower intensity indicate remote couplings

as that of *trans*-crotonaldehyde can be interpreted. Assignments stem from the correlation peaks (the previously mentioned cross peaks) whose internal structure is somewhat irrelevant. As a matter of fact, this structure tends to disappear in practice because of poor digital resolution and peak overlap.

1.3.5 HETERONUCLEAR *J* COUPLINGS. SPECTROSCOPY OF LOW NATURAL ABUNDANCE NUCLEI

In principle, indirect coupling exists between all nuclei possessing a non-zero spin, provided that their relative location within the molecule enables such a coupling to show up. This would imply that a proton spectrum would exhibit numerous splittings due to *J* coupling with all the heteronuclei existing in the molecule considered. Examination of the spectra encountered up to now indicates that this is not often the case.

Let us first consider nuclei whose spin number is greater than 1/2. These nuclei are dubbed *quadrupolar* by the fact that the distribution of nuclear charges is no longer of spherical symmetry. As a consequence, a quadrupolar moment appears which interacts with any electric field gradient existing at the nucleus. This electric field gradient has its origin in the electronic distribution; the interaction between the quadrupolar moment and the electric field gradient actually occurs unless the electronic distribution is of spherical symmetry. It turns that this interaction governs a relaxation mechanism (Chapter 4) whose strength is generally capable of inducing transitions at a frequency much greater than the J coupling. This is, to a certain extent, analogous to the spin decoupling process explained above, however without a secondary radio-frequency field, whose role is played here by the interaction between the quadrupolar moment and the electric field gradient. Apart from the special case of a spherical environment, this situation is generally relevant for all nuclei of spin greater than 1/2, with some exceptions; for instance, deuterium, whose quadrupolar moment is weak enough to make possible the observation of J coupling. Couplings with nitrogen-14 or Br, Cl and I seldom show up (it should be noted that fluorine, the first of the halogens, possesses a spin 1/2).

As far as heteronuclei are concerned, the other category of interest is that of spin 1/2 nuclei of low natural abundance, the isotope most frequently encountered being carbon-13, with a natural abundance of 1.1%. The carbon-12 isotope, whose natural abundance is close to 99%, has a spin zero, which at first sight precludes the detection of J couplings with carbons in the proton spectrum of organic molecules. However, by a thorough examination of the baseline of a proton spectrum, one can notice the presence of signals of weak intensity, called satellites, which arise from ^{1}H–^{13}C couplings. Their intensity with respect to the main signals (sometimes called parents) reflects the carbon-13 natural abundance (Figure 1.35).

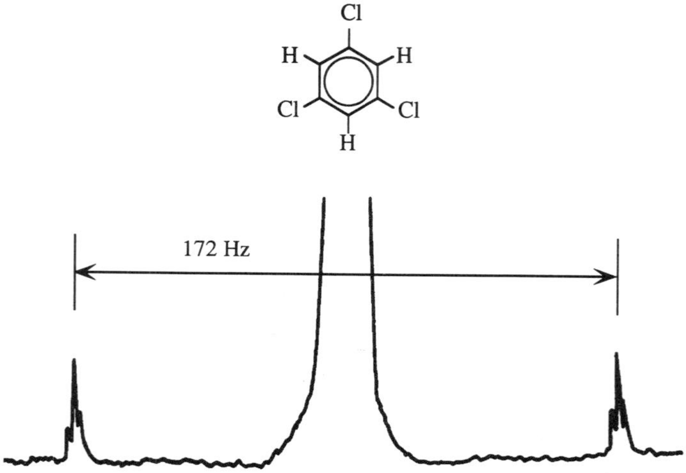

Figure 1.35 ^{13}C satellites in the proton spectrum of 1,3,5-trichlorobenzene

Probabilities are 0.99 and 0.01 for a given site to be occupied by a carbon-12 and a carbon-13, respectively. This means that there exists one molecule in a hundred possessing a carbon-13 at a given site and one in ten thousand possessing two carbon-13 nuclei at two different sites (this stems from the product of probabilities). Clearly, satellites visible in a proton spectrum originate from molecules naturally labeled by one carbon-13 and their global intensity is about 1% of that of the parents. The distance between the two satellites provides the J_{CH} coupling value as shown in Figure 1.35. This example is interesting for another reason: the existence of a carbon-13 breaks the magnetic equivalence of the three protons in the molecule of 1,3,5-trichlorobenzene; as a result, each satellite appears in the form of a triplet which can be interpreted according to the decomposition explained in the previous section, and which therefore yields the proton–proton coupling constant.

$^1J_{CH}$ couplings (for protons directly bonded to the considered carbon) range from 120 Hz up to 200 Hz; typical values are 135 Hz for aliphatic carbons, 160 Hz for aromatic or ethylenic carbons and 200 Hz for acetylenic carbons. $^2J_{CH}$ and $^3J_{CH}$ are of the order of 10 Hz (or smaller). The $^1J_{NH}$ coupling (between a directly bonded proton and nitrogen-15) has a special interest regarding molecules of biological interest (peptides, proteins); its value of 90 Hz is quasi-invariant.

Of course, as soon as the proton spectrum gets a little complicated, the observation of satellites is almost impossible because of numerous overlaps. If one excepts the two dimensional techniques of heteronuclear correlation run in the inverse mode (observation of those proton transitions involving J coupling with the heteronucleus and suppression of the parent signals; techniques to be described later), one is led to the direct observation of the heteronucleus at its own resonance frequency; let us recall that, at a B_0 field of 9.4 T, for which proton resonances occur at around 400 MHz, carbon-13 and nitrogen-15 resonances are observed around 100.6 MHz and 40.5 MHz, respectively. Direct observation of heteronuclei suffers generally from some sensitivity problems due to their low resonance frequency (Chapter 2) and to their usually small natural isotopic abundance. In order partly to circumvent this sensitivity problem, but also to simplify the spectra, one may rely on spin decoupling techniques; in these circumstances, spin decoupling is applied to the totality of the proton transitions in order to cancel any effect of coupling between protons and the considered heteronucleus. For instance, in carbon-13 spectroscopy, each carbon appears in the form of a singlet as shown in Figure 1.36, where the raw ^{13}C spectrum of *trans*-crotonaldehyde is compared with the spectrum obtained under proton decoupling. It can be observed that the raw spectrum (top of Figure 1.36) exhibits some complexity, which originates from J couplings between the considered carbon and directly bonded protons, or possibly with more remote protons.

Of course, in a natural abundance ^{13}C spectrum, ^{13}C–^{13}C couplings could manifest themselves only through satellites, which are difficult to observe although they may allow the establishment of carbon–carbon connectivities (Chapter 5). Hence, only the chemical shift information is clearly visible and provides at first glance an obvious means to determine the number of carbons existing in the molecule under investigation. In a more elaborate way, these carbons can be identified by comparison with data found in chemical shift tables or diagrams

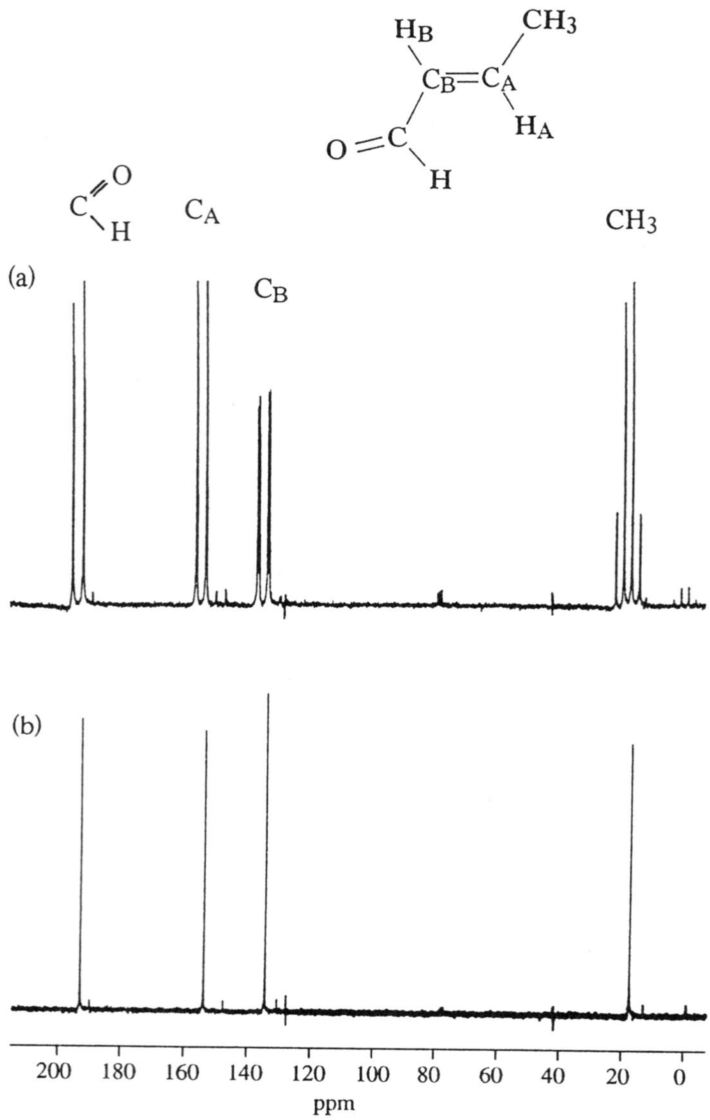

Figure 1.36 (a) Raw ^{13}C spectrum of *trans*-crotonaldehyde; (b) the same spectrum obtained under conditions of total proton decoupling

(Figure 1.10). One must also bear in mind that, without special precautions, relative intensities are not really reliable because of rearrangement of energy level populations. This rearrangement has its origin in the saturation of proton transitions as a consequence of the decoupling field, and can be understood on the basis of the nuclear Overhauser effect considered in Chapter 4. Finally, it can be interestingly mentioned that the isotopic abundance depends, among other things, on the biochemical pathways followed in the course of synthesis of molecules of natural origin. Although significant differences can be detected for carbon-13 or nitrogen-15, this feature is even more pronounced in the case of deuterium: it is possible to differentiate in that way sugars which provide ethanol of alcoholic beverages, and deuterium NMR constitutes at the moment the most reliable tool for detecting chaptalized wines. Another method for studying metabolic pathways rests on the utilization of a precursor labeled with carbon-13 (or nitrogen-15) and an identification, through the NMR spectrum, of the metabolites which show up. This approach is illustrated by the ^{13}C spectrum of Figure 1.37.

We shall end this brief survey of NMR of low abundance nuclei by mentioning two inverse detection techniques (already alluded to above), known under the acronyms of HMQC (heteronuclear multiple quantum correlation) and HSQC (heteronuclear single quantum correlation), which are now in routine use with up-to-date instruments. These two-dimensional techniques, which involve proton detection (thus optimal detection conditions), provide the following items of information (Figure 1.38):

- the proton spectrum by means of the projection onto the f_2 axis;
- the heteronuclear spectrum by means of the projection onto the f_1 axis;
- correlation between these two isotopes (cross peaks in the 2D map) through their mutual coupling (generally one-bond couplings).

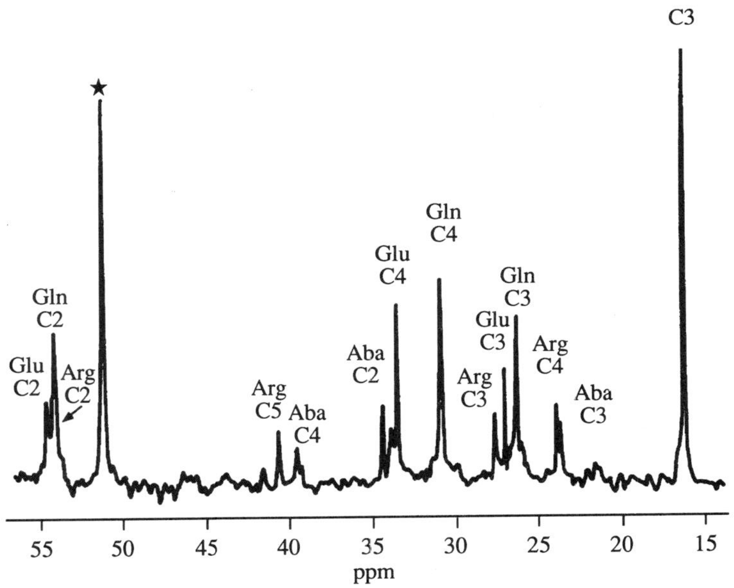

Figure 1.37 ^{13}C spectrum of an extract of mycelium which developed for a period of 5 hours in a medium containing glucose carbon-13 labeled at position 1. Only metabolites formed from labeled glucose show up in the spectrum: Ala (Alanine), Arg (Arginine), Aba (γ-aminobutyrate), Glu (glutamate), Gln (glutamine), * (ethylenediaminetetra-acetate)

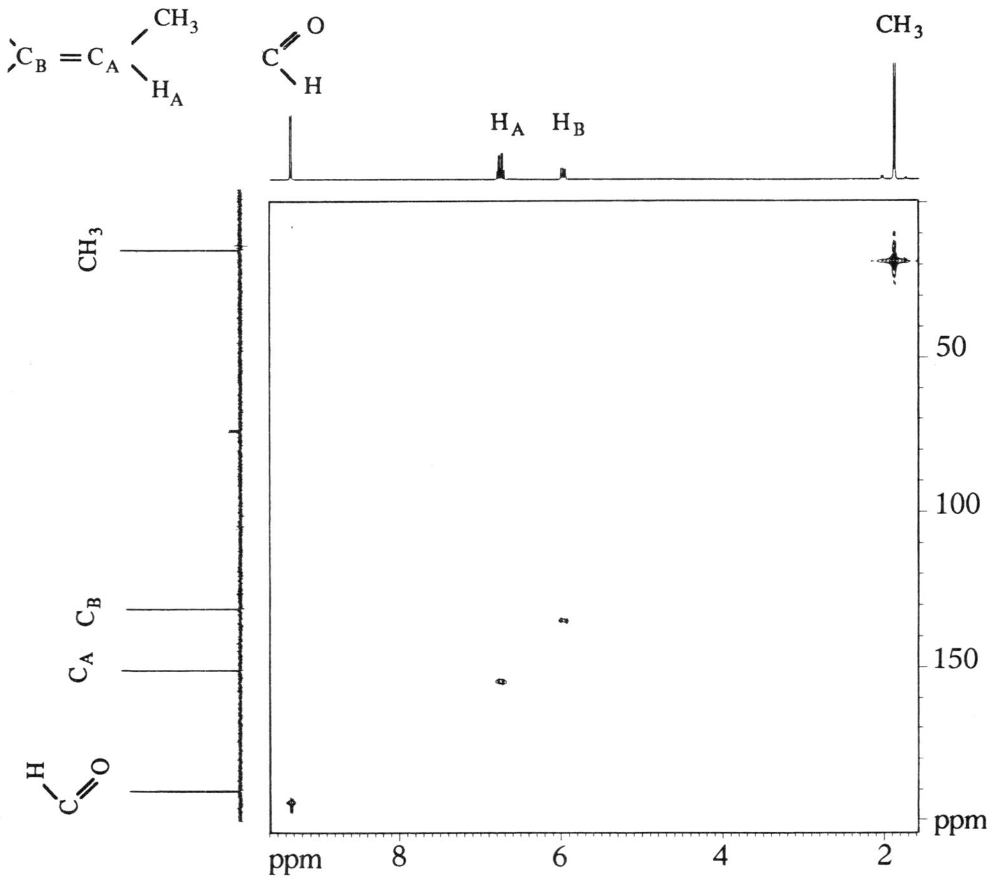

Figure 1.38 $^{1}H-^{13}C$ correlation diagram of *trans*-crotonaldehyde (see Figures 1.36 and 1.34). Spectra on each side of the 2D diagram are the projections of the entire map taken perpendicularly to the considered side; they represent the conventional ^{1}H and ^{13}C spectra respectively. Correlation peaks lead to the identification of protons directly bonded to each carbon

1.3.6 SPECTRAL SIMULATIONS (QUANTUM MECHANICAL APPROACH)

The precise structure of the spectrum corresponding to a system of n spins stems from the knowledge of its hamiltonian eigenvalues and eigenvectors. Because of the chemical shift phenomenon, the Zeeman term, which accounts for the interaction of the n spins with the static magnetic field B_0, must be written as:

$$\hat{\mathcal{H}}_Z = -\sum_{i=1}^{n} \nu_i \hat{I}_z^i \tag{1.17}$$

where $\nu_i = \gamma_i\ (1 - \sigma_i)B_0/2\pi$ stands for the i-th spin resonance frequency. For now, equation (1.17) replaces equation (1.10). This Zeeman term, expressed in Hz, is complemented by a term accounting for indirect coupling (J coupling) within the spin system:

$$\hat{\mathscr{H}}_J = \sum_{i=1, j>i}^{n} J_{ij} \hat{\boldsymbol{I}}_i \hat{\boldsymbol{I}}_j \tag{1.18}$$

The form of $\hat{\mathscr{H}}_J$ could be deduced from a detailed study of the relevant mechanism. (J_{ij} is the coupling constant between spins i and j). It is convenient to express the vectorial operator product as a function of raising and lowering operators (shift operators):

$$\hat{\boldsymbol{I}}_i \hat{\boldsymbol{I}}_j = \hat{I}_z^i \hat{I}_z^j + (\hat{I}_+^i \hat{I}_-^j + \hat{I}_+^i \hat{I}_+^j)/2. \tag{1.19}$$

It is worth recalling in this context that operators relating to different spins commute, regardless of their nature. Moreover, the indirect coupling being *intramolecular* in essence, the spin system includes only those spins belonging to *the same molecule*.

Systems of spin 1/2 nuclei without coupling

The hamiltonian reduces to $\hat{\mathscr{H}}_Z$, which is made of the sum of independent operators, $\nu_i \hat{I}_z^i$, each of them applying to a single spin with α_i or β_i as eigenvectors. Eigenvectors of $\hat{\mathscr{H}}_Z$, labeled by (k), are obtained by writing down all possible products

$$\varphi^{(k)} = \eta_1^{(k)} \eta_2^{(k)} \ldots \eta_i^{(k)} \ldots \eta_n^{(k)} \tag{1.20}$$

with $\eta_i^{(k)} = \alpha_i$ or β_i.

As only two possibilities exist for each η_i, the number of vectors $\varphi^{(k)}$ is 2^n. Those vectors are necessarily normalized because each η_i is itself normalized; they are orthogonal because two distinct $\varphi^{(k)}$ differ by at least one η_i. The set of vectors $\varphi^{(k)}$ therefore constitutes a basis whose elements are sometimes called *simple spin products*.

The eigenvalue $E^{(k)}$ associated with $\varphi^{(k)}$, which represents also the energy of the eigenstate (k), is derived by applying the hamiltonian to $\varphi^{(k)}$:

$$\hat{\mathscr{H}}_Z \varphi^{(k)} = -\sum_i \nu_i \hat{I}_z^i \left(\prod_{i'} \eta_{i'}^{(k)} \right) = -\sum_i \nu_i (\hat{I}_z^i \eta_i^{(k)}) \left(\prod_{i' \neq i} \eta_{i'}^{(k)} \right)$$

This result comes from the fact that $\hat{I}_z^i$ acts only on $\eta_i^{(k)}$ and leaves unchanged any other vector $\eta_{i'}$ for which $i' \neq i$. Since $\hat{I}_z^i \eta_i^{(k)} = (1/2)\eta_i^{(k)}$ or $(-1/2)\eta_i^{(k)}$ for $\eta_i^{(k)} = \alpha_i$ or β_i, respectively, this yields

$$\hat{\mathscr{H}}_Z \varphi^{(k)} = E^{(k)} \varphi^{(k)}$$

where

$$E^{(k)} = -\sum_i \varepsilon_i^{(k)} \nu_i/2 \tag{1.21}$$

with $\varepsilon_i^{(k)} = 1$ if $\eta_i^{(k)} = \alpha_i$ and $\varepsilon_i^{(k)} = -1$ if $\eta_i^{(k)} = \beta_i$.

The energy level diagram of a system of two spin 1/2 nuclei is schematized in Figure 1.39.

Extending the considerations of Section 1.2.2 concerning selection rules, we can evaluate the intensity of a transition occurring between two eigenstates characterized by the eigenvectors $\varphi^{(k)}$ and $\varphi^{(k')}$

$$L_{kk'} = |\langle \varphi^{(k)} | \hat{F}_+ | \varphi^{(k')} \rangle|^2 \tag{1.22}$$

In (1.22), a scaling factor of instrumental origin has been omitted because it is the

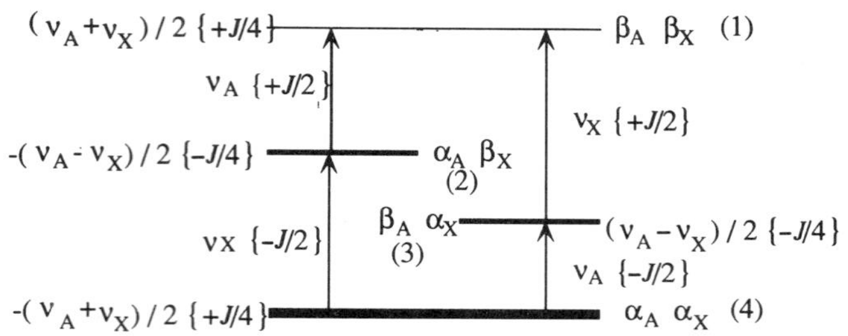

Figure 1.39 Energy level diagram and allowed transitions for a system of two spin 1/2 nuclei. Between braces: modifications of energy values and transition frequencies if the two spins A and X are J coupled (J assumed to be small with respect to $|\nu_A - \nu_X|$)

same for all lines in the spectrum, and the 'total' raising operator, which corresponds to the whole set of spins in the system considered, has been introduced:

$$\hat{F}_+ = \sum_i \hat{I}^i_+. \tag{1.23}$$

Since the set $\varphi^{(k)}$ is a basis, $L_{kk'}$ is non-zero provided that the vector obtained after the application of $\hat{F}_+$ on $\varphi^{(k')}$ is proportional (or involves a term which is proportional) to $\varphi^{(k)}$. Owing to the structure of $\hat{F}_+$, this occurs at the condition that $\varphi^{(k)}$ and $\varphi^{(k')}$ differ by only a single η_i, for instance $\varphi^{(k)} = \eta_1\eta_2 \ldots \alpha_i \ldots \eta_n$ and $\varphi^{(k')} = \eta_1\eta_2 \ldots \beta_i \ldots \eta_n$. With this selection rule, the transition frequency can be calculated by subtracting the energies corresponding to $\varphi^{(k)}$ and $\varphi^{(k')}$ respectively. The result is expressed in Hz because the hamiltonian was at the outset expressed in Hz. We obtain:

$$E^{(k')} - E^{(k)} = \left[-\sum_{i'\neq i} \varepsilon_{i'}^{(k')}\nu_{i'}/2 - (-\nu_i/2)\right] - \left[-\sum_{i'\neq i} \varepsilon_{i'}^{(k)}\nu_{i'}/2 - (\nu_i/2)\right] = \nu_i$$

Thus, a transition at frequency ν_i can be induced between two states which differ only by a spin i. The structure of vectors $\varphi^{(k)}$ and of the operator $\hat{F}_+$ [see (1.22) and (1.23)] leads to the conclusion that transitions of identical intensities occur at frequencies ν_i, ν_i' etc. For instance, an AX two-spin system involves four transitions, two at the frequency ν_A and two at the frequency ν_X (Figure 1.39).

Weakly coupled spin systems

As soon as two (or more) nuclei are coupled, the problem gets more complicated because of the adjunction of $\hat{\mathcal{H}}_J$ to the Zeeman hamiltonian. $\hat{\mathcal{H}}_J$ is very weak with respect to the Zeeman term since, for usual values of the magnetic field B_0, resonance frequencies ν_i are of the order of tens or thousands of MHz whereas coupling constants J are of the order of some Hz or, at most, some tens of Hz. As a consequence, $\hat{\mathcal{H}}_J$ can be considered as a *perturbation*, according to the relevant method frequently used in quantum mechanics. This method allows one to calculate the corrections to apply to the eigenvalues and eigenvectors of the main part of the Hamiltonian, here $\hat{\mathcal{H}}_Z$. These corrections arise from a much smaller term, here $\hat{\mathcal{H}}_J$, and are evaluated by means of an expansion generally limited to second order; thus, we shall retain in $\hat{\mathcal{H}}_J$ only those terms which lead to a non-zero first order and/or second order correction. Referring to the perturbation method (detailed in any textbook on quantum mechanics), the first order correction, $E^{(k)}_{(1)}$ to the eigenvalue $E^{(k)}$ of the Zeeman hamiltonian can be calculated from $\hat{\mathcal{H}}_J$ and from the eigenvector $\varphi^{(k)}$ associated with $E^{(k)}$:

$$E_1^{(k)} = \langle \varphi^{(k)} | \hat{\mathcal{H}}_J | \varphi^{(k)} \rangle$$

Because the vectors $\varphi^{(k)}$ are simple spin products, it is clear that the only term in $\hat{\mathcal{H}}_J$ yielding a non-zero result is $\sum_{i,j>i} J_{ij} \hat{I}^i_z \hat{I}^j_z$. Regarding the second order correction, which can be written as,

$$E_2^{(k)} = \sum_{k' \neq k} \frac{|\langle \varphi^{(k)} | \hat{\mathcal{H}}_J | \varphi^{(k')} \rangle|^2}{E^{(k)} - E^{(k')}}$$

it appears that the complementary part of $\hat{\mathcal{H}}_J$, i.e. $\sum_{i,j>i} J_{ij}\ (\hat{I}^i_+ \hat{I}^j_- + \hat{I}^i_- \hat{I}^j_+)/2$ can contribute to its numerator. This contribution arises from the term involving J_{ij} if the two vectors $\varphi^{(k)}$ and $\varphi^{(k')}$ differ only by η_i and η_j; in that case, and from (1.19), the denominator of $E_2^{(k)}$ is, equal to $\pm(\nu_i - \nu_j)$. It is obvious that whenever $|\nu_i - \nu_j|$ is much greater than J_{ij}, this contribution becomes negligible and that $J_{ij}(\hat{I}^i_+ \hat{I}^j_- + \hat{I}^i_- \hat{I}^j_+)/2$ can be removed from $\hat{\mathcal{H}}_J$. In such a situation, nuclei i and j are said to be *weakly coupled* or to be relevant to the so-called X-*approximation*, a nomenclature according which two nuclei denoted by A and X can be considered as having a large difference in their resonance frequency (with respect to their coupling constant J_{AX}). A further related terminology, *first order spectra*, reminds us that only the first order correction to the energy (in the sense of the perturbation method) is considered. Within such an approximation, the hamiltonian of two weakly coupled nuclei A and X can be written as:

$$\hat{\mathcal{H}} = -\nu_A \hat{I}^A_z - \nu_X \hat{I}^X_z + J_{AX} \hat{I}^A_z \hat{I}^X_z \tag{1.24}$$

It can be noted that the vectors $\varphi^{(k)}$ are still eigenvectors of $\hat{\mathcal{H}}$, defined by (1.24). The extension to systems including more than two spins is straightforward. Eigenvalues are computed without any difficulty by applying the operators $\hat{I}^i_z$ to vectors $\varphi^{(k)}$. They are indicated, as an example, in the diagram 1.39 for the case of a two-spin system. Selection rules are unchanged with respect to a system without coupling. However, transition frequencies become distinct, and for a two-spin system we obtain two doublets, exhibiting a splitting equal to J_{AX}, centered respectively on ν_A and ν_X. This supports the qualitative approach developed in Section 1.3.3, on which it can be reemphasized that it is meaningful only within the context of weakly coupled systems.

Strongly coupled spin systems

We are dealing here with systems for which the difference in resonance frequencies is of the same order of magnitude as their coupling constant. The coupling hamiltonian $\hat{\mathcal{H}}_J$ must be accounted for in its totality and the vectors $\varphi^{(k)}$ cannot any longer be eigenvectors of the full hamiltonian $\hat{\mathcal{H}} = \hat{\mathcal{H}}_Z + \hat{\mathcal{H}}_J$. The most convenient method consists in constructing on an appropriate basis the matrix [H] associated with $\hat{\mathcal{H}}$ and in looking for its eigenvectors and eigenvalues. A given eigenvalue represents the energy of one eigenstate, the corresponding eigenvector of the hamiltonian being obtained by writing down a linear combination of the basis vectors whose coefficients are just the components of the relevant eigenvector of the matrix [H]. The set of vectors $\varphi^{(k)}$ constitutes an adequate basis since it is an orthonormalized system made of the eigenvectors of $\hat{\mathcal{H}}_Z$. We can simplify the description of [H] by relying on the properties of the operator $\hat{F}_z$, associated with the z component of the total spin momentum:

$$\hat{F}_z = \sum_i \hat{I}_z^i \tag{1.25}$$

This latter operator possesses at least two interesting properties. From (1.8) and (1.9), it can be easily shown that it commutes with the total hamiltonian:

$$[\hat{\mathcal{H}}, \hat{F}_z] = 0 \tag{1.26}$$

Moreover, each vector $\varphi^{(k)}$ is an eigenvector of $\hat{F}_z$:

$$\hat{F}_z \varphi^{(k)} = \left(\tfrac{1}{2}\sum_i \varepsilon_i\right)\varphi^{(k)} = M^{(k)}\varphi^{(k)} \tag{1.27}$$

Several vectors $\varphi^{(k)}$ may correspond to the same eigenvalue $M^{(k)}$: for instance, in the case of a two-spin system, the vectors $\alpha_1\beta_2$ and $\beta_1\alpha_2$ are both eigenvectors of $\hat{F}_z$ with 0 as eigenvalue. The (k, k′) element of the matrix [H] is calculated through the scalar product:

$$H_{kk'} = \langle \varphi^{(k)} | \hat{\mathcal{H}} | \varphi^{(k')} \rangle \tag{1.28}$$

Using (1.26) and (1.27) and the hermiticity of $\hat{F}_z$, we may write:

$$\langle \varphi^{(k)} | \hat{\mathcal{H}}\hat{F}_z - \hat{F}_z\hat{\mathcal{H}} | \varphi^{(k')} \rangle = (M^{(k')} - M^{(k)})\langle \varphi^{(k)} | \hat{\mathcal{H}} | \varphi^{(k')} \rangle = 0$$

An immediate consequence is that $H_{kk'}$ is zero, provided that $M^{(k)}$ and $M^{(k')}$ are different. It can therefore be predicted that, if the basis $\{\varphi^{(k)}\}$ is ordered according to the $\hat{F}_z$ eigenvalues, the matrix [H] will be factorized into independent blocks along the main diagonal. This property is illustrated in Figure 1.40, where the [H] matrices of a system of two spin 1/2 nuclei and of a system of three spin 1/2 nuclei are depicted schematically. These matrices are constructed on the basis of simple spin products, with subscripts omitted (this will be done throughout: for instance, $\alpha\beta$ stands for $\alpha_1\beta_2$).

Energies (eigenvalues) and eigenvectors are derived through the diagonalization of each block. Let us denote E_K one of the eigenvalues and by Ψ_K the corresponding eigenfunction; we can write:

$$[H] = \begin{bmatrix} * & 0 & 0 & 0 \\ 0 & * & * & 0 \\ 0 & * & * & 0 \\ 0 & 0 & 0 & * \end{bmatrix} \begin{matrix} \beta\beta & -1 \\ \alpha\beta & 0 \\ \beta\alpha & 0 \\ \alpha\alpha & 1 \end{matrix} \qquad (M)$$

$$[H] = \begin{bmatrix} * & 0 & 0 & 0 & 0 & 0 & 0 & 0 \\ 0 & * & * & * & 0 & 0 & 0 & 0 \\ 0 & * & * & * & 0 & 0 & 0 & 0 \\ 0 & * & * & * & 0 & 0 & 0 & 0 \\ 0 & 0 & 0 & 0 & * & * & * & 0 \\ 0 & 0 & 0 & 0 & * & * & * & 0 \\ 0 & 0 & 0 & 0 & * & * & * & 0 \\ 0 & 0 & 0 & 0 & 0 & 0 & 0 & * \end{bmatrix} \begin{matrix} \beta\beta\beta & -3/2 \\ \beta\beta\alpha & -1/2 \\ \beta\alpha\beta & -1/2 \\ \alpha\beta\beta & -1/2 \\ \alpha\alpha\beta & 1/2 \\ \alpha\beta\alpha & 1/2 \\ \beta\alpha\alpha & 1/2 \\ \alpha\alpha\alpha & 3/2 \end{matrix} \qquad (M)$$

Figure 1.40 Structure of matrices associated with the hamiltonian constructed on a simple spin-product basis for a system of two spin 1/2 nuclei (left) and three spin 1/2 nuclei (right). Asterisks indicate non-zero elements. Columns labeled by M indicate eigenvalues of the operator $\hat{F}_z$

$$\hat{\mathscr{H}}\Psi_K = E_K \Psi_K \quad (1.29)$$

$$\Psi_K = \sum_{k \in M} c_{kK} \varphi^{(k)} \quad (1.30)$$

The coefficients c_{kK} are the eigenvector components, the symbol $\sum_{k \varepsilon M}$ meaning that the summation is limited to those vectors $\varphi^{(k)}$ associated with the eigenvalue M of $\hat{F}_z$. The transition probability between two energy levels E_K and $E_{K'}$ is derived along the same lines as in (1.22):

$$L_{KK'} = |\langle \Psi_K | \hat{F}_+ | \Psi_{K'} \rangle|^2$$

with

$$\hat{F}_+ |\Psi_{K'}\rangle = \sum_{k' \in M'} c_{k'K'} (\hat{F}_+ |\varphi^{(k')}\rangle) \quad (1.31)$$

$L_{KK'}$ is non-zero provided that $\hat{F}_+|\Psi_{K'}\rangle$ is proportional to at least one of the vectors $\varphi^{(k)}$ appearing in the expansion of Ψ_K. As stated above, a necessary condition (but not sufficient) is that $\Delta M = M - M' = 1$. Therefore, and this is the *minimal selection rule*, a transition can only occur between states arising from two consecutive blocks of [H], in other words between states corresponding to two consecutive eigenvalues of $\hat{F}_z$. Then the transition probability can be calculated according to the following expression:

$$L_{KK'} = |\sum_{k \in M} \sum_{k' \in M-1} c_{kK} c_{k'K'} \langle \varphi^{(k)} | \hat{F}_+ | \varphi^{(k')} \rangle|^2 \quad (1.32)$$

As an example, we now detail the calculations leading to the calculation of the spectrum of two strongly coupled spin 1/2 nuclei (an AB spin system according to the usual nomenclature). The hamiltonian has the form:

$$\hat{\mathscr{H}} = -\nu_A \hat{I}_z^A - \nu_B \hat{I}_z^B - J[\hat{I}_z^A \hat{I}_z^B + (\hat{I}_+^A \hat{I}_-^B + \hat{I}_-^A \hat{I}_+^B)/2] \quad (1.33)$$

The (k, k′) element of matrix [H] is deduced from the scalar products $\langle \varphi^{(k)} | \hat{\mathscr{H}} | \varphi^{(k')} \rangle$, where $\varphi^{(k)}$ and $\varphi^{(k')}$ represent the basis functions: $\varphi^{(1)} = \beta\beta$, $\varphi^{(2)} = \alpha\beta$, $\varphi^{(3)} = \beta\alpha$ and $\varphi^{(4)} = \alpha\alpha$ (Figure 1.40). These scalar products can be calculated by relying on the way in which spin operators act [equations (1.4) and (1.7)] and by acknowledging the fact that basis vectors are normalized and orthogonal. Regarding the two extreme blocks of [H], they are of dimension (1, 1), thus trivially in a diagonal form. Hence, the energy is equal to the single element of the considered block, and one obtains the same results as for a first order (weakly coupled) system:

$$E_1 = (\nu_A + \nu_B)/2 + J/4,\ \Psi_1 = \beta\beta \text{ and } E_4 = -(\nu_A + \nu_B)/2 + J/4,\ \Psi_4 = \alpha\alpha$$

The central block is a submatrix of dimension (2, 2)

$$\begin{bmatrix} -(\nu_A - \nu_B)/2 - J/4 & J/2 \\ J/2 & (\nu_A - \nu_B)/2 - J/4 \end{bmatrix}$$

whose eigenvalues and eigenfunctions are calculated in a straightforward manner:

$$E_2 = -[(\nu_A - \nu_B)/2]\sqrt{1 + u^2} - J/4, \qquad \Psi_2 = C\alpha\beta - \sqrt{1 - C^2}\,\beta\alpha$$

$$E_3 = [(\nu_A - \nu_B)/2]\sqrt{1 + u^2} - J/4, \qquad \Psi_3 = \sqrt{1 - C^2}\,\alpha\beta + C\beta\alpha$$

with $u = J/(\nu_A - \nu_B)$ and

$$C = 1\Bigg/\sqrt{1 + \frac{[1 - \sqrt{1 + u^2}]^2}{u^2}}$$

C has purposely been written in a somewhat complicated manner which makes it possible to follow the evolution of eigenvalues and eigenvectors when u bcomes small with respect to unity, that is, when the strongly coupled spin system AB tends toward the weakly coupled spin system AX. In that latter situation, C can be approximated by $1 - u^2/8$ and, if u becomes unity, C tends to unity, leading to the eigenvalues and eigenfunctions of an AX spin system. The transition frequencies and their corresponding intensities (calculated with (1.32)) are gathered in Table 1.5, according to the labeling of energy levels given in Figure 1.39.

It can be observed that the spectrum includes two unsymmetrical doublets (lines within each doublet are not of identical intensities) whose centers are shifted with respect to ν_A and ν_B, respectively, but whose splitting is still equal to J. This latter feature prevails only for a two-spin system and is not of general relevance. The spectrum of an AB case is schematized in Figure 1.41. It can be verified that the two types of distortions (shift with respect to resonance frequencies and intensity modifications) increase with u, that is, when second order effects become more important. These considerations support the qualitative 'roof top effect' mentioned in Section 1.3.4 and in Figure 1.28.

Beyond two strongly coupled spins, one must rely on numerical treatments, as the dimension of some submatrices exceeds two. This can however be avoided if the spin system involves only *two* strongly coupled nuclei or if it includes sets of equivalent nuclei; this latter situation is the subject of the following section.

Table 1.5 Transition frequencies and corresponding intensities for a system of two strongly coupled spin 1/2 nuclei

Transition	Frequency	Intensity
$2 \to 1$	$\nu_A[1 + \sqrt{1 + u^2}]/2 + \nu_B[1 - \sqrt{1 + u^2}]/2 + J/2$	$1 - 2C\sqrt{1 - C^2}$
$3 \to 1$	$\nu_A[1 - \sqrt{1 + u^2}]/2 + \nu_B[1 + \sqrt{1 + u^2}]/2 + J/2$	$1 + 2C\sqrt{1 - C^2}$
$4 \to 2$	$\nu_A[1 - \sqrt{1 + u^2}]/2 + \nu_B[1 + \sqrt{1 + u^2}]/2 - J/2$	$1 - 2C\sqrt{1 - C^2}$
$4 \to 3$	$\nu_A[1 + \sqrt{1 + u^2}]/2 + \nu_B[1 - \sqrt{1 + u^2}]/2 - J/2$	$1 + 2C\sqrt{1 - C^2}$

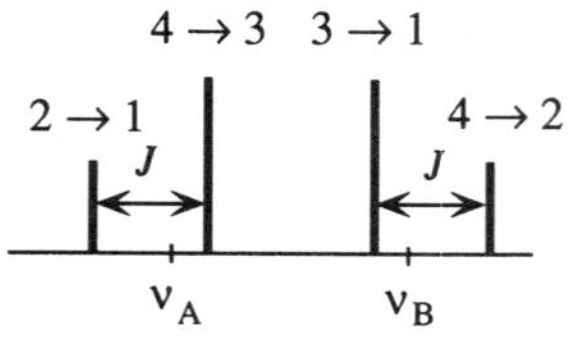

Figure 1.41 Schematic representation of the spectrum whose line frequencies and intensities are given in Table 1.5

Spin systems with equivalence

Considerations of symmetry, stemming from group theory, lead to considerable simplifications in the calculation (simulation) of spectra whenever the considered spin system involves *simple equivalence*, characterized by subset of nuclei possessing the same chemical shift. This is a somewhat sophisticated treatment which is beyond the scope of this introductive presentation and which can be found in specialized textbooks. We shall only consider here spin systems which can be decomposed into subsets of *magnetically equivalent nuclei*. Let us recall that n_A spins are said to constitute a subset A of magnetically equivalent nuclei if $\nu_i = \nu_A$, for $\forall i \in A$, and $J_{ik} = J_{Ak}$, for $\forall i \in A$ and for any spin $k \notin A$. As a whole, the spin system will be denoted by $A_{n_A}B_{n_B}C_{n_C} \ldots$, each of the subsets A, B, C ... possessing the latter properties. Of course, n_A, n_B or n_C etc. may be equal to 1 and the X approximation can possibly be accounted for.

We shall first demonstrate that couplings within subset A are not involved in the spectrum. Let $\hat{\mathcal{H}}_{J_A}$ be the relevant term in the total hamiltonian $\hat{\mathcal{H}} = \hat{\mathcal{H}}_{J_A} + \hat{\mathcal{H}}'$:

$$\hat{\mathcal{H}}_{J_A} = \sum_{i<j, j\in A} J_{ij}[\hat{I}_z^i \hat{I}_z^j + (\hat{I}_+^i \hat{I}_-^j + \hat{I}_-^i \hat{I}_+^j)/2] \tag{1.34a}$$

By reference to commutation rules (1.8) and (1.9), it can be shown that:

$$[\hat{\mathcal{H}}_{J_A}, \hat{\mathcal{H}}] = 0 \tag{1.34b}$$

$$[\hat{\mathcal{H}}_{J_A}, \hat{F}_+] = 0 \tag{1.34c}$$

$\hat{F}_+$ standing for the raising operator of the *whole* spin system. Because of the commutation property (1.34b), it is possible to devise eigenvectors $\psi^{(k)}$, common to $\hat{\mathcal{H}}_{J_A}$ and $\hat{\mathcal{H}}$. As a consequence, the following equation holds:

$$(\hat{\mathcal{H}}' + \hat{\mathcal{H}}_{J_A})\psi^{(k)} = (E'^{(k)} + h^{(k)})\psi^{(k)}$$

where $E'^{(k)}$ and $h^{(k)}$ are the eigenvalues of $\hat{\mathcal{H}}'$ and $\hat{\mathcal{H}}_{J_A}$, respectively. From the hermiticity of $\hat{\mathcal{H}}_{J_A}$ and from (1.34c), we can write:

$$\langle \psi^{(k)} | \hat{\mathcal{H}}_{J_A}\hat{F}_+ - \hat{F}_+\hat{\mathcal{H}}_{J_A} | \psi^{(k')} \rangle = 0$$

$$(h^{(k)} - h^{(k')})\langle \psi^{(k)} | \hat{F}_+ | \psi^{(k')} \rangle = 0$$

Since $\hat{F}_+$ governs the transition probability between states defined by vectors $\psi^{(k)}$ and $\psi^{(k')}$, it turns out that this transition can occur if $h^{(k)} = h^{(k')}$ (this condition is necessary but not sufficient). Consequently, transitions exist only within a subspace defined by a given value of $h^{(k)}$. This latter quantity is subtracted (and disappears) when transition frequencies are calculated (differences between two energy values). $\hat{\mathcal{H}}_{J_A}$ has no influence on the spectrum and can therefore be omitted in the hamiltonian which will be now written as:

$$\hat{\mathcal{H}} = -\sum_A \nu_A \hat{\boldsymbol{I}}_z^A + \sum_{A,B} J_{AB}[\hat{\boldsymbol{I}}_z^A \hat{\boldsymbol{I}}_z^B + (\hat{\boldsymbol{I}}_+^A \hat{\boldsymbol{I}}_-^B + \hat{\boldsymbol{I}}_-^A \hat{\boldsymbol{I}}_+^B)/2] \tag{1.35}$$

Bold-face operators refer to each subset of magnetically equivalent nuclei (for instance $\hat{\boldsymbol{I}}_z^A = \sum_{i\in A} \hat{I}_z^i$), whereas J_{AB} is the coupling constant between two spins of subsets A and B.

We shall now consider the subset of n_A nuclei of spin 1/2 as a *composite particle* whose characteristics will be derived from the properties of the two operators $\hat{\boldsymbol{I}}_z^A$ (which has been defined above) and

$$\hat{\boldsymbol{I}}_A^2 = \left(\sum_{i\in A} \hat{\boldsymbol{I}}_i\right)^2$$

(remember that these latter operators are vectorial quantities). As these operators are actually associated with the total spin momentum of the subset A, they necessarily commute. It is therefore possible to devise a common set of eigenvectors to these two operators which will be denoted by $|\boldsymbol{I}_A, \boldsymbol{M}_A\rangle$ and which satisfy to the following equations by virtue of the general relationships of spin operators [relationships (1.2) and (1.3)]:

$$\hat{I}_z^{A}|\boldsymbol{I}_A, \boldsymbol{M}_A\rangle = \boldsymbol{M}_A|\boldsymbol{I}_A, \boldsymbol{M}_A\rangle$$
$$\hat{I}_A^2|\boldsymbol{I}_A, \boldsymbol{M}_A\rangle = \boldsymbol{I}_A(\boldsymbol{I}_A + 1)|\boldsymbol{I}_A, \boldsymbol{M}_A\rangle \tag{1.36}$$

where $\boldsymbol{I}_A$ is an integer or a halfinteger, the possible values of $\boldsymbol{M}_A$ running between $-\boldsymbol{I}_A$ and $\boldsymbol{I}_A$ by steps of 1. Since the maximum value of $\boldsymbol{M}_A$ is obtained in a situation where all $\hat{I}_z^i$ have the eigenvalue $+1/2$, and since this corresponds to the maximum possible value of $\boldsymbol{I}_A$ (because $-\boldsymbol{I}_A \leqslant \boldsymbol{M}_A \leqslant \boldsymbol{I}_A$), this yields:

$$(\boldsymbol{I}_A)_{\max} = n_A/2$$

which is actually an integer or half an integer according to the parity of n_A. The possible values of $\boldsymbol{I}_A$ can then be listed as indicated below:

$$n_A/2;\ n_A/2 - 1;\ n_A/2 - 2;\ \ldots;\ n_A/2 - r_A;\ \ldots \geqslant 0.$$

From the general theory of angular momentum, it turns out that in the case of a composite particle, all these values must be considered, by contrast with a single particle for which a unique value of I has to be dealt with. Each of the $\boldsymbol{I}_A$ values defines a subspace, and we will show that the total hamiltonian can be factorized according to these subspaces. Referring to relationships (1.8) and (1.9), we can notice that $\hat{I}_A^2$ commutes with $\hat{\mathcal{H}}$ and $\hat{F}_+$. As before, about $\hat{\mathcal{H}}_{J_A}$, we conclude that it is possible to devise eigenvectors shared by $\hat{\mathcal{H}}$ and $\hat{I}_A^2$, and that transitions are forbidden between eigenstates characterized by distinct eigenvalues of $\hat{I}_A^2$. It is thus possible to define as many subspaces as there exist different combinations of $\boldsymbol{I}_A$, $\boldsymbol{I}_B$, $\boldsymbol{I}_C$ .... Each subspace is handled independently from the others; this feature is the key to simplifications which are provided by the composite particle concept when simulating (calculating) NMR spectra involving magnetic equivalence. A last point concerns the degeneracy of each subspace. It can be shown that line intensities, within a given subspace, must be multiplied by the degeneracy factor $g_{I_A,I_B,I_C}\ldots$, given by:

$$g_{I_A,I_B,I_C}\ldots = \prod_A \frac{n_A!(n_A - 2r_A + 1)}{(n_A - r_A + 1)!\,r_A!} \tag{1.37}$$

In practice, one proceeds in the following way: subspaces are first determined by all possible combinations $\boldsymbol{I}_A$, $\boldsymbol{I}_B$, $\boldsymbol{I}_C$ ..., and their degeneracy is calculated via (1.37). This is exemplified in Table 1.6 for spin systems of the A_2B and A_3B_2 types.

Table 1.6 Independent subspaces according to the spin space of A_2B and A_3B_2 systems can be decomposed

System	$\boldsymbol{I}_A$	$\boldsymbol{I}_B$	g
A_2B	1	1/2	1
	0	1/2	1
A_3B_2	3/2	1	1
	3/2	0	1
	1/2	1	2
	1/2	0	2

In a second step, the matrix associated with the hamiltonian is constructed on basis vectors of the type:

$$\varphi^{(k)} = |\boldsymbol{I}_A, \boldsymbol{M}_A; \boldsymbol{I}_B, \boldsymbol{M}_B; \ldots\rangle \tag{1.38}$$

This is simply an extension of the procedure used before for systems without equivalence. Each element of the matrix [H] associated with $\hat{\mathcal{H}}$ can be calculated by means of (1.36) and of:

$$\hat{\boldsymbol{I}}_{\pm}^{A}|\mathbf{I}_A, \boldsymbol{M}_A\rangle = \sqrt{\boldsymbol{I}_A(\boldsymbol{I}_A + 1) - \boldsymbol{M}_A(\boldsymbol{M}_A \pm 1)}\,|\boldsymbol{I}_A, \boldsymbol{M}_A \pm 1\rangle \tag{1.39}$$

This latter expression follows from (1.6) and also yields the transition probabilities by application of (1.32) with $\hat{F}_+ = \sum_A \hat{\boldsymbol{I}}_+^A$.

Moreover, it should be quite clear that the factorization of [H] according to $\hat{F}_z$ eigenvalues is superimposed onto the decomposition into subspaces (Table 1.7) and that the selection rule $\Delta M = 1$ remains valid.

The interest of the composite particle method lies essentially in the reduction of the dimension of matrices which have to be diagonalized. Let us consider for instance the A_2B spin system, which involves three spins; hence the conventional treatment would require to diagonalize two (3, 3) matrices. Thanks to the composite particle decomposition, the matrix of highest rank is of dimension 2. The hamiltonian decomposition (according to subspaces and to $\hat{F}_z$ eigenvalues) is shown in Table 1.7. Also shown in this table are the elements of [H] (diagonal elements are on the same row as the corresponding basis vector). For the sake of completeness, basis vectors are 'translated' into the usual spin functions for which subscripts have again been omitted (for instance, $\beta\beta\beta$ meaning $\beta_A\beta_{A'}\beta_B$).

The whole calculation involves the diagonalization of a (2, 2) matrix. Although this can be performed without any difficulty, we shall limit ourselves to an AX_2 system (such that $|\nu_A - \nu_X| \gg J$) for which the basis vectors are also the hamiltonian eigenvectors, the eigenvalues being identical to the relevant diagonal elements. The energy diagram of an AX_2 system and the corresponding stick spectrum are shown in Figure 1.42. It may be noted that the multiplet structure predicted by the intuitive approach of Section 1.3.4 is effectively retrieved.

Table 1.7 Factorization of the matrix associated with the hamiltonian of an A_2B system according to $\hat{F}_z$ eigenvalues and to the decomposition into two subspaces

Subspace	$\hat{F}_z$ eigenvalues	Basis vectors	[H] elements
$\boldsymbol{I}_A = 1$; $\boldsymbol{I}_B = 1/2$	$-3/2$	(1): $\|1, -1; 1/2, -1/2\rangle = \beta\beta\beta$	$\nu_A + \nu_B/2 + J/2$
	$-1/2$	(2): $\|1, -1; 1/2, 1/2\rangle = \beta\beta\alpha$	$\nu_A - \nu_B/2 - J/2$
	$-1/2$	(3): $\|1, 0; 1/2, -1/2\rangle = (\alpha\beta + \beta\alpha)\beta/\sqrt{2}$	$\nu_B/2$
			$J/\sqrt{2}$ (off-diagonal)
	$1/2$	(4): $\|1, 0; 1/2, 1/2\rangle = (\alpha\beta + \beta\alpha)\alpha/\sqrt{2}$	$-\nu_B/2$
	$1/2$	(5): $\|1, 1; 1/2, -1/2\rangle = \alpha\alpha\beta$	$-\nu_A + \nu_B/2 - J/2$
			$J/\sqrt{2}$ (off-diagonal)
	$3/2$	(6): $\|1, 1; 1/2, 1/2\rangle = \alpha\alpha\alpha$	$-\nu_A - \nu_B/2 + J/2$
$\boldsymbol{I}_A = 0$; $\boldsymbol{I}_B = 1/2$	$-1/2$	(7): $\|0, 0; 1/2, -1/2\rangle = (\alpha\beta - \beta\alpha)\beta/\sqrt{2}$	$\nu_B/2$
	$1/2$	(8): $\|0, 0; 1/2, 1/2\rangle = (\alpha\beta - \beta\alpha)\alpha/\sqrt{2}$	$-\nu_B/2$

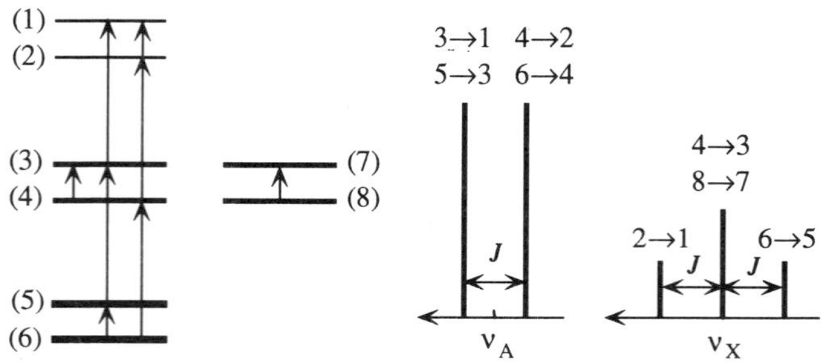

Figure 1.42 Energy diagram, allowed transitions and schematic representation of the spectrum of an A_2X spin system. Refer to Table 1.7 for the labeling of energy levels

1.4 Nuclear Magnetic Resonance in Anisotropic Media

1.4.1 ANISOTROPIC MEDIA AS DEFINED WITH RESPECT TO NMR. THE TENSIONAL NATURE OF INTERACTIONS

Within a liquid, the environment of a given molecule is isotropic *on an average*, meaning that it is the same regardless of the direction considered. Time scales involved in NMR spectroscopy imply that the notion of isotropy, intimately related to that of average, must be stated by reference to interactions undergone by nuclear spins (these interactions manifest themselves either by the shielding effect or by splittings due to coupling with other spins). In order to make this point more clear, let us denote by a and b two orientations that a molecule can adopt. Let T_a and T_b be the values of an interaction T for each of these orientations. T_a and T_b should be different because a preferential orientation exists, that of the induction B_0. T_a and T_b can be expressed in frequency units; if the frequency of exchange between a and b is much larger than $|T_a - T_b|$, it is conceivable that the average of T must be considered instead of its individual values for each orientation. This average indeed prevails in non-viscous liquids, since the frequency of molecular reorientation is of the order of 10^{12} Hz. It can therefore be stated that one is dealing with an *isotropic medium* if any quantity involved in an NMR spectrum is averaged over all possible orientations; this leads to the *two following* conditions.

- In the course of molecular reorientation, all molecular motions must be very fast with respect to the variation of interactions when a molecule switches from one orientation to the other.
- All molecular orientations are allowed to occur with an equal probability.
- On the contrary, it will be decided that one is dealing with an anisotropic medium if one or both conditions do not hold. The two examples given below serve to illustrate this statement.
- Regarding a solid sample (either in the form of a monocrystal or of a powder), the realization of the first condition is quite exceptional, whereas the second condition is very unlikely to occur.

- Organized systems, such as liquid crystals or certain parts of biological tissues, possess, at least locally, one or several preferential directions, so that the second condition cannot be satisfied.

The concept of anisotropy reveals the tensional character of all interactions involved in nuclear magnetic resonance: the shielding effect (through the Zeeman interaction), indirect coupling (J coupling), direct coupling (dipolar coupling), and quadrupolar coupling. In order to be convinced by the tensorial character of any interaction and to understand its manifestations in an anisotropic medium, we shall first consider the shielding effect. A molecule is generally anisotropic by itself (we shall exclude highly symmetrical molecular entities, such as methane). Let us denote by (x, y, z) a molecular frame and let us suppose that the molecule is subjected to a magnetic field B_x oriented along x. The shielding of the magnetic field due to electronic distribution is, in general, anisotropic within a molecule, so that, at the level of a given nuclear spin, the magnetic field $\boldsymbol{B}'$ originating from the applied field B_x possesses three components:

$$B'_x = (1 - \sigma_{xx})B_x$$

$$B'_y = \sigma_{yx}B_x$$

$$B'_z = \sigma_{zx}B_x$$

Writing $(1 - \sigma_{xx})$ rather than σ_{xx} is somewhat formal, and reflects the fact that the shielding effect is very weak and that B_x remains the major component of $\boldsymbol{B}'$.

Now, if a molecule is subjected to a magnetic field of arbitrary direction, represented by its three components B_x, B_y and B_z, the components of the field $\boldsymbol{B}'$, at the level of the considered nuclear spin, are related to those of $\boldsymbol{B}$ by:

$$B'_x = (1 - \sigma_{xx})B_x + \sigma_{xy}B_y + \sigma_{xz}B_z$$

$$B'_y = \sigma_{yx}B_x + (1 - \sigma_{yy})B_y + \sigma_{yz}B_z$$

$$B'_z = \sigma_{zx}B_x + \sigma_{zy}B_y + (1 - \sigma_{zz})B_z$$

or, in matrix form, as:

$$\begin{bmatrix} B'_x \\ B'_y \\ B'_z \end{bmatrix} = \begin{bmatrix} (1 - \sigma_{xx}) & \sigma_{xy} & \sigma_{xz} \\ \sigma_{yx} & (1 - \sigma_{yy}) & \sigma_{yz} \\ \sigma_{zx} & \sigma_{zy} & (1-\sigma_{zz}) \end{bmatrix} \begin{bmatrix} B_x \\ B_y \\ B_z \end{bmatrix} \tag{1.40}$$

Actually, the shielding effect involves nine coefficients instead of the simple shielding constant which prevails on an average in the liquid phase. These nine coefficients constitute a *tensor of rank* 2, associated with the (3, 3) matrix of equation (1.40). For similar reasons, any other interaction affecting nuclear spins should also be tensorial in nature.

1.4.2 INTERACTIONS UNDERGONE BY NUCLEAR SPINS IN AN ANISOTROPIC MEDIUM

So as to figure out the consequences of the tensorial character of the interactions affecting nuclear spins, they must be mapped out and their hamiltonian must be defined. The total hamiltonian governing a spin system can be written according to the following expansion (expressed in Hz, as is usual throughout this book):

$$\hat{\mathcal{H}} = \hat{\mathcal{H}}_Z + \hat{\mathcal{H}}_J + \hat{\mathcal{H}}_D + \hat{\mathcal{H}}_Q \tag{1.41}$$

$\hat{\mathcal{H}}_Z$ is the Zeeman hamiltonian, eventually including shielding effects, whose principal part is:

$$\hat{\mathcal{H}}_0 = -\sum_i \frac{\gamma_i B_0}{2\pi} \hat{I}_z^i$$

$\hat{\mathcal{H}}_0$ is also the major term of $\hat{\mathcal{H}}$, provided that B_0 has a sufficiently large value. This is the case for most usual spectrometers, with some marginal exceptions concerning quadrupolar nuclei possessing large quadrupolar moments such that $\hat{\mathcal{H}}_Q$ may become comparable with $\hat{\mathcal{H}}_0$. $\hat{\mathcal{H}}_J$ and $\hat{\mathcal{H}}_D$ are respectively the indirect coupling hamiltonian (interaction via bonding electrons) and the direct coupling hamiltonian (interaction between the magnetic moments associated with nuclear spins). $\hat{\mathcal{H}}_Q$ is associated with the quadrupolar interaction and relates to nuclei of spin greater than 1/2.

In the following, we will assume that $\hat{\mathcal{H}}_0$ is effectively the dominant term of $\hat{\mathcal{H}}$ and, as before, we shall only retain in the rest of $\hat{\mathcal{H}}$ those terms which lead to a non-zero result at the outset of a perturbation calculation (first order and/or second order). These terms will be called *secular*. In this way, the Zeeman hamiltonian, including the tensor defined in (1.40), reduces to:

$$\hat{\mathcal{H}}_Z = -\sum_i \frac{\gamma i B_0}{2\pi}(1 - \sigma_{ZZ}^i)\hat{I}_z^i \tag{1.42}$$

For a fixed molecule, the shielding effect on nucleus i acts through the quantity $(1 - \sigma_{ZZ}^i)$, where Z is the axis of the laboratory frame collinear with $\boldsymbol{B}_0$.

$\hat{\mathcal{H}}_J$ is expressed as $\boldsymbol{I}_i\, J_{ij}\, \boldsymbol{I}_j$ (J is the indirect coupling tensor); restricted to its secular terms, it can be written as:

$$\hat{\mathcal{H}}_J = \sum_{i<j}[J_{ZZ}^{ij}\hat{I}_z^i\hat{I}_z^j + \tfrac{1}{4}(J_{XX}^{ij} + J_{YY}^{ij})(\hat{I}_+^i\hat{I}_-^j + \hat{I}_-^i\hat{I}_+^j)] \tag{1.43}$$

The second term becomes non-secular if $|\nu_i - \nu_j|$ is much greater than any $J_{\alpha\beta}^{ij}$; (with α, β = X, Y or Z). Thus, the dominant term of (1.43) is the one involving J_{ZZ}^{ij}; the second one, whose manifestation is difficult to detect in an anisotropic medium, is generally disregarded.

The energy accounting for the interaction between two magnetic moments $\boldsymbol{\mu}_i$ and $\boldsymbol{\mu}_j$ (associated with the nuclear spins i and j) is classically expressed as

$$\frac{\boldsymbol{\mu}_i\boldsymbol{\mu}_j}{r_{ij}^3} - \frac{3(\boldsymbol{\mu}_i\boldsymbol{r}_{ij})(\boldsymbol{\mu}_j\boldsymbol{r}_{ij})}{r_{ij}^5}$$

where $\boldsymbol{r}_{ij}$ is the vector joining the two magnetic moments μ_i and μ_j. The hamiltonian is obtained by transforming the classical quantities into their associated operators and dropping non-secular terms:

$$\hat{\mathcal{H}}_D = \sum_{i<j} 2D_{ZZ}^{ij}[(\hat{I}_z^i\hat{I}_z^i - \tfrac{1}{4}(\hat{I}_+^i\hat{I}_-^j + \hat{I}_-^i\hat{I}_+^j)] \tag{1.44}$$

(The factor of 2 has been included for compatibility with the usual notation). Again, the second term in (1.44) becomes non-secular if $|\nu_i - \nu_j| \gg D_{ZZ}^{ij}$. The dipolar hamiltonian is thus directly proportional to the tensorial element D^{ij}_{ZZ}, denoted simply by D_{ij}. In Hz, it is expressed as:

$$D_{ij} = -\left(\frac{\mu_0}{4\pi}\right)\frac{\gamma_i\gamma_j h}{8\pi^2 r_{ij}^3}(3\cos^2\theta_{ij} - 1) \quad (1.45)$$

where θ_{ij} is the angle between $\boldsymbol{r}_{ij}$ and $\boldsymbol{B}_0$. It can be noticed that $\cos^2\theta_{ij}$, which may be regarded as the product of the Z component of $\boldsymbol{r}_{ij}$ by itself, is actually a tensorial element. The factor $\mu_0/4\pi$, where μ_0 is the magnetic permeability of vacuum, has been introduced in conformity with MKSA units.

Finally, the secular form of the quadrupolar hamiltonian for nuclei of spin I_i larger than 1/2 can be expressed as:

$$\mathcal{H}_Q = \sum_i \frac{eQ^i}{4hI^i(2I^i - 1)} V^i_{ZZ}[3(\hat{I}^i_z)^2 - \hat{I}^2_i] \quad (1.46)$$

The magnitude of the quadrupolar interaction is therefore proportional to the following quantity (in Hz)

$$C_{Qi} = \frac{eQ^i V^i_{ZZ}}{4hI^i(2I^i - 1)} \quad (1.47)$$

where (eQ^i) is a constant, named the quadrupolar moment, depending only on the isotope considered and where V^i_{ZZ} stands for the (Z, Z) element of the electric field gradient tensor; in other words, $V^i_{ZZ} = \partial^2 V^i/\partial Z^2$, V^i being the electrical potential originating from the distribution of electrical charges which obeys the Laplace relationship:

$$\frac{\partial^2 V^i}{\partial X^2} + \frac{\partial^2 V^i}{\partial Y^2} + \frac{\partial^2 V^i}{\partial Z^2} = 0.$$

1.4.3 INTERACTIONS EXPRESSED IN A MOLECULAR FRAME. AVERAGING EFFECTS. PREFERENTIAL DIRECTION

The tensorial element T_{ZZ} involved in each of the interactions considered above can eventually be exploited for structural purposes provided it is expressed in a molecular frame. It is appropriate to choose a molecular frame (x, y, z) which diagonalizes the relevant tensor, hence, such that:

$$t_{xy} = t_{yx} = t_{xz} = t_{zx} = t_{xy} = t_{yx} = 0$$

where $t_{\alpha\beta}$ stands for a tensorial element in such a frame (note that lower case letters are used to differentiate the molecular frame from the laboratory frame). Such a molecular frame is called the principal axis system. A second rank tensor transforms actually in the same way as a (3, 3) matrix; this can be written as $[T] = [U^{-1}][t][U]$, [U] being the transformation matrix for switching from the molecular frame (x, y, z) to the laboratory frame (X, Y, Z). It can be recalled that columns of U are made of the components of the unitary vectors, which defines the X, Y, Z axes in the frame (x, y, z) and that $U^{-1}_{ij} = U_{ji}$ (this property comes from the cartesian nature of both frames). According to the conventions of Figure 1.43, we thus obtain:

$$T_{ZZ} = \cos^2\theta_x t_{xx} + \cos^2\theta_y t_{yy} + \cos^2\theta_z t_{zz} \quad (1.48)$$

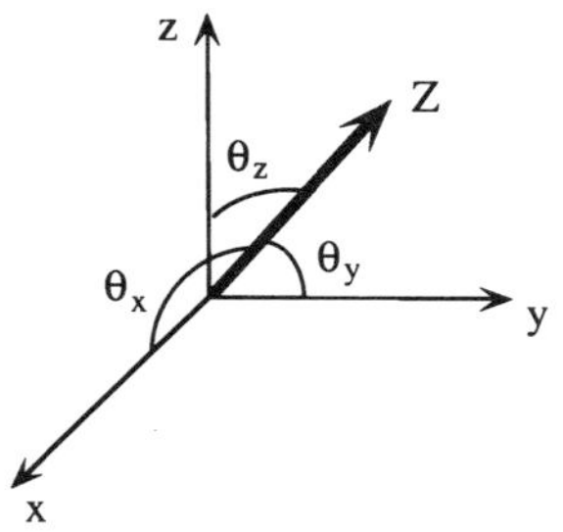

Figure 1.43 The orientation of the fixed Z axis with respect to a molecular frame (x, y, z)

Single crystal

In principle, the elements t_{xx}, t_{yy} and t_{zz} are derived from measurements performed at several orientations of the single crystal with respect to $\boldsymbol{B}_0$, so as to modify $\cos^2\theta_x$, $\cos^2\theta_y$ and $\cos^2\theta_z$. This method is widely used to determine the elements of the shielding tensor.

Average with equal weight over all orientations (isotropic medium)

The measured quantity is:

$$\langle T_{ZZ}\rangle = \langle\cos^2\theta_x\rangle t_{xx} + \langle\cos^2\theta_y\rangle t_{yy} + \langle\cos^2\theta_z\rangle t_{zz} \tag{1.49}$$

If one is dealing with a tensor of axial symmetry ($t_{xx} = t_{yy}$), this can be simplified to

$$\langle T_{ZZ}\rangle = \tfrac{1}{3}Tr(t) + \frac{\Delta t}{3}\langle 3\cos^2\theta_z - 1\rangle \tag{1.50}$$

with $\Delta t = t_{zz} - t_{xx}$.

If all possible orientations share the same probability, then

$$\langle\cos^2\theta_x\rangle = \langle\cos^2\theta_y\rangle = \langle\cos^2\theta_z\rangle = 1/3$$

and, regardless of the tensor symmetry properties, $\langle T_{ZZ}\rangle = (1/3)Tr(t)$, which is also $(1/3)Tr(T)$, since the trace of a matrix is invariant (this property means that the trace is the same regardless of the frame in which the matrix is constructed). (The same result holds for $(\langle T_{XX}\rangle + \langle T_{YY}\rangle)/2$, this latter property being used later). In an isotropic medium, one therefore has access to a single quantity which is the third of the trace of the tensor associated with the considered interaction. This statement is of prime importance for the shielding tensor and for the indirect coupling tensor which are indeed involved in liquid state NMR spectra. Concerning the two other interactions, namely the direct coupling and the quadrupolar interaction, their trace is zero, and therefore do not contribute to NMR spectrum of isotropic media.

Average with respect to preferential direction

We shall assume that the condition of the fast reorientation, which prevails in an isotropic medium, holds here only with respect to a preferential direction denoted by D. This can occur in a natural way in the case of liquid crystals (D is the phase director) or can be artificially created by fast sample spinning. In order to account for the particularities of such a situation in evaluating $\langle T_{ZZ}\rangle$, we must first go from the laboratory frame (X, Y, Z) to another frame (X′, Y′, Z′), whose axis Z′ coincides with D (Figure 1.44).

This is performed with a transformation matrix, which can be constructed as indicated before. This yields:

$$\langle T_{ZZ}\rangle = \frac{\sin^2\psi}{2}Tr(T) + \tfrac{1}{2}(3\cos^2\psi - 1)\langle T_{Z'Z'}\rangle - \sin\psi\cos\psi[\langle T_{X'Z'}\rangle + \langle T_{Z'X'}\rangle] \tag{1.51}$$

where an axial symmetry with respect to D has been assumed; as a further consequence, the following relationships hold

$$\langle T_{X'X'}\rangle = \langle T_{Y'Y'}\rangle = \tfrac{1}{2}(Tr(T) - \langle T_{Z'Z'}\rangle).$$

We then go to a molecular frame (x, y, z), in which all tensors to be considered are not necessarily diagonal (this is because we may have to deal with several interactions, each of them possessing a specific principal axis system). Let [U′] be the transform matrix such that $[T'] = [U']^{-1}[t][U']$ where each column of [U′] contains the components of unitary vectors of the axes X′, Y′, Z′ expressed in the molecular frame (x, y, z). One has:

$$\langle T_{Z'Z'}\rangle = \sum_{i,j}\langle u'_{i3}u'_{j3}\rangle t_{ij}$$

$$\langle T_{X'Z'}\rangle = \langle T_{Z'X'}\rangle = \sum_{i,j}\langle u'_{i1}u'_{j3}\rangle t_{ij}$$

We shall further assume that D is apolar, meaning that D behaves in the same fashion with respect to +Z′ or to −Z′. As a consequence, $u'_{j3}\ u'_{j3}$ and $u'_{i1}(-u'_{j3})$ correspond to the same probability, since u'_{j3} represents the projection of the unitary vector of D onto the molecular axis j ($j = 1, 2, 3$ stands for x, y and z respectively). From these properties, it can be seen that $\langle T_{X'Z'}\rangle$ and $\langle T_{Z'X'}\rangle$ are zero. In replacing u'_{13}, u'_{23} and u'_{33} by the cosine of angles defined in Figure 1.45, we can introduce the so-called Saupe matrix. This is a (3, 3) matrix, whose trace is zero, and whose each element is given by:

$$S_{\alpha\beta} = \tfrac{1}{2}\langle 3\cos\theta_\alpha\cos\theta_\beta - \delta_{\alpha\beta}\rangle \tag{1.52}$$

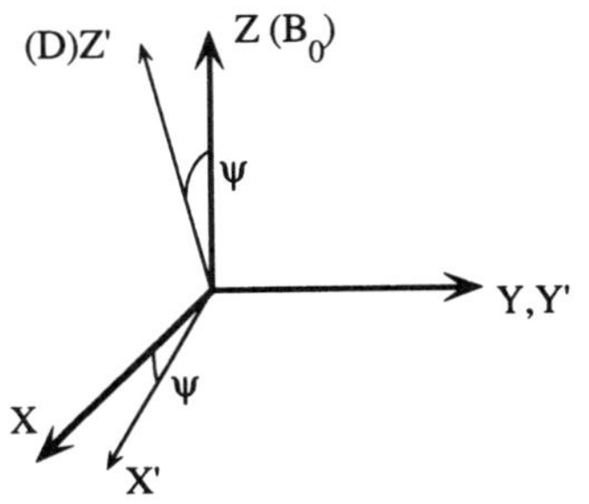

Figure 1.44 Definition of the frame X′, Y′, Z′ (for which Z′ coincides with D), with respect to the usual laboratory frame (X, Y, Z) for which Z coincides with the static field B_0

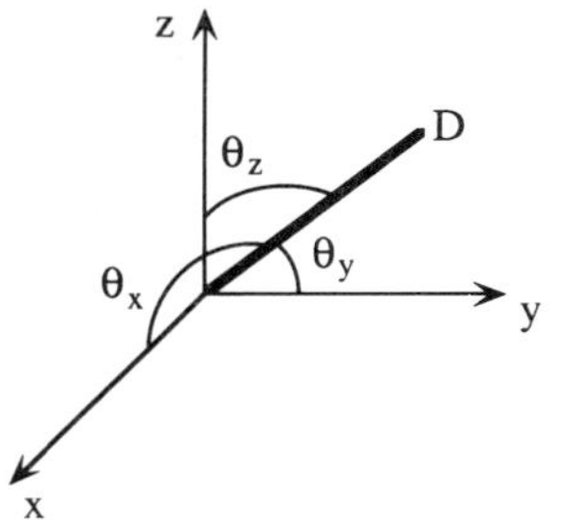

Figure 1.45 Definition of the preferential direction D with respect to a molecular frame (x, y, z)

where α and β stand for x, y or z. Diagonal elements of S are usually called order parameters or orientation parameters. $\langle T_{ZZ} \rangle$ can then be recast as:

$$\langle T_{ZZ} \rangle = \frac{1}{3}\left[Tr(T) + (3\cos^2\psi - 1)\sum_{\alpha,\beta} S_{\alpha\beta}t_{\alpha\beta}\right] \tag{1.53}$$

The same result holds for $1/2(\langle T_{XX} \rangle + \langle T_{YY} \rangle)$, which is involved only in the indirect coupling tensor. It can be noted that an isotropic medium is just a special case of (1.53) for which $S_{\alpha\beta} = 0$ for any α or β. There is another important consequence of equation (1.53); in order to retrieve properties analogous to those of an isotropic medium, $(3\cos^2\psi - 1)$ can be set to zero, which amounts to orienting the preferential direction D at an angle of 54.74° with respect to B_0. This particular value of ψ is dubbed *magic angle*.

1.4.4 STRUCTURE OF SPECTRA IN AN ANISOTROPIC MEDIUM WITH A PEFERENTIAL DIRECTION. SYSTEMS INVOLVING A FINITE NUMBER OF NUCLEI

We shall consider here a sample made of N independent systems (N being very large), each system including n nuclei. In addition, we shall assume that each system behaves in the same manner with respect to the preferential direction. This would be the case for a single crystal in which n interacting nuclei could define a system which would not interact with another identical system. For instance, the system would include nuclear spins of a given molecule, which would necessarily interact, but whose interaction with those of a neighboring molecule is negligibly small because the dipolar coupling depends on $1/r^3$ (r is the distance between two nuclei). This is also the case of oriented molecules within a liquid crystal: owing to fast translational motions, *intermolecular* dipolar interactions are averaged out.

Spectra are simulated either according to the multiplet construction rules for weakly coupled systems or, if the weak coupling approximation cannot be used, by calculating the hamiltonian eigenvalues and eigenvectors. An additional complication arises from dipolar and quadrupolar terms, which were not involved in an isotropic medium. Forthcoming calculations, which involve the orientation matrix, apply also to a single crystal. In that latter case, ψ must be set to zero [equation (1.51)] and the elements $S_{\alpha\beta} = 1/2(3\cos\theta_\alpha \cos\theta_\beta - \delta_{\alpha\beta})$ must be considered as defining the transformation from the laboratory frame (X, Y, Z) to the molecular frame. In a general way and with the help of formula (1.53), which expresses $\langle T_{ZZ} \rangle$ as a function of the elements $S_{\alpha\beta}$, the different terms of the hamiltonian [relations (1.42)–(1.46)] can be written as:

$$\hat{\mathcal{H}}_Z = -\sum_i \nu_i \hat{I}^i_z \tag{1.54}$$

with

$$\nu_i = \frac{\gamma_i B_0}{2\pi}\left[1 - \sigma^i_0 + \tfrac{1}{2}(3\cos^2\psi - 1)\tfrac{2}{3}\sum_{\alpha,\beta} S_{\alpha\beta}\sigma^i_{\alpha\beta}\right] \tag{1.55}$$

where σ^i_0 is the shielding constant in the isotropic medium;

$$\hat{\mathcal{H}}_J = \sum_{i<j}\{J_{ij}[\hat{I}^i_z\hat{I}^j_z + \tfrac{1}{2}(\hat{I}^i_+\hat{I}^j_- + \hat{I}^i_-\hat{I}^j_+] + J^{ij}_{\text{aniso}}[\hat{I}^i_z\hat{I}^j_z - \tfrac{1}{4}(\hat{I}^i_+\hat{I}^j_- + \hat{I}^i_-\hat{I}^j_+)]\} \tag{1.56}$$

J_{ij} being the value of the indirect coupling in the isotropic medium and

$$J^{ij}_{\text{aniso}} = \tfrac{1}{2}(3\cos^2\psi - 1)\tfrac{2}{3}\sum_{\alpha,\beta} S_{\alpha\beta}J^{ij}_{\alpha\beta} \tag{1.57}$$

$$\hat{\mathcal{H}}_D = \sum_{i<j} 2D_{ij}[\hat{I}^i_z\hat{I}^j_z - \tfrac{1}{4}(\hat{I}^i_+\hat{I}^j_- + \hat{I}^i_-\hat{I}^j_+)] \tag{1.58}$$

with

$$D_{ij} = -\left(\frac{\mu_0}{4\pi}\right)\tfrac{1}{2}(3\cos^2\psi - 1)\frac{\gamma_i\gamma_j h}{r^s_{ij}}\sum_{\alpha,\beta} S_{\alpha\beta}x_\alpha x_\beta \tag{1.59}$$

x_α is one of the components of $\boldsymbol{r}_{ij}$ in the molecular frame;

$$\hat{\mathcal{H}}_Q = \sum_i C_{Qi}[3(\hat{I}^i_z)^2 - \hat{\boldsymbol{I}}^2_i] \tag{1.60}$$

$$C_{Qi} = \frac{eQ_i}{4hI^i(2I^i - 1)}\tfrac{1}{2}(3\cos^2\psi - 1)\tfrac{2}{3}\sum_{\alpha,\beta} S_{\alpha\beta}V^i_{\alpha\beta} \tag{1.61}$$

$V^i_{\alpha\beta}$ is an element of the field gradient tensor expressed in the molecular frame.

Systems of spin 1/2 nuclei

By gathering and organizing (1.54), (1.56) and (1.58), we may write the total hamiltonian as:

$$\hat{\mathcal{H}} = -\sum_i \nu_i\hat{I}^i_z + \sum_{i<j} J_{ij}[\hat{I}^i_z\hat{I}^j_z + \tfrac{1}{2}(\hat{I}^i_+\hat{I}^j_- + \hat{I}^i_-\hat{I}^j_+)] + \sum_{i<j}\mathcal{D}_{ij}[\hat{I}^i_z\hat{I}^j_z - \tfrac{1}{4}(\hat{I}^i_+\hat{I}^j_- + \hat{I}^i_-\hat{I}^j_+)] \tag{1.62}$$

where $\mathcal{D}_{ij} = 2D_{ij} + J^{ij}_{\text{aniso}}$.

Whenever the system is weakly coupled ($|\nu_i - \nu_j|$ much greater than any J_{ij} or $\mathcal{D}_{ij}$) the hamiltonian simplifies to:

$$\hat{\mathcal{H}} = -\sum_i \nu_i\hat{I}^i_z + \sum_{i<j}(J_{ij} + \mathcal{D}_{ij})\hat{I}^i_z\hat{I}^j_z \tag{1.63}$$

Therefore a 'first order' spectrum of spin 1/2 nuclei differs from its counterpart in an isotropic medium only by the multiplet splittings; $J + 2D$ is substituted for J (provided that J^{ij}_{aniso} can be neglected, as this is usually the case). Comparison between spectra in isotropic and anisotropic media (Figure 1.46) yields direct couplings and hence the

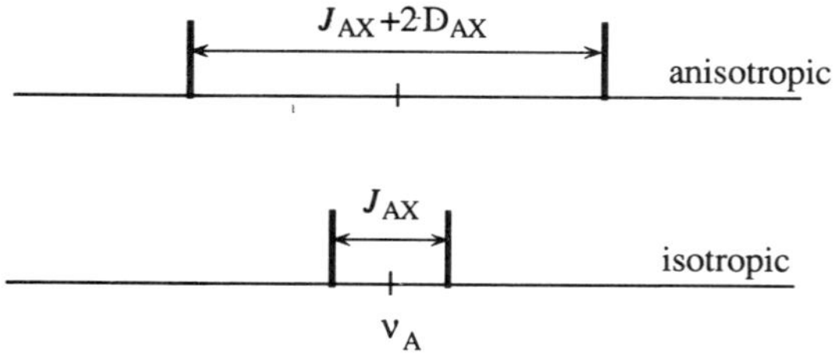

Figure 1.46 The A part of the AX spectrum (two spin 1/2 nuclei) in isotropic and anisotropic media

molecular geometry if order parameters $S_{\alpha\beta}$ are known; this is because these latter parameters enter the expressions of direct couplings. Conversely, ratios of either internuclear distances or angles are available without any additional hypothesis.

From (1.62), it is relatively easy to show that the spectrum of two magnetically equivalent nuclei (A_2 systems) shows up in the form of a doublet of splitting $3D_{AA}$ where D_{AA} refers to the definition given in (1.59) (Figure 1.47).

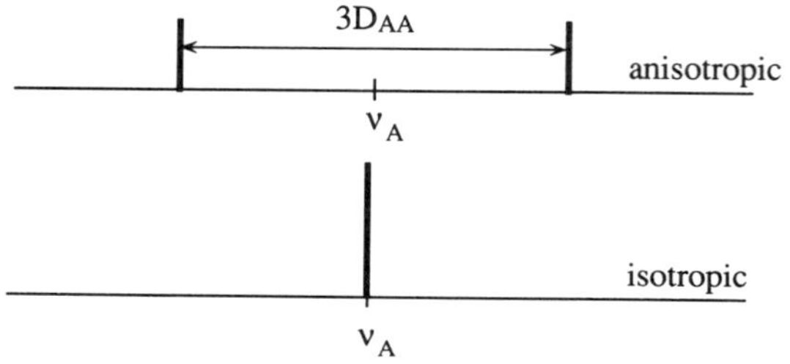

Figure 1.47 Spectra of two magnetically equivalent spin 1/2 nuclei in isotropic and anisotropic media

Quadrupolar nuclei

In order to avoid useless complications, we shall limit ourselves to a system involving a single nucleus of spin I, since the consideration of additional interactions (direct or indirect couplings, most of the time negligible with respect to the quadrupolar interactions) does not involve any particular difficulty. From (1.54) and (1.60), the hamiltonian is of the form

$$\mathscr{H} = -\nu\hat{I}_Z + C_Q[3(\hat{I}_z)^2 - \hat{I}^2]$$

The eigenvectors of this hamiltonian can be denoted as $|I, m\rangle$, with $-I \leqslant m \leqslant I$, corresponding to the eigenvalues $-m\nu + 3C_Q m^2 - C_Q I\ (I+1)$. The usual selection rule $|\Delta m| = 1$ holds and the intensity of the transition $|I, m\rangle \rightarrow |I, m-1\rangle$ is proportional to $|\langle I, m_+|\hat{I}_+|I, m-1\rangle|^2 = (I - m + 1)\ (I + m)$.

Since, in the coupling term of the above hamiltonian, the spin operator $\hat{I}$ acts through its square, lines are displayed symmetrically with respect to ν. Thus, the spectrum of a quadrupolar nucleus shows up in the form of a multiplet of $2I$ lines centered on ν, of splitting $6C_Q$ (the value of C_Q being given by equation (1.61)). The spectrum of a nucleus of spin 1, such as deuterium (Figure 1.48), is a doublet, and the

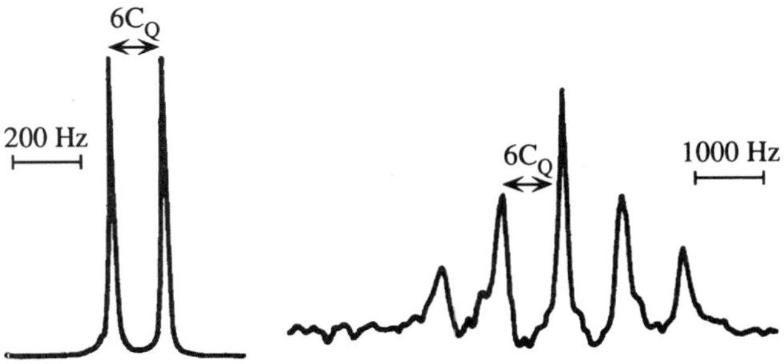

Figure 1.48 Deuterium (left) and oxygen-17 (right) spectra of heavy water in a lyotropic liquid crystal macroscopically oriented

spectrum of a spin 3/2 nucleus is a triplet whose intensities are in the ratio 3:4:3, and the spectrum of a nucleus of spin 5/2 is a quintet whose intensities are in the ratios 5:8:9:8:5 (Figure 1.48).

1.4.5 STRUCTURE OF SPECTRA IN ANISOTROPIC MEDIA WITHOUT PREFERENTIAL DIRECTION (POWDER SPECTRA)

We shall consider only two simple cases: (i) a single spin 1/2 nucleus subjected to the sole Zeeman term with so-called chemical shift anisotropy (due to the shielding tensor), for which couplings are negligible or have been removed by application of a decoupling radio-frequency field; (ii) two spin 1/2 nuclei, magnetically equivalent, with negligible chemical shift anisotropy but interacting by dipolar or direct coupling (the spectrum of a spin 1 nucleus with an axially symmetric field gradient tensor is formally equivalent to that of the latter system; this equivalence rests on the substitution of the quadrupolar interaction by the dipolar interaction).

Chemical shift anisotropy

Let us consider the molecular frame (x, y, z) in which the shielding tensor is diagonal, and let us denote by θ_x, θ_y and θ_z the angles which define the orientation of $\boldsymbol{B}_0$ with respect to this frame (Figure 1.49).

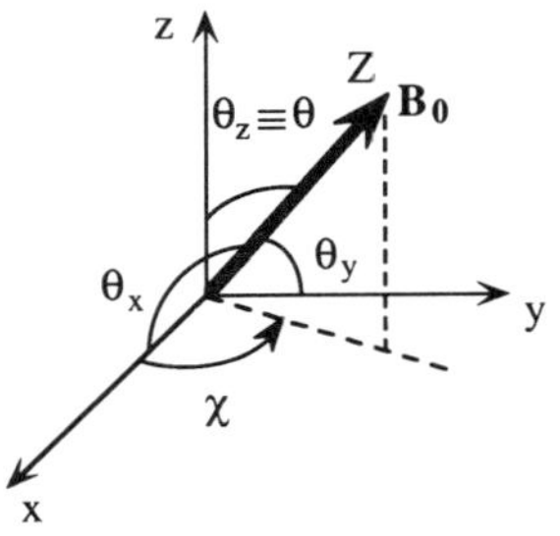

Figure 1.49 Orientation of $\boldsymbol{B}_0$ with respect to a molecular frame in which the shielding tensor is diagonal

The resonance frequency for a given molecular orientation is derived from (1.42)

$$\nu = \frac{\gamma B_0}{2\pi}(1 - \sigma_{ZZ}) = \frac{\gamma B_0}{2\pi}[1 - \cos^2\theta_x\sigma_{xx} - \cos^2\theta_y\sigma_{yy} - \cos^2\theta_z\sigma_{zz}] \quad (1.64)$$

According to the usual convention, the molecular axes are chosen so that $|\sigma_{zz}| \geqslant |\sigma_{yy}| \geqslant |\sigma_{xx}|$. In (1.64) σ_{ZZ} has been expressed in the molecular frame by applying the transformation procedures described above (formula (1.48)). The 'powder spectrum' is simulated by adding the transitions which correspond to all possible orientations within the sample, each of them being supposed to occur with equal probability. It can be mentioned that one is dealing here with an ensemble average and not a time average.

We shall first consider the special case of an axial shielding tensor with the following notations: $\sigma_{zz} = \sigma_{\|}$; $\sigma_{xx} = \sigma_{yy} = \sigma_{\perp}$ and $\Delta\sigma = \sigma_{\|} - \sigma_{\perp}$; when inserted in (1.64), this yields the simplified expression (with $\theta \equiv \theta_z$):

$$\nu = \frac{\gamma B_0}{2\pi}(1 - \sigma_{\perp} - \Delta\sigma\cos^2\theta). \quad (1.65)$$

Since $\cos^2\theta$ lies in the interval [0,1], the spectrum ranges from $(\gamma B_0/2\pi)$ $(1 - \sigma_{\perp})$ to $(\gamma B_0/2\pi)$ $(1 - \sigma_{\|})$. The relevant shape function $S(\nu)$ is such that:

$$S(\nu)\,d\nu = dN$$

where dN stands for the number of molecules giving rise to a resonance within the interval $d\nu$ centered on ν. Writing dN in the form

$$\frac{dN}{d\theta}\frac{d\theta}{d\nu}d\nu,$$

we can express $S(\nu)$ as:

$$S(\nu) = \frac{dN}{d\theta}\frac{d\theta}{d\nu}$$

Now, dN can be written $dN = P(\theta, \chi)\sin\theta\,d\theta\,d\chi$, where $P(\theta, \chi)$ is the probability of having an orientation defined by θ and χ within the solid angle $\sin\theta\,d\theta\,d\chi$. Since all orientations share the same probability, one has $P(\theta, \chi) \equiv P$ and $P \equiv 1/4\pi$ owing to the fact that:

$$\int_0^{\pi}\int_0^{2\pi} P\sin\theta\,d\theta\,d\chi = 1$$

This leads to:

$$\frac{dN}{d\theta} = \sin\theta\int_0^{2\pi} P(\theta, \chi)\,d\chi = \frac{\sin\theta}{2}$$

As far as $d\theta/d\nu$ is concerned, it can be expressed by differentiating (1.65).

Finally, substituting θ by ν through (1.65), we obtain:

$$S(\nu) = \tfrac{1}{2}\sqrt{\frac{\pi\,\Delta\sigma}{2\gamma B_0}\frac{1}{\nu - (\gamma B_0/2\pi)(1 - \sigma_{\perp})}} \quad \text{for } \sigma_{\perp} > \sigma_{\|}$$

$$S(\nu) = \tfrac{1}{2}\sqrt{\frac{\pi\,\Delta\sigma}{2\gamma B_0}\frac{1}{(\gamma B_0/2\pi)(1 - \sigma_{\perp}) - \nu}} \quad \text{for } \sigma_{\perp} < \sigma_{\|} \quad (1.66)$$

This theoretical function goes to infinity for $\nu = \gamma B_0/2\pi$ $(1 - \sigma_{\perp})$, as shown on Figure 1.50 (plotted for the case $\sigma_{\perp} < \sigma_{\|}$). An experimental spectrum, obviously, does not go to infinity. In order to simulate the actual lineshape, we must take the product of

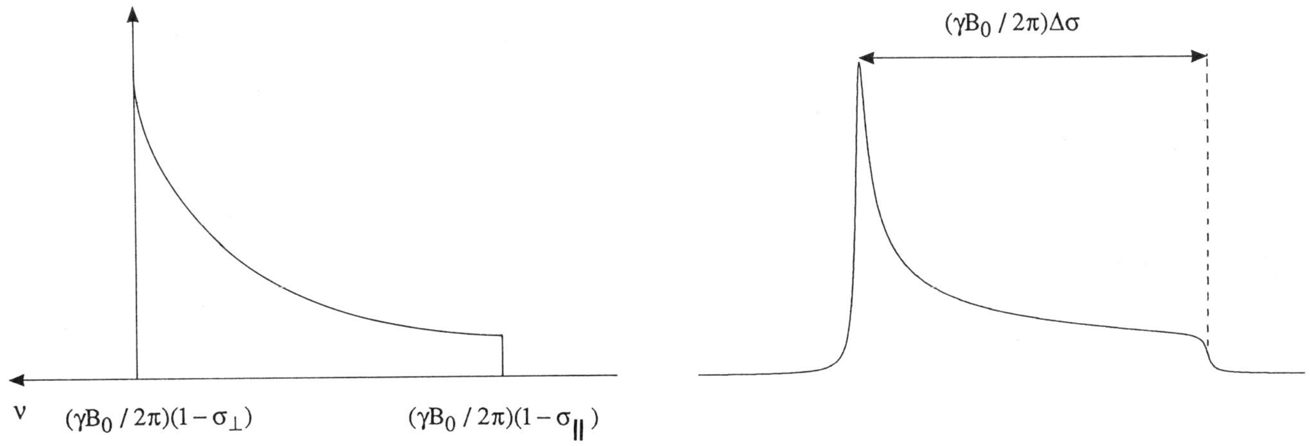

Figure 1.50 Left, theoretical lineshape function $S(\nu)$ from formula (1.66) for a shielding tensor of axial symmetry; right, simulated spectrum $\mathcal{S}(\nu)$ obtained by calculating the convolution product of $S(\nu)$ with a gaussian function

convolution of the theoretical function $S(\nu)$ by a broadening function which accounts for transverse relaxation and for all additional interactions undergone by the considered nucleus. Generally a gaussian function is retained; if β stands for its standard deviation, the actual line shape $S(\nu)$ can be calculated by:

$$\mathcal{S}(\nu) = \int_{-\infty}^{+\infty} S(h)\frac{1}{\beta\sqrt{2\pi}}\exp\left[-\frac{(\nu - h)^2}{2\beta^2}\right]\mathrm{d}h \qquad (1.67)$$

The spectrum therefore directly provides $\Delta\sigma$ (Figure 1.50) which constitutes a major structural piece of information. In the general case, when symmetry is lacking in the shielding tensor, similar calculations lead to a more complicated lineshape which nevertheless yields the three principal elements of the tensor (Figure 1.51).

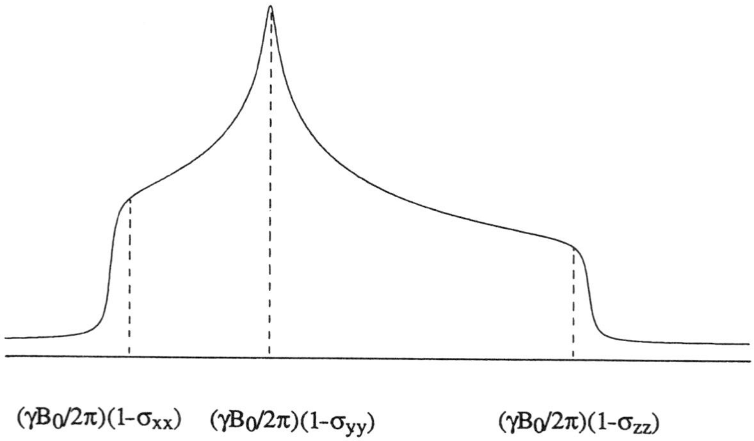

Figure 1.51 Powder spectrum for a shielding tensor without symmetry

Dipolar coupling between two spin 1/2 nuclei. Quadrupolar coupling of a spin 1 nucleus possessing an axially symmetric field gradient tensor

In both cases, for a given molecular orientation defined by $\theta_z \equiv \theta$ (Figure 1.49), two transitions are generated at frequencies $\nu' = \nu_n - u(3\cos^2\theta - 1)$ and $\nu'' = \nu_n + u(3\cos^2\theta - 1)$, ν_n being the resonance frequency (Zeeman term). When dealing with the dipolar coupling between two equivalent spin 1/2 nuclei separated by a distance r, and z being defined as the direction joining the nuclei, u is given by $(\mu_0/4\pi)3\gamma^2 h/(16\pi^2 r^3)$, whereas u becomes $(\mu_0/4\pi)\gamma_A\gamma_X h/(8\pi^2 r^3)$ for two distinct nuclei of respective gyromagnetic ratio γ_A and γ_X [see equation (1.45)]. In the case of a spin 1 nucleus, z represents the principal axis of the field gradient tensor in relation with the assumed axial symmetry ($V_{xx} = V_{yy}$) and u can be written as $(3/4h)\mathrm{e}QV_{zz}$ [relationship (1.46)]. If all possible orientations are considered, an approach similar to that of the previous paragraph leads to the theoretical lineshape functions for the two transitions at frequencies ν' and ν'' (taking the frequency origin at the resonance, i.e. $\nu_n = 0$):

$$S'(\nu') = \frac{1}{4\sqrt{3u}\sqrt{u - \nu'}}$$

$$S''(\nu'') = \frac{1}{4\sqrt{3u}\sqrt{u + \nu''}}$$

ν' lies in the interval $[-2u; u]$ and ν'' in $[-u; 2u]$. Summing up the two contributions yields:

$$
\begin{aligned}
S(\nu) &= \frac{1}{4\sqrt{3u}\sqrt{u-\nu}} && -2u \leqslant \nu < -u \\
S(\nu) &= \frac{1}{4\sqrt{3u}}\left(\frac{1}{\sqrt{u-\nu}} + \frac{1}{\sqrt{u+\nu}}\right) && -u \leqslant \nu < u \qquad (1.68) \\
S(\nu) &= \frac{1}{4\sqrt{3u}\sqrt{u+\nu}} && u \leqslant \nu < 2u
\end{aligned}
$$

The actual spectrum (Figure 1.52), obtained by convoluting $S(\nu)$ with an appropriate broadening function [see equation (1.67)], is termed the Pake doublet. The splitting is equal to $2u$ and its measurement leads to the internuclear distance or the electric field gradient.

This type of spectrum also prevails for a sample made of an infinity of small domains, each possessing a preferential direction, these directions being distributed randomly within the sample (e.g. a liquid crystal non-macroscopically oriented). In that case, u must be interpreted by taking into account a factor related to the order parameter of the considered molecule inside each domain.

The situation described above, for which it is possible to define the system as including either two isolated spin 1/2 nuclei or a single spin 1 nucleus, is seldom encountered. Very often, mutual dipolar couplings are such that the system includes the totality of spins within the sample. The spectrum exhibits broad lines whose shapes

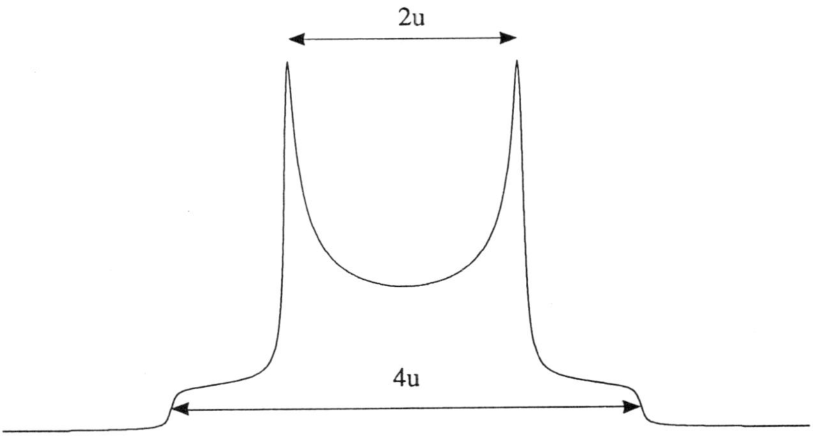

Figure 1.52 The main splitting ($2u$) is equal to:

- $\dfrac{\mu_0}{4\pi}\dfrac{3\gamma^2 h}{8\pi^2 r^3}$ in the case of two equivalent spin 1/2 nuclei;
- $\dfrac{\mu_0}{4\pi}\dfrac{\gamma_A \gamma_X h}{4\pi^2 r^3}$ in the case of two distinct spin 1/2 nuclei;
- $\dfrac{3}{2h} eQV_{zz}$ in the case of a spin 1 nucleus with an axially symmetric field gradient tensor

can nevertheless be tentatively analysed, yielding limited structural information. This aspect of solid state NMR will not be considered further here.

1.4.6 HIGH RESOLUTION SPECTROSCOPY OF POWDER SAMPLES

In order to remove the line broadening due to chemical shift anisotropy or to dipolar couplings, a first method would consist in artificially creating a preferred direction D by fast sample spinning. It has been noted in Section 1.4.3 that a high resolution spectrum is retrieved (similar to a liquid phase spectrum) provided that D makes the so called magic angle with the static magnetic field direction and that the rotation frequency is much larger than the interactions to be removed. It turns out that the largest rotation frequency available (of the order of 20 kHz) is still far too small to suppress the effects of dipolar or quadrupolar couplings (of the order of several tens or hundreds of kHz). Various methods are currently in use for alleviating this problem. For homonuclear systems, somewhat complicated pulse sequences lead, by averaging effects, to the cancellation of the dipolar hamiltonian.

Heteronuclear systems involving a rare spin (i.e. of low natural abundance), such as carbon-13, are especially attractive. The low natural abundance is such that any carbon–carbon coupling evidently cannot be taken into consideration. There thus remains only two causes of broadening: chemical shift *aniso*tropy and direct (dipolar) couplings with the abundant spins (generally protons). The latter are simply annihilated by spin decoupling, which however requires a strong irradiation at the proton frequency as the proton spectrum occupies a large spectral range because of the homonuclear dipolar couplings. Concerning chemical shift anisotropy, owing to its relatively weak broadening effects, it can be removed by rotation at the magic angle, (magic angle spinning, MAS). If the spinning rate is too low, satellite lines will appear on each side of the principal peak, at frequencies which are integer multiples of the spinning rate. This technique is illustrated in Figures 1.53 and 1.54 by spectra of a powder sample of hexamethylbenzene. It can be observed that a true high resolution spectrum is obtained, however with the drawback of containing only the chemical shift information. The overall sensitivity of the experiment is generally improved by polarization transfer from the abundant nuclei to the rare spin (CP: cross polarization).

Further remarks about the 'magic angle' spinning procedure

When sample spinning is sufficiently fast with respect to the interactions undergone by nuclear spins, the approach of Section 1.4.3. [equations (1.49) to (1.51)], which concerned organized systems such as liquid crystals, can be transposed in a straightforward manner:

- D becomes the rotation axis, with an effective axial symmetry created by fast spinning;

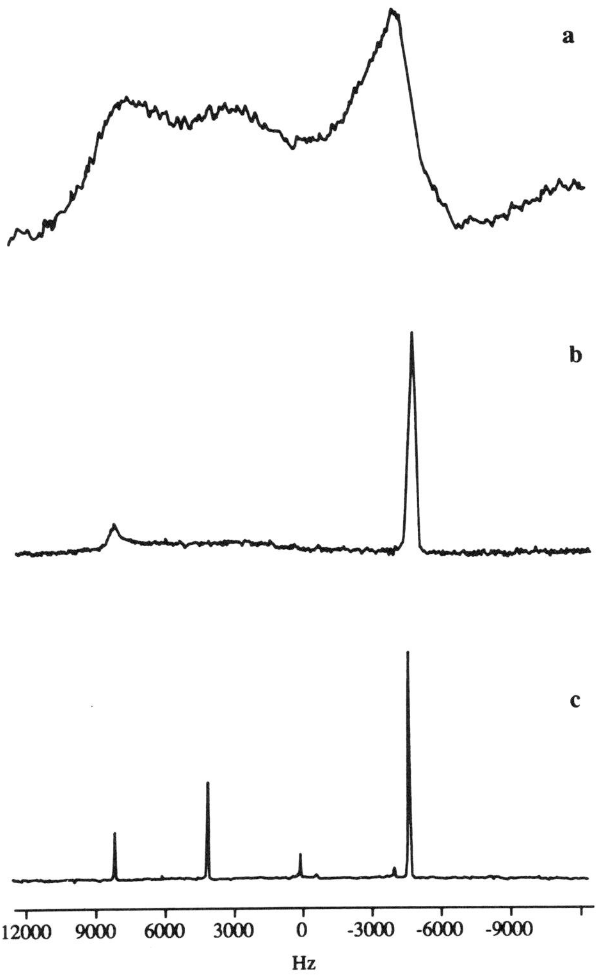

Figure 1.53 ^{13}C spectra (75 MHz) of a powder of hexamethylbenzene obtained by cross polarization (CP; transfer of proton magnetization toward carbon-13 by means of radio-frequency fields applied simultaneously at proton and carbon frequencies) in order to improve the sensitivity of the experiment. (a): raw spectrum; (b): with proton decoupling to remove the effect of $^{13}C - {}^{1}H$ dipolar couplings; (c): as (b) with in addition magic angle sample spinning (MAS) at a frequency of 4 kHz. This latter technique removes the chemical shift anisotropy patterns (Figures 1.50 and 1.51). Residual sidebands, regularly spaced according to the rotation frequency, originate from the fact that the rotation frequency is still too small with respect to the considered interaction. They are especially noticeable for aromatic carbons which possess a large chemical shift anisotropy (200 ppm)

- In the expression of $\langle T_{X'Z'} \rangle$, u'_{i1} represents the cosine of the angle between the i axis of the molecular frame and the X′ axis of a laboratory frame whose axis Z′ coincides with D. The probability relative to u'_{i1} is the same as for $-u'_{i1}$. Consequently, $\langle u'_{i1} u'_{i3} \rangle$ is averaged out.

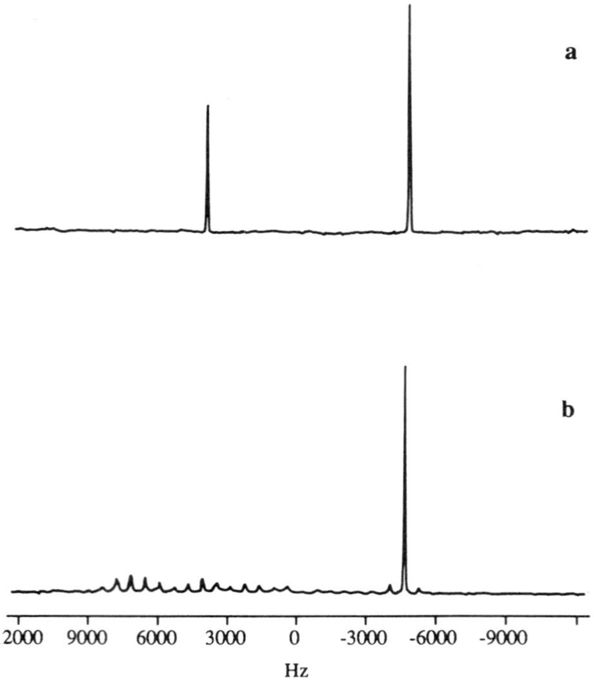

Figure 1.54 (a): Spectrum obtained in the same conditions as for Figure 1.53(c) with the application of a special multipulse sequence designed for suppressing rotational sidebands. (b): Same as Figure 1.53(c) with a low spinning rate (700 Hz)

Whenever the spinning rate of the sample cannot be considered as very large with respect to anisotropic interactions, the principle of time average, symbolized by $\langle\ \rangle$, no longer applies. Hence, the observed spectrum results from the superposition of the individual spectra of the microcrystallites constituting the sample (such a calculation has been performed for a static sample in Section 1.4.5). The lack of effective axial symmetry makes it mandatory to account for $T_{Z'Z'}$, $T_{X'X'}$, $T_{X'Z'}$ and $T_{Z'X'}$ in the calculation of T_{ZZ}. Since $T_{X'X'}$ depends on $\cos^2(2\pi\nu_r t)$ (where ν_r is the spinning rate) and $T_{Z'X'}$ and $T_{X'Z'}$ depend on $\cos(2\pi\nu_r t)$, rotational sidebands at $2\nu_r$, ν_r and their multiples are expected. The rotational sideband pattern resembles the powder spectrum when ν_r tends to zero, as shown by the envelope of sidebands in Figure 1.54b.

Finally, an important application of magic angle spinning deserves to be mentioned. It concerns quadrupolar nuclei of half integer spin ($I = 3/2, 5/2, \ldots$). The large value of quadrupolar interactions generally makes the whole spectrum impossible to observe with the exception of the $|I, 1/2\rangle \rightarrow |I, -1/2\rangle$ transition, whose frequency is precisely the resonance frequency of the considered quadrupolar nucleus (Zeeman term). It turns out that the relevant line is relatively narrow under magic angle spinning conditions, which cancel first order quadrupolar effects. The residual broadening is attributable to second order effects which cannot be removed by MAS (Figure 1.55).

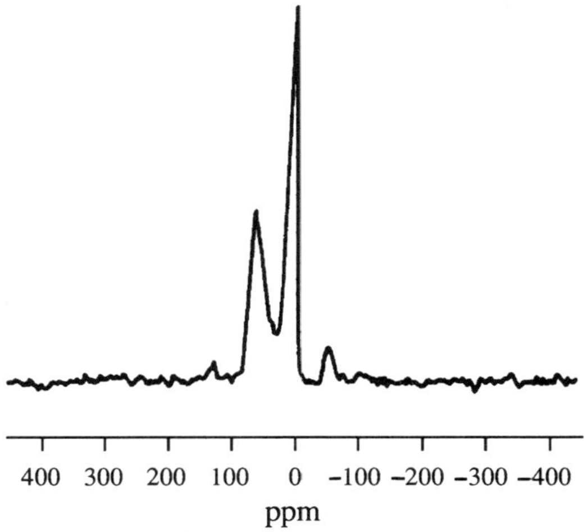

Figure 1.55 ^{27}Al MAS spectrum (spinning rate: 4 kHz; Larmor frequency: 78.5 MHz) of the powder of an aluminosilicate compound. This spectrum permits the discrimination and the quantitation of aluminum atoms in a tetrahedral site (left peak) and in an octahedral site (right peak). Signals of weak intensity are rotational sidebands

Bibliography

General textbooks

J. A. Pople, W. G. Schneider and H. J. Bernstein: *High Resolution Nuclear Magnetic Resonance*, McGraw-Hill, New York, 1959

J. W. Emsley, J. Feeney and L. H. Sutcliffe: *High Resolution Nuclear Magnetic Resonance Spectroscopy*, Pergamon Press, Oxford, 1965

A. Carrington and A. D. McLachlan: *Introduction to Magnetic Resonance*, Harper and Row, London, 1967

M. L. Martin, J.-J. Delpuech and G. J. Martin: *Practical NMR Spectroscopy*, Heyden, London, 1980

R. K. Harris: *Nuclear Magnetic Resonance Spectroscopy. A Physicochemical View*, Pitman, 1983

J. W. Akitt: *NMR and Chemistry, an Introduction to Modern NMR Spectroscopy*, 3rd ed., Chapman and Hall, New York, 1992

E. Breitmeir: *Structure Elucidation by NMR Inorganic Chemistry. A Practical guide*, Wiley, Chichester, 1993

J. K. M. Sanders and B. K. Hunter: *Modern NMR Spectroscopy. A Guide for Chemists*, 2nd ed., Oxford University Press, Oxford, 1993

H. Günther: *NMR Spectroscopy. A Practical Approach. Basic Principles, Concepts and Applications in chemistry*, 2nd ed., Wiley, Chichester, 1995

Multinuclear Spectroscopy

R. K. Harris and B. E. Mann (Eds.): *NMR and the Periodic Table*, Academic Press, London, 1978

C. Brevard and P. Granger: *Handbook of High Resolution Multinuclear NMR*, Wiley, New York, 1981

G. C. Levy and R. L. Lichter: *Nitrogen-15 Nuclear Magnetic Resonance Spectroscopy*, Wiley, New York, 1979

G. C. Levy, R. L. Lichter and G. L. Nelson: *Carbon-13 Nuclear Magnetic Resonance Spectroscopy*, 2nd ed., Wiley, New York, 1980

G. J. Martin, M. L. Martin and J. P. Gouesnard: *^{15}N NMR Spectroscopy*, *NMR—Basic Principles and Progress*, Vol. 18, Springer-Verlag, Berlin, 1981

J. Mason, Ed.: *Multinuclear NMR*, Plenum Press, New York, 1987

P. Granger and R. K. Harris (Eds.): *Multinuclear Magnetic Resonance in Liquids and Solids—Chemical Applications*, Kluwer, Dordrecht, 1990

Spectral simulation

P. L. Corio: *Structure of High Resolution NMR Spectra*, Academic Press, New York, 1966

R. J. Abraham: *The Analysis of High Resolution NMR Spectra*, Elsevier, Amsterdam, 1971

NMR in anisotropic media

A. Abragam: *The Principles of Nuclear Magnetism*, Clarendon Press, Oxford, 1961, Chapters 4 and 7

J. W. Emsley and J. C. Lindon: *NMR Spectroscopy using Liquid Crystal Solvents*, Pergamon Press, Oxford, 1975

M. Mehring: *High Resolution NMR in Solids*, Springer-Verlag, Berlin, 1983

J. W. Emsley (Ed.): *Nuclear Magnetic Resonance of Liquid Crystals*, Reidel, Dordrecht, 1983

C. A. Fyfe: *Solid State NMR for Chemists*, C.F.C. Press, Guelph, 1983

R. Y. Dong: *Nuclear Magnetic Resonance of Liquid Crystals*, Springer-Verlag, New York, 1994

NMR of biomolecules

O. Jardetzky and G. C. K. Roberts: *NMR in Molecular Biology*, Academic Press, New York, 1981

D. G. Gadian: *NMR and its Applications to Living Systems*, 2nd ed., Oxford University Press, Oxford, 1995

K. Wüthrich: *NMR of Proteins and Nucleic Acids*, Wiley, New York, 1986

I. Bertini and C. Luchinat: *NMR of Paramagnetic Molecules in Biological Systems*, Benjamin Cummings, Menlo Park, 1986

I. Bertini, H. Molinari and N. Niccolai (Eds.): *NMR and Biomolecular Structure*, VCH, Weinheim, 1991

2 BASIC MATHEMATICS AND PHYSICS OF NMR

2.1 The Phenomenological Approach; the Bloch Equations

2.1.1 THERMAL EQUILIBRIUM MAGNETIZATION

The concept of *macroscopic* nuclear magnetization is related to the whole set of systems existing within the sample under investigation (*cf*. the Avogadro number). Until now, we have essentially considered *one* system involving the nuclear spins of one molecule and for the sake of simplicity, it has usually been assumed that this elementary system involves a single nucleus of spin 1/2 (e.g. the proton of the chloroform molecule). It must however be borne in mind that any experiment is performed on a sample which necessarily includes a very great number of replicas of the elementary system, and that the measured quantity is macroscopic in nature. In nuclear magnetic resonance, this quantity is the nuclear magnetization which results from the polarization of nuclear spins under the application of the static magnetic field B_0. Hence a statistical treatment must be applied to the ensemble of the systems defined above, each of them, in the simple case assumed here, being characterized by two states $|\alpha\rangle$ and $|\beta\rangle$ of energy $-(\gamma\hbar B_0/2)$ and $(\gamma\hbar B_0/2)$, respectively, as shown in Figure 2.1.

Let P_α and P_β represent the probabilities for *one* system to be in the α or β state. If one is interested in all the systems in the sample, P_α and P_β become the relative populations of the relevant energy levels. These probabilities are evidently related to each other by $P_\alpha + P_\beta = 1$ and further obey the Boltzmann law

$$P_\alpha/P_\beta = \exp\left[(\gamma\hbar B_0)/(k_B T)\right]$$

where k_B is the Boltzmann constant ($k_B = 1.3806 \times 10^{-23}\ \mathrm{J\,K^{-1}}$) and T is the absolute temperature in kelvin. For typical values of B_0 and T (the so-called 'high temperature approximation') the exponential can be adequately represented by the first order term of its expansion, as follows:

$$\exp\left[(\gamma\hbar B_0)/(k_B T)\right] \approx 1 + (\gamma\hbar B_0)/(k_B T)$$

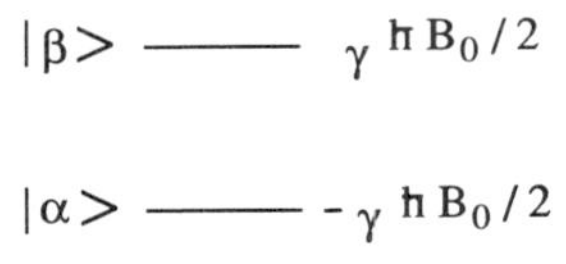

Figure 2.1 Energy diagram for a spin 1/2 nucleus in the presence of a static magnetic field B_0

As a consequence, $P_\alpha - P_\beta$ can be safely approximated by $(\gamma\hbar B_0)/(2k_BT)$ and the magnetization per volume unit, denoted by M_0, can be calculated from $M_0 = \mathcal{N}(P_\alpha - P_\beta)\mu$, where $\mathcal{N}$ is the number of systems per unit volume and μ is the nuclear magnetic moment for one system whose absolute value turns out to be the same for both energy levels. Because $\boldsymbol{\mu} = \gamma\hbar\boldsymbol{I}$ (see section 1.1), one has $\mu = \gamma\hbar/2$ for the present case of a spin 1/2 nucleus and

$$M_0 = \mathcal{N}\frac{\gamma^2\hbar^2 B_0}{4k_B T} \tag{2.1}$$

which is an expression of the Curie law, easily transposed to an isotope of arbitrary spin number I:

$$M_0 = \mathcal{N}\frac{\gamma^2\hbar^2 B_0 I(I+1)}{3k_B T} \tag{2.2}$$

2.1.2 BEHAVIOR OF NUCLEAR MAGNETIZATION IN THE PRESENCE OF STATIC INDUCTION

From now on, instead of dealing with the magnetic field vector $\boldsymbol{H}$, we shall be concerned with the magnetic induction because we are discussing a macroscopic sample in which the magnetic susceptibility of the medium plays a role. This quantity, generally denoted by χ, is dimensionless, positive for a paramagnetic substance, negative for most substances which are diamagnetic, and zero for vacuum.

The magnetic induction, which we have until now dubbed incorrectly as the magnetic field, is related to the magnetic field applied externally to the sample by the relationship

$$\boldsymbol{B} = \mu_0(1 + \chi)\boldsymbol{H}$$

where μ_0 is the permeability of vacuum, whose value is $4\pi \times 10^{-7}$ in the MKSA system and 4π in the cgs system. However, we shall be using interchangeably the term 'field' or induction to mean the vector $\boldsymbol{B}$, even though field is not strictly correct from the standpoint of electromagnetic theory. We shall now proceed by looking at the behavior of an elementary magnetic moment $\boldsymbol{m}$, in the presence of a magnetic induction $\boldsymbol{B}$, neglecting for the moment any other time-dependent mechanisms (especially relaxation phenomena, which will be introduced later). The basic laws of electromagnetism tell us that $\boldsymbol{m}$ is subjected to a torque equal to $\boldsymbol{m} \wedge \boldsymbol{B}$ ($\wedge$ indicates the vector product) and, by virtue of the angular momentum theorem, one has $\mathrm{d}\boldsymbol{L}/\mathrm{d}t = \boldsymbol{m} \wedge \boldsymbol{B}$. As the magnetic moment is proportional to the angular momentum ($\boldsymbol{m} = \gamma\boldsymbol{L}$), this yields $\mathrm{d}\boldsymbol{m}/\mathrm{d}t = \gamma\boldsymbol{m} \wedge \boldsymbol{B}$. Taking into account all the nuclear magnetic moments in the sample yields

$$\frac{\mathrm{d}\boldsymbol{M}}{\mathrm{d}t} = \gamma\boldsymbol{M} \wedge \boldsymbol{B} \tag{2.3a}$$

which can be broken down into expressions for each of the components of $\boldsymbol{M}$ (with the direction of $\boldsymbol{B}$ defining the z axis)

$$\frac{\mathrm{d}M_x}{\mathrm{d}t} = \gamma B M_y \tag{2.3b}$$

$$\frac{dM_y}{dt} = -\gamma B M_x \tag{2.3c}$$

$$\frac{dM_z}{dt} = 0. \tag{2.3d}$$

These are the most general equations obtained from classical mechanics, which indeed apply to macroscopic magnetization but which should not hold for a single spin whose behavior is actually governed by quantum mechanics. They will be first solved for the case where B is constant and denoted by B_0. At thermal equilibrium, the nuclear magnetization is given by (2.2) and its direction is along the z axis. There is no reason for components along x or y to exist, since the probability for having m_x is equal to the probability for having $-m_x$, owing to the fact that z is the only preferential direction. Without any external perturbation, and by virtue of equations (2.3b–d), the nuclear magnetization remains invariant: $M_x(t) = M_y(t) = 0$ and $M_z(t) = M_0$. On the other hand, if we assume that, at the outset, the nuclear magnetization has been tipped from its equilibrium configuration through an angle α about the x axis (for instance, by means of a radio-frequency pulse, as explained later), equations (2.3) tell us how M_x, M_y and M_z will evolve, provided that we know their initial values, which here can be expressed as

$$\begin{aligned} M_x(0) &= 0 \\ M_y(0) &= M_0 \sin\alpha \\ M_z(0) &= M_0 \cos\alpha \end{aligned} \tag{2.4}$$

It can be noted that there is no loss of generality in considering the initial magnetization in the (yz) plane, as z is the only preferential direction. Solving equations (2.3) with the above initial conditions (2.4) is straightforward and yields

$$\begin{aligned} M_x(t) &= M_0 \sin\alpha \sin(\omega_0 t) \\ M_y(t) &= M_0 \sin\alpha \cos(\omega_0 t) \\ M_z(t) &= M_0 \cos\alpha \end{aligned} \tag{2.5}$$

with $\omega_0 = \gamma B_0$.

Thus it can be seen that the nuclear magnetization undergoes a precessional motion (Figure 2.2) at the Larmor frequency $\nu_0 = \omega_0/2\pi = \gamma B_0/2\pi$, as already mentioned in Chapter 1. The angular velocity should actually be written $-\omega_0$ since, in the case of a positive gyromagnetic ratio, precession takes place in a clockwise fashion, which is the opposite of the normal trigonometric rotational convention.

2.1.3 EFFECT OF AN ALTERNATING MAGNETIC FIELD IN A DIRECTION PERPENDICULAR TO THE STATIC MAGNETIC FIELD. DEFINITION OF THE ROTATING FRAME. RADIO-FREQUENCY (r.f.) PULSES

We now consider the effect of applying an alternating magnetic induction (or 'field') expressed as $2B_1 \cos(\omega_r t)$, with $B_1 \ll B_0$, polarized along, say, the x axis

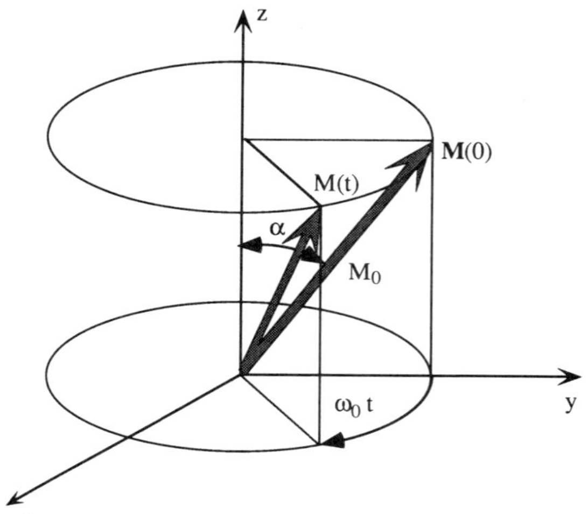

Figure 2.2 Precession of nuclear magnetization as described by equations (2.5)

of the laboratory frame. This means that this field is perpendicular to B_0, a necessary, although not sufficient, condition for inducing transitions among the energy level system of nuclear spins (see Section 1.2.2); hence, this represents a way of acting on nuclear magnetization. This alternative magnetic field can be formally decomposed into two *rotating* magnetic fields, one with the angular velocity $+\omega_r$, the other with the angular velocity $-\omega_r$ (Figure 2.3). Owing to that decomposition, it is reasonable to consider that the magnetic field rotating in the same sense as the nuclear precession is the one able to act on nuclear magnetization. From a pragmatic point of view, one can state that two vectors rotating in

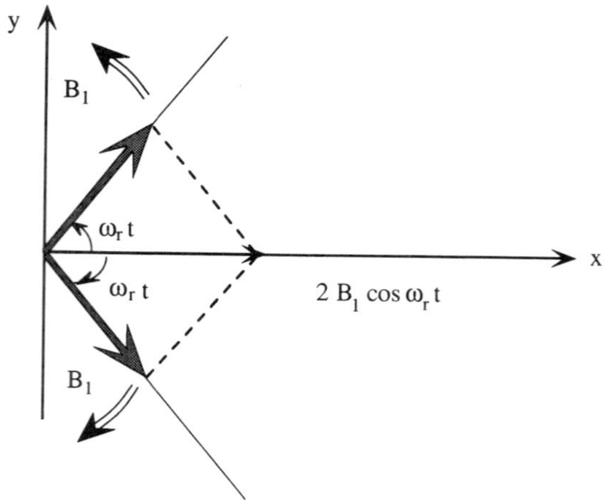

Figure 2.3 Decomposition of an alternative magnetic field polarized along x into two rotating magnetic fields

opposite directions do not 'talk' with each other and that their mutual interaction is consequently negligible.

To account for the NMR experiment, we can thus consider a static field $\boldsymbol{B}_0$ oriented along the z direction of the laboratory frame and an induction of amplitude $\boldsymbol{B}_1$ rotating in the xy plane of the same frame at an angular velocity $-\omega_r$. In order to handle this problem, it is convenient to define a frame (x', y', z) which rotates around z at the angular velocity $-\omega_r$ (Figure 2.4), so that $\boldsymbol{B}_1$ appears stationary, e.g. along the x' direction. Let us also define $\boldsymbol{i}'$, $\boldsymbol{j}'$ and $\boldsymbol{k}$ as the unit vectors along the x', y' and z directions respectively.

Expanding the nuclear magnetization as $\boldsymbol{M} = M_{x'}\boldsymbol{i}' + M_{y'}\boldsymbol{j}' + M_z\boldsymbol{k}$, we obtain:

$$\frac{\mathrm{d}\boldsymbol{M}}{\mathrm{d}t} = M_{x'}\frac{\mathrm{d}\boldsymbol{i}'}{\mathrm{d}t} + M_{y'}\frac{\mathrm{d}\boldsymbol{j}'}{\mathrm{d}t} + M_z\frac{\mathrm{d}\boldsymbol{k}}{\mathrm{d}t} + \left(\frac{\mathrm{d}M_{x'}}{\mathrm{d}t}\boldsymbol{i}' + \frac{\mathrm{d}M_{y'}}{\mathrm{d}t}\boldsymbol{j}' + \frac{\mathrm{d}M_z}{\mathrm{d}t}\boldsymbol{k}\right)$$

Denoting by $\delta\boldsymbol{M}/\delta t$ the expression between parentheses, which represents the derivative of $\boldsymbol{M}$ with respect to time viewed from the rotating frame, and introducing the instantaneous rotation vector $\boldsymbol{\Omega}$ of components $(0, 0, -\omega_r)$, which is such that $\mathrm{d}\boldsymbol{i}'/\mathrm{d}t = \boldsymbol{\Omega} \wedge \boldsymbol{i}'$ (and a similar relation for $\boldsymbol{j}'$), we can write:

$$\frac{\mathrm{d}\boldsymbol{M}}{\mathrm{d}t} = \boldsymbol{\Omega} \wedge \boldsymbol{M} + \frac{\delta\boldsymbol{M}}{\delta t} \tag{2.6}$$

This equation is representative of the fact that the absolute velocity is equal to the leading velocity plus the relative velocity. Referring to the expression of $(\mathrm{d}\boldsymbol{M}/\mathrm{d}t)$ in (2.3), we arrive at

$$\frac{\delta\boldsymbol{M}}{\delta t} = \boldsymbol{M} \wedge \gamma\boldsymbol{B}_{\mathrm{eff}} \tag{2.7}$$

with $\boldsymbol{B}_{\mathrm{eff}} = (B_0 - \omega_r/\gamma)\boldsymbol{k} + B_1\boldsymbol{i}'$.

We thus notice that the equation of motion pertaining to nuclear magnetization, viewed from the rotating frame, is similar to the one which prevails in the laboratory frame, provided that we replace $\boldsymbol{B}$ by an effective induction $\boldsymbol{B}_{\mathrm{eff}}$ defined as indicated in Figure 2.5. From now on, we shall systematically proceed in the rotating frame and omit the primes.

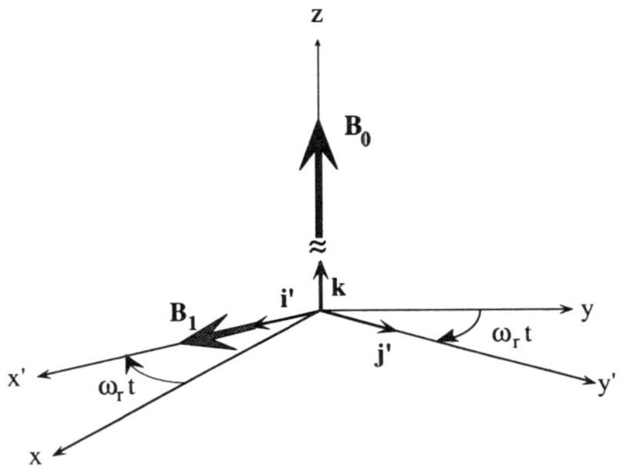

Figure 2.4 Definition of the rotating frame in which both $\boldsymbol{B}_0$ and $\boldsymbol{B}_1$ appear stationary

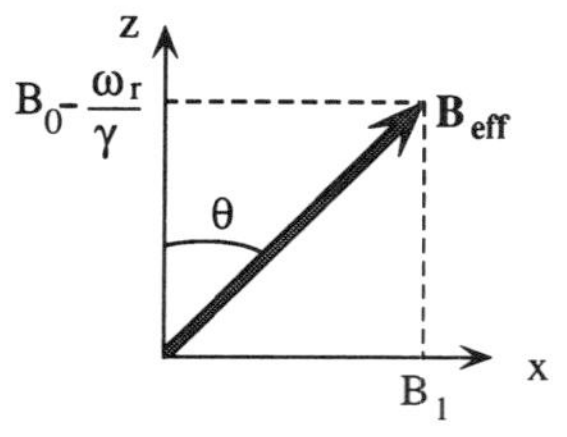

Figure 2.5 Effective magnetic field

Through equation (2.7) and its resemblance to equation (2.3), we are in a position to predict the behavior of the nuclear magnetization in the presence of both a static magnetic field B_0 and an alternating magnetic field B_1 of frequency $\omega_r/2\pi$. In the frame rotating about z at an angular velocity ω_r, the nuclear magnetization precesses around $\boldsymbol{B}_{\text{eff}}$ (this particular precession in a rotating frame is usually dubbed nutation) at the frequency

$$\nu_{\text{nut}} = \frac{\gamma B_{\text{eff}}}{2\pi} = \frac{\sqrt{\gamma^2 B_1^2 + (\gamma B_0 - \omega_r)^2}}{2\pi} \tag{2.8}$$

$\boldsymbol{B}_{\text{eff}}$ making with z an angle θ such that

$$\tan\theta = \frac{\gamma B_1}{\gamma B_0 - \omega_r}$$

A widespread use of (2.8) is in the context of application of *radio-frequency pulses*, which imply that the frequency of the B_1 field ($\omega_r/2\pi$) is in the vicinity of the resonance frequencies of interest (that is, around $\gamma B_0/2\pi$). It is because the latter frequencies are of the order of tens or hundreds of MHz that one generally uses the terminology radio-frequency, or simply r.f. Moreover, if we assume that the amplitude of B_1 is sufficiently large so that $\gamma B_1 \gg |\gamma B_0 - \omega_r|$, B_{eff} coincides with B_1 and the nutation frequency simply becomes equal to $\gamma B_1/2\pi$. It is thus possible to rotate nuclear magnetization at will. For instance, starting from equilibrium (magnetization $\boldsymbol{M}_0$ along z), we can apply a field B_1 (coinciding with the x axis of the rotating frame) for a duration equal to $\tau_{\pi/2}$ such that $\gamma B_1 \tau_{\pi/2} = \pi/2$ and, in this way, rotate the magnetization vector into the y direction of the rotating frame (Figure 2.6). This is what is usually called a $\pi/2$ (or 90°) pulse.

In practice, the amplitude of B_1 is chosen in such a way that the duration of a $\pi/2$ pulse does not exceed some tens of microseconds (typically 10 μs or less) so that the angle θ between the B_{eff} direction and the z axis remains close to 90° for all individual magnetizations of a spectrum covering the usual spectral widths. In that way, unwanted 'off-resonance' effects, arising from a resonance frequency

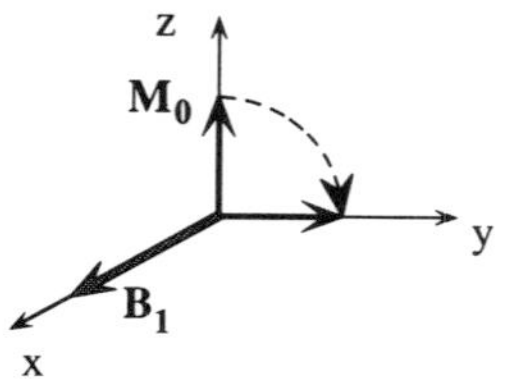

Figure 2.6 Motion of the nuclear magnetization in the rotating frame, starting from its equilibrium position, under the application of a $\pi/2$ pulse

remote with respect to ν_r, are avoided. Once nuclear magnetization is located in the xy plane, it is a simple matter to detect it after having switched off the B_1 field (it is for this reason that the xy plane is sometimes called the measuring plane). From that moment, only B_0 holds, $\boldsymbol{B}_{\text{eff}}$ coincides with $(B_0 - \omega_r/\gamma)\mathbf{k}$ and precession takes place in the xy plane (of the rotating frame) at the frequency $\nu_r - \nu_0 = (\omega_r - \gamma B_0)/2\pi$ (or, evidently, at the frequency $\gamma B_0/2\pi$ in the laboratory frame).

As already mentioned in Chapter 1, nuclear precession leads to an induced electromotive force which is established within the dedicated coil of the NMR probe, the same coil being generally employed for both transmit and receive operations (Figure 2.7).

In fact, the NMR signal is *demodulated* with respect to ν_r which, by reference to radio engineering techniques, is called the *carrier frequency*. The major consequence of this detection scheme is that the receive operations take place also in the *rotating frame*. Returning to transmit operations, it is clear that a rotation of any angle can be produced by an r.f. pulse of appropriate duration, as the flip angle is directly proportional to this duration. In particular, the complete inversion of the nuclear magnetization (π pulse) is widely used in the determination of relaxation times (Chapter 4) or more generally in multipulse sequences. It is also easier to calibrate pulse durations by means of a π pulse which corresponds in principle to signal cancellation rather than by a $\pi/2$ pulse which corresponds to a flat maximum. Moreover, it is possible to select the axis around which rotation occurs by modifying the r.f. phase. If the frequency source output is (arbitrarily) identified with the x axis of the rotating frame, it suffices to modify its phase by 90° to operate along the y axis, by 180° to operate along $-x$, by 270° to operate along

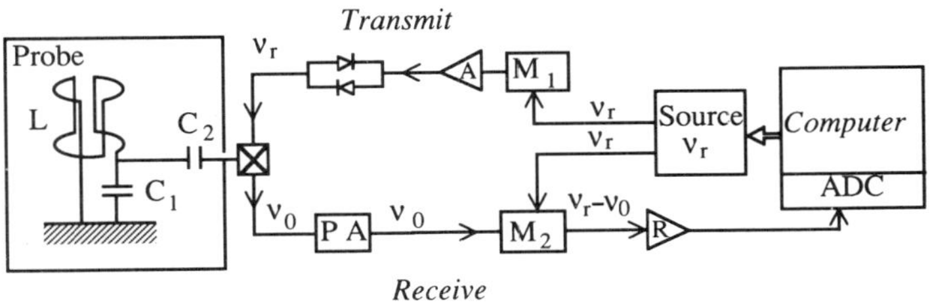

Figure 2.7 Block diagram of a pulse NMR spectrometer. The source (frequency synthesizer) provides a sine wave whose frequency ν_r is close to the nuclear resonance frequency ν_0. This r.f. wave is shaped into pulsed form by means of the mixer M_1 and subsequently amplified (power amplifier A delivering several tens or hundreds of watts) and applied to the probe, which essentially comprises a resonant circuit LC_1 (with the tuning condition $4\pi^2\nu_r^2LC_1 \approx 1$); the capacity C_2 is used for matching the whole circuit (LC_1, C_2) to a pure resistive impedance of 50 Ω. L is the self-induction coefficient of the coil surrounding the sample, which is generally of the saddle type, for a vertical cryomagnet and a vertical sample tube (when other configurations are used, a solenoid is preferable). The purpose of the crossed diodes in the transmitter circuit is to isolate the latter during signal reception. The NMR signal is detected through the same circuit LC_1C_2. The cross is a symbol for diode-based circuitry aimed at commuting and protecting the preamplifier (PA). The NMR signal, at frequency ν_0, is demodulated with respect to ν_r (by means of the mixer M_2) and thus shifted to a low frequency signal $\nu_r - \nu_0$, which is amplified (amplifier R), digitized in an analog-to-digital-converter (ADC) and processed by a computer

$-y$ and, in a general way, by an angle φ to operate along an axis making an angle φ with x.

All these considerations constitute the basis of multipulse experiments, first devised for measuring relaxation times and later for selecting particular spectroscopic information or establishing various types of correlations in the context of multidimensional spectroscopy (Chapter 5).

2.1.4 THE BLOCH EQUATIONS. EVOLUTION OF NUCLEAR MAGNETIZATION IN THE ABSENCE OF RADIO-FREQUENCY FIELDS

It is obvious that equations (2.3) and (2.6) are not correct, as they would predict an endless motion of nuclear magnetization when it has been tipped from its equilibrium configuration. In order to account for the necessary return to thermal equilibrium, damping terms must be introduced, just in the same way as friction forces gradually attenuate the oscillatory motion of a pendulum.

Two distinct relaxation times, T_1, concerning the longitudinal magnetization M_z, and T_2, concerning the transverse components $M_{x,y}$, will account for the various time-dependent interactions which are undergone by nuclear spins (especially the magnetic dipole–dipole interactions) and which are responsible for the damping phenomena alluded to above. Equations (2.3) and (2.7) must be modified by including the terms $-M_{x,y}/T_2$ for the transverse magnetization and the $(M_0 - M_z)/T_1$ for the longitudinal magnetization. This leads to the famous Bloch equations which describe the behavior of nuclear magnetization in the laboratory frame and also in the rotating frame:

$$\frac{\mathrm{d}M_{x,y}}{\mathrm{d}t} = \gamma(\boldsymbol{M} \wedge \boldsymbol{B})_{x,y} - \frac{M_{x,y}}{T_2} \tag{2.9a}$$

$$\frac{\mathrm{d}M_z}{\mathrm{d}t} = \gamma(\boldsymbol{M} \wedge \boldsymbol{B})_z + \frac{(M_0 - M_z)}{T_1} \tag{2.9b}$$

The introduction of relaxation times amounts to allowing nuclear magnetization to return to its equilibrium configuration, according to first order kinetic equations with time constants T_2 and T_1 for the transverse and longitudinal components, respectively. Relaxation times are of the order of a second or tens of seconds if they originate from the dipolar interaction mentioned above. They can however be shortened by several orders of magnitude in the case of efficient relaxation mechanisms (for instance, the quadrupolar interaction). Relaxation times depend also on molecular motions, and especially on molecular reorientation. When molecular reorientation slows down, the longitudinal relaxation first decreases, goes through a minimum value and then increases again (in solids, T_1 may become fairly long) whereas the transverse relaxation time decreases continuously. These points will be examined in more detail in Chapter 4. Except for selective pulses, the duration of r.f. pulses is generally negligible with respect to T_1 and T_2 [equations (2.7) and (2.8) are thus sufficient for treating their effects] and the Bloch equations are merely used for handling the intervals separating these pulses. However, selective pulses are becoming more and more important in the strategy of multipulse NMR experiments and there is a real need to account for relaxation phenomena during an interval of finite duration involving the application of a r.f. field. It turns out that the Bloch equations can be solved analytically. A complete derivation is given in the appendix to this chapter.

T_1 is associated with the recovery of longitudinal magnetization toward its equilibrium value, whereas T_2 is associated with the disappearance of transverse magnetization. This can be viewed through the following example: assume that, at $t = 0$, nuclear magnetization has been taken along the x axis of the rotating frame and that no r.f. field is applied afterwards. Thus the only field acting on the nuclear magnetization is B_0, and the Bloch equation in the rotating frame can be written as

$$\frac{dM_x}{dt} = -(1/T_2)M_x + (\gamma B_0 - \omega_r)M_y$$

$$\frac{dM_y}{dt} = -(\gamma B_0 - \omega_r)M_x - (1/T_2)M_y$$

$$\frac{dM_z}{dt} = (1/T_1)(M_0 - M_z)$$

These three differential equations are more easily solved by introducing the complex transverse magnetization $M_t = M_x + iM_y$; $\nu' = \nu_r - \nu_0$ being the precession frequency in the rotating frame (actually, ν' would be negative with the usual conventions), they reduce to:

$$\frac{dM_t}{dt} = -(1/T_2 - 2i\pi\nu')M_t$$

$$\frac{dM_z}{dt} = (1/T_1)(M_0 - M_z)$$

With the initial conditions $M_t(0) = M_0$ and $M_z(0) = 0$, we obtain

$$M_t = M_0 \exp(-t/T_2) \exp(2i\pi\nu' t) \tag{2.10a}$$

Hence

$$M_x = M_0 \exp(-t/T_2) \cos(2\pi\nu' t) \tag{2.10b}$$

$$M_y = M_0 \exp(-t/T_2) \sin(2\pi\nu' t) \tag{2.10c}$$

$$M_z = M_0[1 - \exp(-t/T_1)] \tag{2.10d}$$

In (2.10b, c, d), the meaning of T_1 and T_2 appears clearly: when t goes to infinity, M_z recovers to the magnetization equilibrium value (in fact, for $t = 5T_1$, $M_z = 0.993M_0$), while any transverse magnetization disappears. It can be noticed that both processes are exponential and independent of each other with, however, $T_2 \leqslant T_1$ (see Chapter 4). Very often, the notation corresponding to relaxation rates, $R_1 = 1/T_1$ and $R_2 = 1/T_2$, will be used. The transverse relaxation rate R_2 reflects the loss of coherence (or the defocusing) of the spins belonging to the sample. It must be appended by a contribution arising from the spatial inhomogeneity of B_0 which produces a distribution of the precession frequency ν'. This distribution enhances the coherence loss in the xy plane; therefore R_2 must be replaced by $R_2^* = R_2 + (R_2)_{\text{inhomogeneity}}$. As a result, the effective relaxation T_2^* is shorter than T_2.

The cosine and sine factors in (2.10b) and (2.10c) represent nuclear precession in the rotating frame. Hence, the NMR signal appears as a damped cosine and sine function (Figure 2.8). For a spectrum containing several resonance lines, the magnetization vector for each of these is described by a separate equation of the form of (2.10a). The result is therefore a superposition of damped sine (or cosine) functions. This set of damped sine (or cosine) functions cannot be interpreted directly. For retrieving the conventional spectrum, one can apply to the complex signal M_t a mathematical operation known under the name of Fourier transformation

$$\mathrm{FT}(M_t) = \int_0^\infty M_t(t)\exp(-2i\pi\nu t)\,\mathrm{d}t \tag{2.11}$$

which will be considered in more detail in the next chapter. Evaluation of the integral in (2.11) leads to

$$\mathrm{FT}(M_t) = \frac{M_0 T_2^*}{1 + 4\pi^2 T_2^{*2}(\nu - \nu')^2} - i\frac{M_0 T_2^{*2} 2\pi(\nu - \nu')}{1 + 4\pi^2 T_2^{*2}(\nu - \nu')^2} \tag{2.12}$$

The first term (which is the real part of the Fourier transform) corresponds to a Lorentzian line with a maximum equal to $M_0 T_2^*$ occurring at $\nu = \nu'$ and a linewidth at half height equal to $1/\pi T_2^*$ (Figure 2.9).

It can finally be noted that, for a pulse flip angle of α smaller than $\pi/2$, similar results hold, with, however, a factor equal to $\sin\alpha$ in front of the right hand sides of (2.10b) and (2.10c) and a factor equal to $\cos\alpha$ in front of the right hand side of (2.10d).

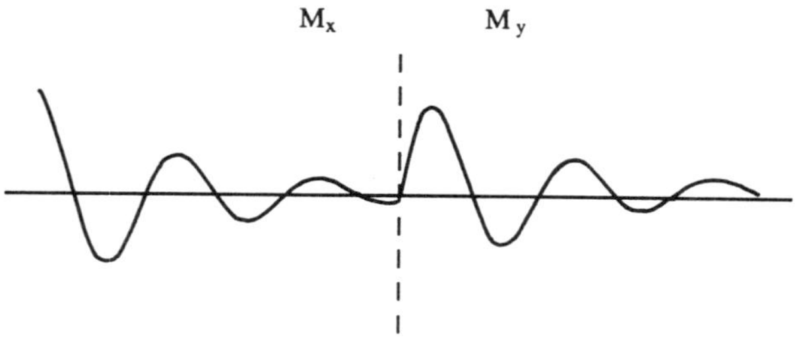

Figure 2.8 Temporal evolution of the two transverse components of nuclear magnetization

2.1.5 DETECTION OF THE NMR SIGNAL BY THE CONTINUOUS WAVE METHOD

This is the mode of detection of early NMR spectrometers. Employed till the 1970s, it is nowadays almost completely abandoned (except at the level of the lock channel, the field–frequency stabilization device) and will therefore be only briefly mentioned. It makes use of a low-amplitude B_1 field, applied continuously, whose

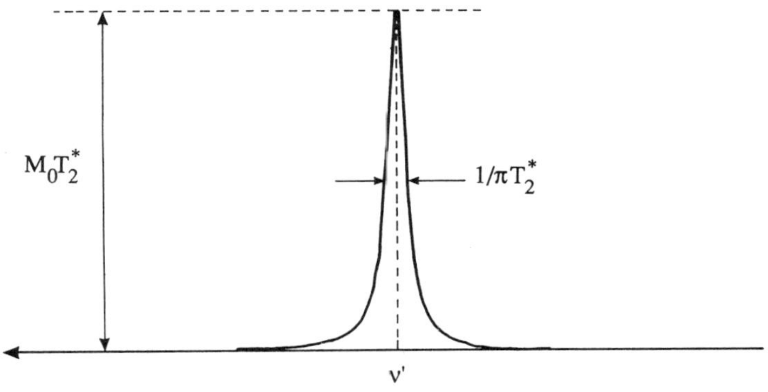

Figure 2.9 A Lorentzian line (the normal line shape for NMR spectra)

frequency ν_r is swept very slowly so that the various resonances in the spectrum appear successively. (From the Larmor equation, slowly sweeping B_0 can be seen to be an equivalent procedure). Under conditions of slow sweep, the system may be considered as quasi-stationary; this can be formulated as $\mathrm{d}\boldsymbol{M}/\mathrm{d}t = 0$. The differential equations (2.9) reduce to a simple set of linear equations:

$$M_x - (\gamma B_0 - \omega_r)T_2^* M_y = 0$$

$$(\gamma B_0 - \omega_r)T_2^* M_x + M_y - \gamma B_1 T_2^* M_z = 0$$

$$\gamma B_1 T_1 M_y + M_z = M_0.$$

These equations allow us to calculate M_y, which is the magnetization component effectively measured because it is in quadrature with respect to the transmitter assumed to be applied along the x axis of the rotating frame. We obtain

$$M_y = \frac{M_0 T_2^{*2} \gamma B_1}{1 + 4\pi^2 T_2^{*2}(\nu_r - \nu_0)^2 + \gamma^2 B_1^2 T_1 T_2^*}. \tag{2.13}$$

If the radio-frequency field is sufficiently weak, then $\gamma^2 B_1^2 T_1 T_2 \ll 1$, and (2.13) shows up again in the form of a Lorentzian signal; this indicates a strong relationship between the two detection modes: continuous wave and pulse mode followed by a Fourier transform. The advantages of the latter method will be detailed in the next chapter.

2.2 Quantum Mechanical Approach. Introduction to the Density Operator

The previous section almost ignores quantum mechanics, which should however always be employed because nuclear spins are microscopic entities. Yet the Bloch equations appear to be quite adequate in many situations. On the other hand, the quantum mechanical treatment presented in Chapter 1 describes perfectly the features of an NMR spectrum although it does not seem *a priori* to be able to account for the temporal evolution of the spin system under radio-frequency pulses or during free precession intervals. This failure arises from the non-recognition of the *ensemble* concept. As a matter of fact, this latter concept must be associated with the consideration of the different states pertaining to *a single* system, as they are derived from the application of the quantum mechanics principles. Fortunately, there exists a tool capable of reconciling both these concepts, which is called the *density operator*. Of course, this tool must be defined and the way in which it evolves must be rationalized. This will be achieved through the Liouville–von Neumann equation.

Before going into the details of this approach, it may be worth delineating the domain of application of the Bloch equations, or equivalently of equations (2.7), which up to now have enabled us to predict the behavior of nuclear magnetization under the application of a radio-frequency pulse. The answer may be formulated in a simple way: these equations originate from classical mechanics and concern *vectorial and macroscopic* quantities. *Therefore, provided that the spin system can be treated as a vector (for instance, the nuclear magnetization whose components are* M_x, M_y *and* M_z*), precession and nutation motions can be adequately described by the Bloch equations*. Regarding spin relaxation, we may conceive that the two time constants (the longitudinal and transverse relaxation times), introduced on a purely phenomenological basis, can suffer some exceptions. This point will be discussed in Chapter 4.

The latter considerations lead us to resort to the Bloch equations for the three components of nuclear magnetization, in particular for the case of a system involving a single spin. However, in situations where the state of a spin system can no longer be described by these three components, it is likely that Bloch equations will fail in predicting its evolution. An example frequently encountered is that of an antiphase doublet schematized in Figure 2.10. This configuration represents a mandatory stage in most coherence transfer experiments (Chapter 5) and is achieved through an appropriate succession of events (r.f. pulses and precession intervals).

It is quite obvious that the three magnetization components of A cannot by any means describe this state. This is because I_x^A, which is substituted for M_x as far as spin A is concerned, can be associated with the A doublet only if the two lines are

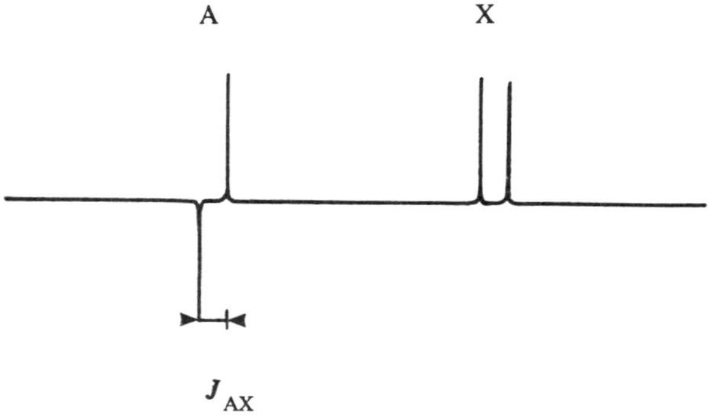

Figure 2.10 The A antiphase doublet for a weakly J-coupled spin system (two spins 1/2 A and X, the indirect coupling is denoted by J_{AX})

of the same sign. Whenever they are of different sign, we shall see (Section 2.2.6) that a property relating to X must be introduced leading to the quantity $2I_x^A I_z^X$ which represents the A antiphase doublet and will permit its evolution to be followed. There exist in fact many other situations which require an approach more sophisticated than the one based on Bloch equations. The density operator approach will prove to be universal in reaching such goals.

2.2.1 DEFINITION OF THE DENSITY MATRIX AND OF THE DENSITY OPERATOR. EVOLUTION EQUATIONS

We shall proceed here in an unusual way by first defining a matrix and subsequently looking for the properties of the operator with which this matrix is associated. The density matrix is defined for an arbitrary state of the system, described by a vector ψ which can be expanded on an appropriate basis. This basis will be naturally taken as the set of the eigenvectors of the static hamiltonian $\hat{\mathcal{H}}_0$. Let us recall that $\hat{\mathcal{H}}_0$ is time-independent and leads to the energy levels of the system (eigenvalues) and to the relevant eigenvectors ψ_K. Because ψ, in the most general case, is time-dependent, the temporal dependence should appear in the expansion coefficients as indicated by

$$\psi(t) = \sum_{K=1}^{2^n} c_K(t)\psi_K \tag{2.14}$$

The summation over K extends to 2^n, which is the number of eigenstates in a system of n spin 1/2 nuclei. The (K, L) element of the density matrix is defined as

$$\sigma_{KL} = \overline{c_K c_L^*} \tag{2.15}$$

where the bar denotes an average over the ensemble of systems included in the sample. The following properties can be derived from this definition:

(1) The density matrix, and consequently the density operator, associated with the density matrix, are hermitian.

(2) The Kth diagonal element can be written

$$\sigma_{KK} = \overline{|c_K|^2} = p_K \tag{2.16}$$

and represents the population of the Kth energy level, owing to the fact that in (2.14) c_K represents the contribution of the eigenstate ψ_K, and that the concept of probability (or population) is related to the square of the modulus of the relevant vector. As a consequence, the trace of σ is equal to 1 (it is recalled that the trace of a matrix is obtained by summing all its diagonal elements).

(3) A non-zero value of an off-diagonal element σ_{KL} ($K \neq L$) reveals a coherence between eigenstates K and L. This coherence of course does not hold at thermal equilibrium since, in that case, the relative phases pertaining to eigenstates K and L are randomly distributed (it can be recalled that any vector ψ can be multiplied by an arbitrary phase factor $e^{i\varphi}$, because its only physical significance stems from the square of its modulus $|\psi|^2$ which indicates the probability of the relevant state). An external perturbation (e.g. a radio-frequency pulse) can, however, create a phase relationship and induce coherence. The latter will be classified according to the difference in $\hat{F}_z$ eigenvalues ($\hat{F}_z = \sum_{i=1}^{n} \hat{I}_z^i$). In that way, for $|(F_z)_{KK} - (F_z)_{LL}| = 0, 1, 2, 3 \ldots$, σ_{KL} represents zero quantum, one quantum,

two quanta, three quanta... coherences, respectively (one quantum coherences are the only ones physically observable).

(4) The density operator evolves according to the Liouville–von Neumann equation

$$\mathrm{d}\hat{\sigma}/\mathrm{d}t = 2i\pi[\hat{\sigma}, \hat{\mathcal{H}}(t)] \tag{2.17}$$

where $\hat{\mathcal{H}}(t)$, expressed in Hz, is the *full* hamiltonian of the system including all time-dependent terms which arise from radio-frequency fields or from interactions modulated by molecular motions (capable of inducing relaxation phenomena). Owing to its importance, equation (2.17) deserves a proof, which is given now. Any quantum system governed by the time-dependent hamiltonian $\hat{\mathcal{H}}(t)$ obeys the Schrödinger equation (which can be considered as universal in quantum mechanics):

$$\hat{\mathcal{H}}(t)\psi = -\frac{1}{2i\pi}\frac{\partial\psi}{\partial t}$$

This equation can be applied to a vector ψ describing an arbitrary state of the system. Inserting the expansion (2.14) in the above equation, one obtains

$$\hat{\mathcal{H}}(t)\sum_{\mathrm{K}'} c_{\mathrm{K}'}\psi_{\mathrm{K}'} = -\frac{1}{2i\pi}\sum_{\mathrm{K}'}\frac{\mathrm{d}c_{\mathrm{K}'}}{\mathrm{d}t}\psi_{\mathrm{K}'}$$

Taking the scalar product to the left by ψ_{K} and using the orthogonality and normalization properties of the set of eigenvectors $\{\psi_{\mathrm{K}}\}$, we can derive a simple expression for $\mathrm{d}c_{\mathrm{K}}/\mathrm{d}t$:

$$\frac{\mathrm{d}c_{\mathrm{K}}}{\mathrm{d}t} = -2i\pi\sum_{\mathrm{K}'} c_{\mathrm{K}'}\mathrm{H}_{\mathrm{KK}'}$$

where $\mathrm{H}_{\mathrm{KK}'} = \langle\psi_{\mathrm{K}}|\hat{\mathcal{H}}(t)\psi_{\mathrm{K}'}\rangle$. In a similar way, we obtain for $\mathrm{d}c_{\mathrm{L}}^*/\mathrm{d}t$:

$$\frac{\mathrm{d}c_{\mathrm{L}}^*}{\mathrm{d}t} = 2i\pi\sum_{\mathrm{K}'} c_{\mathrm{K}'}^*\mathrm{H}_{\mathrm{K}'\mathrm{L}}$$

The two previous relationships lead to the expression of $\mathrm{d}\sigma_{\mathrm{KL}}/\mathrm{d}t$ (which can be expanded as $(\mathrm{d}c_{\mathrm{K}}/\mathrm{d}t)c_{\mathrm{L}}^* + c_{\mathrm{K}}(\mathrm{d}c_{\mathrm{L}}^*/\mathrm{d}t)$):

$$\frac{\mathrm{d}\sigma_{\mathrm{KL}}}{\mathrm{d}t} = -2i\pi\sum_{\mathrm{K}'}(\mathrm{H}_{\mathrm{KK}'}\sigma_{\mathrm{K}'\mathrm{L}} - \sigma_{\mathrm{KK}'}\mathrm{H}_{\mathrm{K}'\mathrm{L}})$$

The right-hand side of the latter equation can be expressed as the (K, L) element of the matrix associated with $\hat{\sigma}\hat{\mathcal{H}} - \hat{\mathcal{H}}\hat{\sigma}$

$$\frac{\mathrm{d}\sigma_{\mathrm{KL}}}{\mathrm{d}t} = 2i\pi(\sigma\mathrm{H} - \mathrm{H}\sigma)_{\mathrm{KL}}$$

which is none other than equation (2.17) in a matricial form and completes the proof of the Liouville–von Neumann equation.

The ensemble average of any quantity G (eventually determined through an experimental procedure) can be evaluated from a knowledge of the density operator and through the following trace relationship

$$\langle G(t)\rangle = Tr(\hat{\sigma}\hat{G}) \tag{2.18}$$

where $\hat{G}$ is the operator associated with the relevant quantity. (In fact $\langle G(t)\rangle$

should be denoted by $\overline{\langle \hat{G}(t) \rangle}$, since we are dealing with both an ensemble average and an expectation value).

The proof of (2.18) is straightforward, and is obtained through the equations given below, where $\langle \hat{G}(t) \rangle$ is calculated for an arbitrary state described by the vector Ψ:

$$\overline{\langle \hat{G}(t) \rangle} = \overline{\langle \Psi | \hat{G} | \Psi \rangle} = \overline{\left\langle \sum_{\mathrm{L}} c_{\mathrm{L}} \psi_{\mathrm{L}} | \hat{G} | \sum_{\mathrm{K}} c_{\mathrm{K}} \psi_{\mathrm{K}} \right\rangle}$$

$$= \sum_{\mathrm{L,K}} \overline{c_{\mathrm{L}}^* c_{\mathrm{K}}} \langle \psi_{\mathrm{L}} | \hat{G} | \psi_{\mathrm{K}} \rangle = \sum_{\mathrm{L,K}} \sigma_{\mathrm{KL}} G_{\mathrm{LK}}$$

$$= \sum_{\mathrm{K}} (\sigma \mathrm{G})_{\mathrm{KK}} = Tr(\sigma \mathrm{G})$$

In spite of their abstract character, equations (2.18) and (2.17) constitute the basis of all rigorous treatments of the temporal evolution of the ensemble of spin systems. However, this approach is meaningful only as long as we have at hand initial conditions pertaining to the density operator. In particular, the density operator at thermal equilibrium, denoted by $\hat{\sigma}_{\mathrm{eq}}$, must be known. It can be evaluated through the usual rules of statistical mechanics applied to the Kth diagonal element of the density matrix:

$$\sigma_{\mathrm{KK}}^{\mathrm{eq}} = p_{\mathrm{K}}^{\mathrm{eq}} = \exp(-E_{\mathrm{K}}/\mathrm{k_B} T)/Z$$

where Z is the partition function, equal to $\sum_{\mathrm{K}} \exp(-E_{\mathrm{K}}/\mathrm{k_B} T)$, and which can be approximated by

$$Z \approx \sum_{\mathrm{K}} (1 - E_{\mathrm{K}}/\mathrm{k_B} T).$$

according to the so-called high-temperature hypothesis. Since $\sum_{\mathrm{K}} E_{\mathrm{K}} = 0$ (because, when a static magnetic induction is applied, energy level splittings are equally distributed between negative and positive values), Z reduces to the number Ne of energy levels of the system (when dealing with n nuclei of spin 1/2, Ne is equal to 2^n) and $\sigma_{\mathrm{KK}}^{\mathrm{eq}}$ can be written as

$$\sigma_{\mathrm{KK}}^{\mathrm{eq}} \approx (1 - E_{\mathrm{K}}/\mathrm{k_B} T)/Ne$$

As far as the calculation of $\sigma_{\mathrm{KK}}^{\mathrm{eq}}$ is concerned, we may safely retain only the Zeeman term in the static hamiltonian (terms relating to direct or indirect couplings modify $E_{\mathrm{K}}/\mathrm{k_B} T$ by a tiny amount), and E_{K} can be approximated as

$$E_{\mathrm{K}} \approx \left\langle \psi_{\mathrm{K}} | - \sum_j \gamma_j \hbar \hat{I}_z^j | \psi_{\mathrm{K}} \right\rangle$$

which leads to the expression of the density operator at thermal equilibrium ($\hat{E}$ stands for the identity operator)

$$\hat{\sigma}_{\mathrm{eq}} = \frac{1}{N_e} \hat{E} + \frac{\hbar}{N_e \mathrm{k_B} T} \sum_j \gamma_j \hat{I}_z^j \tag{2.19}$$

For convenience, we shall write $\hat{\sigma}_{\mathrm{eq}}$ as

$$\hat{\sigma}_{\mathrm{eq}} = P_0 \hat{E} + \sum_j \Delta_j \hat{I}_z^j \tag{2.20a}$$

where P_0 and Δ_j are easily deduced from the comparison of (2.19) and (2.20). For a homonuclear system, the expression of $\hat{\sigma}_{eq}$ is even simpler:

$$(\hat{\sigma}_{eq})_{\text{homonucl.syst.}} = P_0\hat{E} + \Delta\hat{F}_z \tag{2.20b}$$

where $\hat{F}_z$ is the operator associated with the z component of the total spin momentum (Chapter 1). In that case, Δ represents the population difference of energy levels corresponding to consecutive eigenvalues of $\hat{F}_z$. The meaning of Δ, and of Δ_j, is clarified in the two examples given in Figure 2.11.

The possibility of expressing the density operator in terms of spin operators must be emphasized [see equations (2.19) and (2.20)] and we shall see in the next section that such an expansion, for an arbitrary state, is always possible. This expansion can involve the so-called product operators, which provides a better insight into the evolution of a spin system either under the application of r.f. pulses or in the course of free precession intervals.

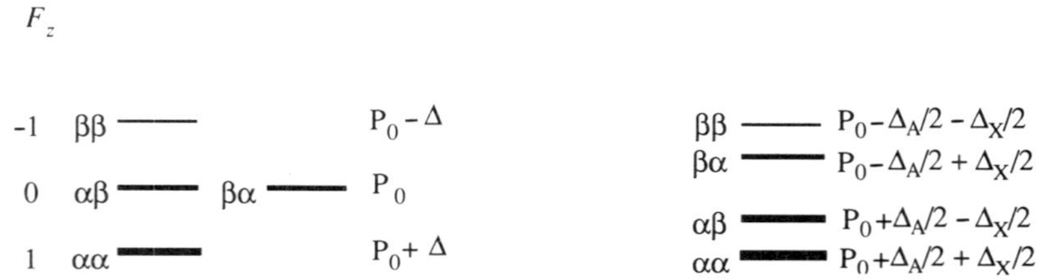

Figure 2.11 Energy level populations of a system of two spin 1/2 nuclei. Left: an homonuclear system. Right: a AX heteronuclear system with $\Delta_A > \Delta_X$ (A being for instance a proton and X a carbon-13)

2.2.2 EXPANSION OF THE DENSITY OPERATOR ON A BASIS OF PRODUCT OPERATORS (SPIN 1/2 SYSTEMS)

A system including n nuclei of spin 1/2 possesses 2^n eigenstates, hence the density matrix involves $(2^n \times 2^n) = 2^{2n}$ elements. The dimension of the *Liouville space* is thus 2^{2n} and requires the density operator to be expanded on a basis of 2^{2n} operators, constructed from $\hat{I}_x$, $\hat{I}_y$ and $\hat{I}_z$. We shall now demonstrate that it is feasible, in a general way, to devise 2^{2n} operators, orthogonal and normalized, appearing as products of spin operators and written as

$$\hat{U}_r = N_r \prod_{j=1}^{n} \hat{S}_j^{(r)} \tag{2.21}$$

where $\hat{S}_j^{(r)} = \hat{E}_j$, $\hat{I}_x^j$, $\hat{I}_y^i$ or $\hat{I}_z^j$, the subscript j referring to one of the spins and N_r being a normalization factor which will be calculated according to procedures indicated below. Furthermore, each $\hat{U}_r$ is supposed to be different from the other by at least one $\hat{S}_j$; this ensures the independence of the product operators $\hat{U}_r$, whose number is indeed 2^{2n} because there exist four possible operators for each spin, hence $4^n = 2^{2n}$ different ways of designing the $\hat{U}_r$ operators. It remains to demonstrate that the set of $\hat{U}_r$ operators is normalized and orthogonal. These two latter features can be defined according to trace properties

$$Tr(\hat{U}_r\hat{U}_S) = \delta_{rs} \tag{2.22}$$

It may be worth recalling that the trace of an operator is identical to the trace of the matrix with which it is associated, the trace of a matrix being independent of the basis used for constructing this matrix. A well-known example is provided by the Pauli matrices which apply to a single spin 1/2 and which are constructed on the $|\alpha\rangle$, $|\beta\rangle$ basis

$$I_x = \frac{1}{2}\begin{bmatrix} 0 & 1 \\ 1 & 0 \end{bmatrix} \quad I_y = \frac{i}{2}\begin{bmatrix} 0 & -1 \\ 1 & 0 \end{bmatrix} \quad I_z = \frac{1}{2}\begin{bmatrix} 1 & 0 \\ 0 & -1 \end{bmatrix}$$

By simple matrix products, the following trace relationships can be derived:

$$Tr(\hat{E}) = 2;\ Tr(\hat{I}_u) = 0;\ Tr(\hat{I}_u^2) = 1/2;\ Tr(\hat{I}_u\hat{I}_v) = 0 \tag{2.23}$$

with $u \neq v$ and $u, v = x, y$ or z.

The trace of the product $\hat{U}_r\hat{U}_s$ can be written in the form of a product of traces:

$$Tr(\hat{U}_r\hat{U}_s) = N_rN_s\prod_{j=1}^{n} Tr(\hat{S}_j^{(r)}\hat{S}_j^{(s)})$$

As there exists at least one spin j′ by which the two operators $\hat{U}_r$ and $\hat{U}_s$ differ, $Tr(\hat{S}_{j'}^{(r)}\hat{S}_{j'}^{(s)})$ is necessarily zero by virtue of the last relationship in (2.23). Consequently, $\hat{U}_r$ and $\hat{U}_s$ are orthogonal. Again, normalization coefficients can be evaluated through relationships (2.23). We obtain

$$Tr(\hat{U}_r^2) = N_r^2\prod_{j=1}^{n} Tr[\hat{S}_j^{(r)2}] = N_r^2 2^{e_r}(1/2)^{n-e_r}$$

where e_r denotes the number of identity operators $\hat{E}_j$ in $\hat{U}_r$.

Finally, for a system of n spin 1/2 nuclei, any operator $\hat{U}_r$ can be expressed as

$$\hat{U}_r = \frac{1}{2^{e_r - n/2}}\prod_{j=1}^{n}\hat{S}_j^{(r)}$$

where $\hat{S}_j^{(r)} = \hat{E}_j$, $\hat{I}_x^j$, $\hat{I}_y^j$ or $\hat{I}_z^j$ and e_r is the number of identity operators in $\hat{U}_r$. As has been shown above, the set of $\hat{U}_r$ operators constitutes a complete orthonormalized basis upon which any operator, and especially the density operator, can be expanded.

We shall consider here only two spin systems: (i) a single spin 1/2 (dimensionality of the relevant Liouville space: 4); (ii) two spins 1/2 (dimensionality of the relevant Liouville space: 16), which will be analyzed in the next section. Extension to more complicated spin systems is straightforward and is only a matter of calculation. As far as a single spin 1/2 system is concerned, the basis $\{\hat{U}_r\}$ is obvious: $\hat{E}/2$, $\hat{I}_x$, $\hat{I}_y$ and $\hat{I}_z$. Owing to the vectorial nature of this basis, Bloch equations apply.

Quite generally, it will always be possible to write the density operator representing an arbitrary vector in the form

$$\hat{\sigma} = \sum_{r=1}^{2^{2n}} c_r\hat{U}_r \tag{2.24}$$

The corresponding state is actually defined by the coefficients c_r. The interest of (2.24) lies in the ability to follow the evolution of $\hat{\sigma}$. This will be done from the knowledge of the evolution of each $\hat{U}_r$, which can be calculated once and for all (see Sections 2.2.4 and 2.2.6). Likewise, it will prove possible to determine the evolution of any quantity $\langle G\rangle$ through the equation (2.18):

$$\langle G\rangle = Tr(\hat{\sigma}\hat{G}) = \sum_{r=1}^{2^{2n}} c_r Tr(\hat{U}_r\hat{G}).$$

2.2.3 BASIS OF PRODUCT OPERATORS OF WEAKLY COUPLED SYSTEM OF TWO SPIN 1/2 NUCLEI. PHYSICAL MEANING

Our aim is here to construct the set of 16 operators according to the rules stated in the preceding section. Moreover, for visualization purposes, we shall try to devise vectorial representations and describe the way of in which they must be treated. One of these representations concerns the four transitions of a weakly coupled AX spin system. It is schematized in Figures 2.12a and 2.12b (vectors A_1, A_2, X_1 and X_2) and is consistently justified by the form of the matrix associated with each of the relevant product operators. One must however bear in mind that simple vector rotations do not necessarily account for spin dynamics. There exist circumstances where another representation must be put forward: it consists of associating with each product operator a 'global' vector shown as double arrows in Figures 2.12a and 2.12b according to the relevant coherence order. As indicated below, we shall resort to one or the other representation depending on the type of event to be handled.

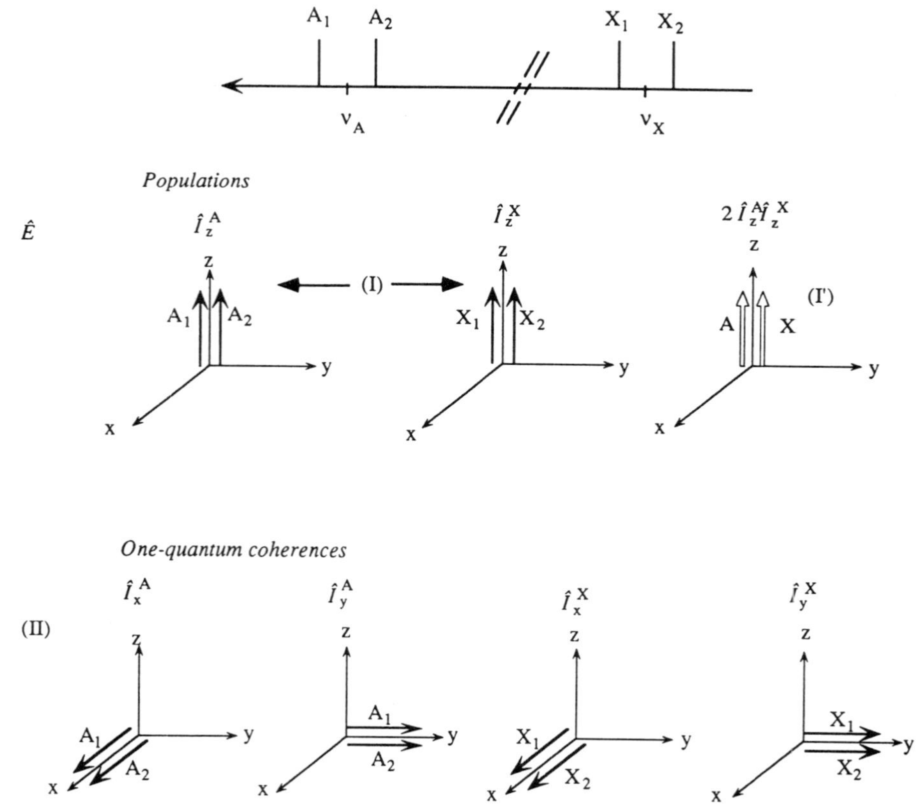

Figure 2.12a Vectorial representations of product operators for a weakly coupled system of two spin 1/2 nuclei. Classes are defined by roman numbers (here, classes (I), (I′) and (II)); each element in a given class transform according to similar rules. Simple arrows are associated with the different lines in the conventional spectrum. Double arrows refer to the 'global' representation

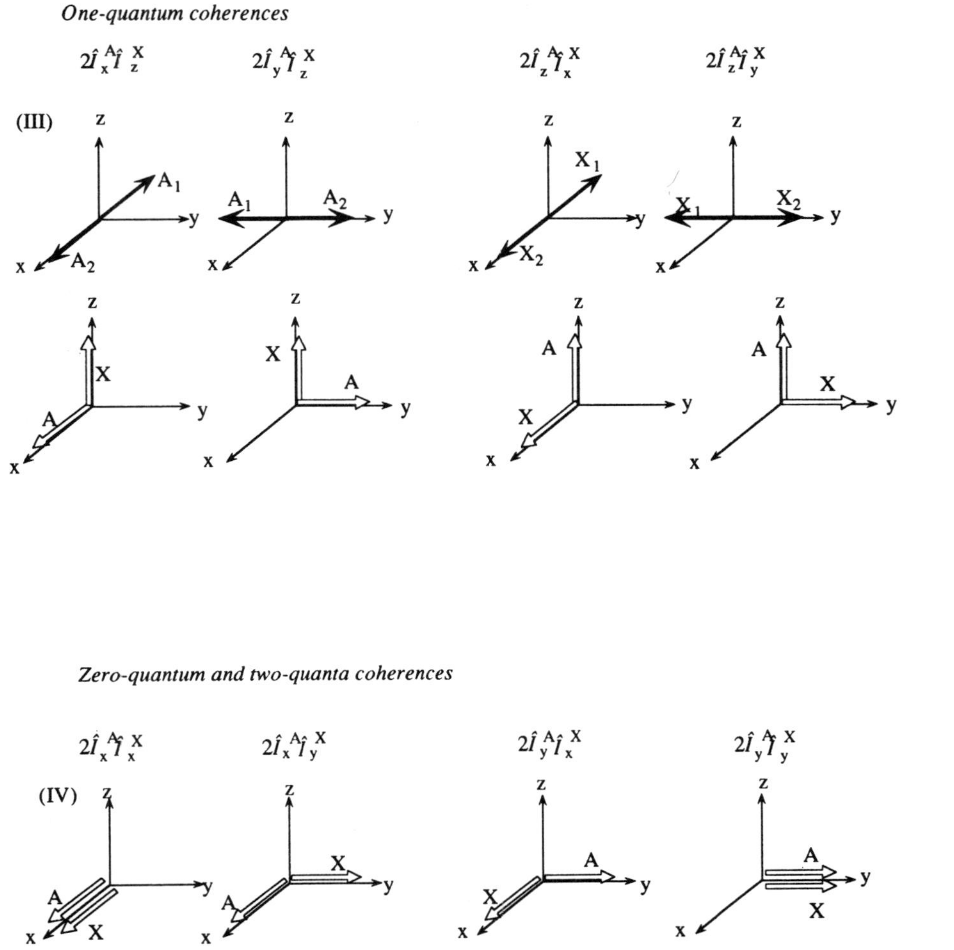

Figure 2.12b Classes (III) and (IV). See Figure 2.12a for the meaning of the different symbols

These representations have been constructed with the help of the following guidelines:

(1) Populations. The density matrix is in a diagonal form. Class (I) involves operators concerning a single spin, albeit coupled to another one, and can be represented by the vectors associated with the transitions pertaining to that spin. Class (I′) corresponds to the product operator $2\hat{I}_z^A\hat{I}_z^X$ and is necessarily represented by A and X 'global' vectors (double arrow) pointing toward z.

(2) One-quantum coherences. The density matrix involves only off-diagonal elements σ_{KL}, such that $|(F_z)_K - (F_z)_L| = 1$. Again, single spin operators are represented by vectors associated with individual transitions. The case of product operators (involving two spins) is more delicate: they can be described by 'global' vectors oriented along the relevant directions or by vectors associated with individual

transitions (because such a representation can be considered here), owing to the fact that one-quantum coherences are directly observable. The circumstances under which one or the other representation is more appropriate will be discussed later. For the moment, let us consider the example of the product operator $2\hat{I}_x^A\hat{I}_z^X$ and let us devise the relevant representation involving vectors associated with individual transitions. Relying on density matrix calculations, we can easily show that this product operator corresponds to vectors A_1 and A_2 in an antiphase configuration, along $-x$ and $+x$, respectively. In a more intuitive way, we can notice that A magnetization along x, accounted for by I_x^A, is affected by I_z^X (in the product operator). Since A_2 is associated with the eigenvalue $+1/2$ of I_z^X (Figure 1.39), A_2 must be oriented along the $+x$. Conversely, A_1 being associated with the eigenvalue $-1/2$ of I_z^X, it must be oriented along $-x$.

(3) Zero-quantum and two-quanta coherences. The corresponding density matrix elements are off-diagonal. They are denoted by σ_{KL}, and fulfill the condition $|(F_z)_K - (F_z)_L| = 0$ or 2 for zero-quantum or two-quanta coherences, respectively. Since no directly observable transitions can be defined in such cases, the only possible representation is that of 'global' vectors. It can be noted that pure zero-quantum or two-quanta coherences do not match product operators but, rather, are linear combinations of the latter:

- zero-quantum: $$2\hat{I}_x^A\hat{I}_x^X + 2\hat{I}_y^A\hat{I}_y^X = \hat{I}_+^A\hat{I}_-^X + \hat{I}_-^A\hat{I}_+^X$$ $$2\hat{I}_y^A\hat{I}_x^X - 2\hat{I}_x^A\hat{I}_y^X = i(\hat{I}_-^A\hat{I}_+^X - \hat{I}_+^A\hat{I}_-^X)$$

(the product of *one* raising operator by *one* lowering operator corresponds to no change in the $\hat{F}_z$ eigenvalues, and thus the terminology of zero-quantum coherences)

- two-quanta: $$2\hat{I}_x^A\hat{I}_x^X - 2\hat{I}_y^A\hat{I}_y^X = \hat{I}_+^A\hat{I}_+^X + \hat{I}_-^A\hat{I}_-^X$$ $$2\hat{I}_y^A\hat{I}_x^X + 2\hat{I}_x^A\hat{I}_y^X = i(\hat{I}_-^A\hat{I}_-^X - \hat{I}_+^A\hat{I}_+^X)$$

(the product of *two* raising or of *two* lowering operators corresponds to a change of ± 2 in the $\hat{F}_z$ eigenvalues, and thus the terminology of two-quanta coherences).

2.2.4 EVOLUTION DURING A FREE PRECESSION PERIOD (TIME-INDEPENDENT HAMILTONIAN)

We shall assume here that the hamiltonian $\hat{\mathcal{H}}(t)$ in equation (2.17) is restricted to the static hamiltonian $\hat{\mathcal{H}}_0$. This means that relaxation phenomena are disregarded and that no r.f. pulse is applied during the relevant period. The solution to equation (2.17) may be written as

$$\hat{\sigma}(t) = \exp(-2i\pi\hat{\mathcal{H}}_0 t)\hat{\sigma}(0)\exp(2i\pi\hat{\mathcal{H}}_0 t) \tag{2.25}$$

which can be easily formulated for a system involving a single spin 1/2 since, in that case, the hamiltonian (in Hz) is the simple form $\hat{\mathcal{H}}_0 = -\nu\hat{I}_z$

$$\hat{\sigma}(t) = \exp(2i\pi\nu t\hat{I}_z)\hat{\sigma}(0)\exp(-2i\pi\nu t\hat{I}_z) \tag{2.26}$$

where $\hat{\sigma}(0)$ stands for the density operator at the beginning of the free precession period. Since, in a general way, $\hat{\sigma}$ can be expanded on the basis defined above (which includes $\hat{E}$, $\hat{I}_x$, $\hat{I}_y$ and $\hat{I}_z$), it is mandatory to determine the effect of the exponential operators of (2.26) upon $\hat{I}_x$, $\hat{I}_y$ and $\hat{I}_z$. Such an operator is defined by the expansion of the exponential function

$$\exp(\theta \hat{I}_z) = \hat{E} + \theta \hat{I}_z + \frac{\theta^2}{2!}(\hat{I}_z)^2 + \frac{\theta^3}{3!}(\hat{I}_z)^3 + \ldots$$

and we must calculate $f = \exp(i\theta \hat{I}_z)\hat{I}_x \exp(-i\theta \hat{I}_z)$. Deriving f with respect to θ and recognizing that $\hat{I}_z$ commutes with $\exp(i\theta \hat{I}_z)$, we obtain

$$\frac{\mathrm{d}f}{\mathrm{d}\theta} = i\exp(i\theta \hat{I}_z)(\hat{I}_z\hat{I}_x - \hat{I}_x\hat{I}_z)\exp(-i\theta \hat{I}_z)$$

which, by virtue of the commutation rules given in (1.8), yields

$$\frac{\mathrm{d}f}{\mathrm{d}\theta} = -\exp(i\theta \hat{I}_z)\hat{I}_y \exp(-i\theta \hat{I}_z)$$

By a further differentiation with respect to θ and again applying the commutation rules of spin operators, we obtain

$$\frac{\mathrm{d}^2 f}{\mathrm{d}\theta^2} = -i\exp(i\theta \hat{I}_z)(\hat{I}_z\hat{I}_y - \hat{I}_y\hat{I}_z)\exp(-i\theta \hat{I}_z) = -f$$

which can also be written as $\mathrm{d}^2 f/\mathrm{d}\theta^2 + f = 0$. The solution of this differential equation has the form $f = a\cos\theta + b\sin\theta$. a and b are determined from $f(0) = \hat{I}_x$ and $(\mathrm{d}f/\mathrm{d}\theta)_{\theta=0} = -\hat{I}_y$, respectively. This provides the first of the relationships listed below (2.27), the second being derived along the same lines and the third from the evident commutation of $\hat{I}_z$ with $\exp(i\theta \hat{I}_z)$

$$\begin{aligned} \exp(i\theta \hat{I}_z)\hat{I}_x \exp(-i\theta \hat{I}_z) &= \cos\theta \hat{I}_x - \sin\theta \hat{I}_y \\ \exp(i\theta \hat{I}_z)\hat{I}_y \exp(-i\theta \hat{I}_z) &= \cos\theta \hat{I}_y + \sin\theta \hat{I}_x \\ \exp(i\theta \hat{I}_z)\hat{I}_z \exp(-i\theta \hat{I}_z) &= \hat{I}_z \end{aligned} \tag{2.27}$$

These relationships prove to be of primary importance, and further relationships involving $\hat{I}_x$ and $\hat{I}_y$ can be deduced from (2.27) by x, y, z permutations. An essential feature emerges from the examination of (2.27): *everything behaves as if the operators $\hat{I}_x$ and $\hat{I}_y$ have been rotated clockwise (this is the sense of precession) by an angle θ around the z axis*. As expected, we retrieve results consistent with Bloch equations.

The behavior of a two-spin system, even weakly coupled, may be quite different. For such a system, the hamiltonian $\mathcal{H}_0 = -\nu_A \hat{I}_z^A - \nu_X \hat{I}_z^X + J\hat{I}_z^A\hat{I}_z^X$ will interfere in the exponential operators of equation (2.25) without leading necessarily to simple rotations. However, the evolution of product operators still rest on relationships (2.27); since the density operator $\hat{\sigma}$ is expanded on the basis of product operators, the relevant calculations are required. These calculations are better handled with the help of the particular properties of spin 1/2 operators, which can be deduced from the Pauli matrices

$$\hat{I}_x^2 = \hat{I}_y^2 = \hat{I}_z^2 = \hat{E}/4 \tag{2.28}$$

$$\hat{I}_x\hat{I}_y = -\hat{I}_y\hat{I}_x = i\hat{I}_z/2 \tag{2.29}$$

Relations homologous to (2.29) can be derived by changing x into y, y into z, z into x and so on. There exist other useful relationships which again stem directly from the Pauli matrices:

$$\cos(2u\hat{I}_z) = (\cos u)\hat{E} \tag{2.30a}$$

$$\sin(2u\hat{I}_z) = 2(\sin u)\hat{I}_z \tag{2.30b}$$

In the latter equations, u must be free from any operator relevant to the considered spin but may include operators pertaining to another spin. Moreover, because the different terms contained in $\hat{\mathcal{H}}_0$ commute, $\exp(i\hat{\mathcal{H}}_0 t)$ can be written as a product of exponential operators by virtue of the fact that if the operators $\hat{A}$ and $\hat{B}$ commute, then $\exp(\hat{A} + \hat{B}) = \exp(\hat{A})\exp(\hat{B})$. Finally, we shall use the trivial property that operators relevent to different spins commute. As a matter of example, let us consider the evolution of $\hat{I}_x^A$; this rests on the evaluation of

$$\exp(-2i\pi\hat{\mathcal{H}}_0 t)\,\hat{I}_x^A\,\exp(2i\pi\hat{\mathcal{H}}_0 t)$$

which, by virtue of the above mentioned properties, can be expressed as:

$$\exp(2i\pi\nu_A\hat{I}_z^A t)\exp(-2i\pi J\hat{I}_z^A\hat{I}_z^X t)\hat{I}_x^A\exp(2i\pi J\hat{I}_z^A\hat{I}_z^X t)\exp(-2i\pi\nu_A\hat{I}_z^A t).$$

The first step is to calculate $\exp(-2i\pi J\hat{I}_z^A\hat{I}_z^X t)\,\hat{I}_x^A\,\exp(2i\pi J\hat{I}_z^A\hat{I}_z^X t)$. From (2.30a) and (2.30b), it is easy to show that

$$\exp(-2i\pi J\hat{I}_z^A\hat{I}_z^X t) = \cos(\pi J t/2)\hat{E} - 2i\sin(\pi J t/2)(2\hat{I}_z^A\hat{I}_z^X)$$

and, using (2.29), we arrive at

$$\exp(-2i\pi J\hat{I}_z^A\hat{I}_z^X t)\,\hat{I}_x^A\,\exp(2i\pi J\hat{I}_z^A\hat{I}_z^X t) = \cos(\pi J t)\hat{I}_x^A + \sin(\pi J t)(2\hat{I}_y^A\hat{I}_z^X)$$

This corresponds to the superposition of two states: one represented by $\hat{I}_x^A$ (the A doublet along the x axis of the rotating frame) weighted by $\cos(\pi J t)$, the other represented by $2\hat{I}_y^A\hat{I}_z^X$ (the A doublet in an *antiphase* configuration along y), weighted by $\sin(\pi J t)$. This amounts to the rotation of A_1 and A_2 by an angle of $\pm\pi J t$ as indicated in Figure 2.13. The remainder of the calculation concerns evolution according to the precession frequency of nucleus A:

$$\exp(2i\pi\nu_A\hat{I}_z^A t)[\cos(\pi J t)\hat{I}_x^A + \sin(\pi J t)(2\hat{I}_y^A\hat{I}_z^X)]\exp(-2i\pi\nu_A\hat{I}_z^A t)$$

$$= \cos(\pi J t)\cos(2\pi\nu_A t)\hat{I}_x^A - \cos(\pi J t)\sin(2\pi\nu_A t)\hat{I}_y^A$$

$$+ \sin(\pi J t)\sin(2\pi\nu_A t)(2\hat{I}_x^A\hat{I}_z^X) + \sin(\pi J t)\cos(2\pi\nu_A t)(2\hat{I}_y^A\hat{I}_z^X)$$

This result indicates a global precession by an angle $2\pi\nu_A t$; hence, the global configuration shown in the right-hand side of Figure 2.13 has to be rotated (clockwise) by an angle of $2\pi\nu_A t$. Finally, and consistently with the conventional picture, *the vector A_1 precesses at a frequency equal to $\nu_A + J/2$ whereas the vector A_2 precesses at a frequency equal to $\nu_A - J/2$.*

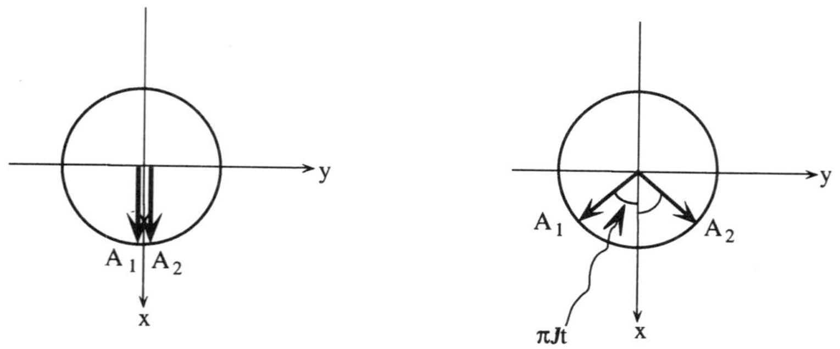

Figure 2.13 Evolution of $\hat{I}_x^A$, represented by A_1 and A_2 vectors, under the indirect coupling hamiltonian

From similar calculations, the following rules, which are valid only for a *free precession interval*, can be established.

(1) Quantities in classes (I) and (I′) *are invariant* because they involve only $\hat{I}_z$ operators.

(2) Quantities in classes (II) and (III) *evolve 'classically'*, that is to say, according to a model for which a vector is associated with each transition. For instance, as has just been demonstrated, A_1 and A_2 precess at frequencies $\nu_A + J/2$ and $\nu_A - J/2$, respectively. It is therefore easy to evaluate, by simple geometrical considerations, the way in which $\hat{I}_x^A$, $\hat{I}_y^A$, $2\hat{I}_x^A\hat{I}_z^X$ and $2\hat{I}_y^A\hat{I}_z^X$ are transformed, and also the homologous operators obtained by interchanging A and X. The two examples shown in Figure 2.14 illustrate the method.

Owing to the fact that ν_A and ν_X should be understood as the resonance frequencies viewed from the rotating frame, we shall assume for simplicity (and without loss of generality) that A is *on-resonance* (in other words, the absolute resonance frequency of A coincides with the carrier frequency and $\nu_A = 0$). We shall determine the fate of $\hat{I}_x^A$ and of $2\hat{I}_x^A\hat{I}_z^X$ subsequently to a free precession interval of duration equal to $1/2J$. The answer is directly given by the diagrams of Figure 2.14. A_1 and A_2 rotate at an angular velocity equal to πJ, clockwise and anticlockwise, respectively. After an interval of duration $1/2J$, each of these vectors will have rotated by $\pm\pi/2$. As a result $\hat{I}_x^A$ is transformed into $2\hat{I}_y^A\hat{I}_z^X$ whereas $2\hat{I}_x^A\hat{I}_z^X$ is transformed into $\hat{I}_y^A$. Were the interval duration to be different from $1/2J$, a linear combination of these quantities is obtained, whose structure can be deduced from simple geometrical arguments.

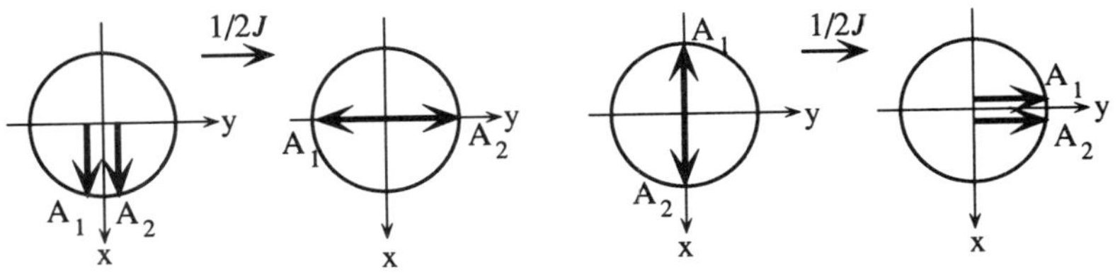

Figure 2.14 Evolution of operators $\hat{I}_x^A$ (left) and $2\hat{I}_x^A\hat{I}_z^X$ (right) subsequently to a free precession period of duration equal to $1/2J$. A is assumed to be on resonance ($\nu_A = 0$)

(3) *Quantities in class (IV) evolve according to the precession of the relevant 'global' vectors A and X at frequencies ν_A and ν_X, respectively*. Let us recall that the 'global' vectors have been introduced in Figures 2.12 and are symbolized by a double arrow (for instance, $2\hat{I}_x^A\hat{I}_x^X$ is represented by two such 'global' A and X vectors oriented along the x axis of the rotating frame). The 'global' vector A precesses at frequency ν_A and the 'global' vector X precesses at frequency ν_X, regardless of the J coupling value. This property arises from the expansion of equation (2.25) in which $\hat{\sigma}(0)$ is replaced by one of the operators of class (IV) and rests on the property that the relevant product operators commute with the coupling term $J\hat{I}_z^A\hat{I}_z^X$ in the hamiltonian. The example of the evolution of

$2\hat{I}_x^A\hat{I}_x^X$ is shown in Figure 2.15 and leads to the following result, which stems from the projections of 'global' vectors over the x and y axes:

$$2\hat{I}_x^A\hat{I}_x^X\cos(2\pi\nu_A t)\cos(2\pi\nu_X t) + 2\hat{I}_y^A\hat{I}_y^X\sin(2\pi\nu_A t)\sin(2\pi\nu_X t) - 2\hat{I}_x^A\hat{I}_y^X\cos(2\pi\nu_A t)\sin(2\pi\nu_X t) - 2\hat{I}_y^A\hat{I}_x^X\sin(2\pi\nu_A t)\cos(2\pi\nu_X t)$$

The previous examples corroborate a general property according to which the final states can always be expressed on the basis of product operators. This method can thus be considered as universal provided that one is dealing with a weakly coupled system; it has the ability to lead to the fate of a given state in the course of a free precession period. Finally, we note that classes (II) and (III) can mix together (an in-phase doublet may transform into an antiphase doublet) whereas class (IV) remains exclusive (a product operator belonging to class (IV) is invariably transformed into product operators of that class).

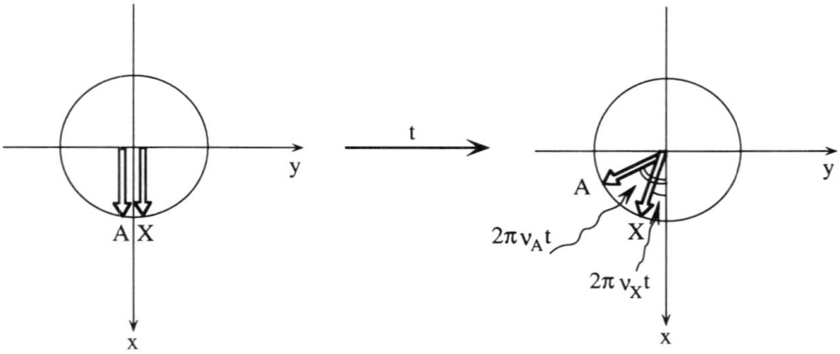

Figure 2.15 Evolution of the product operator $2\hat{I}_x^A\hat{I}_x^X$, represented by the 'global' A and X vectors

2.2.5 EVOLUTION UNDER THE APPLICATION OF A RADIO-FREQUENCY PULSE

Before considering the effect of a radio-frequency pulse, we are going to assess mathematically the use of the density operator in the rotating frame(s) and derive an equation similar to (2.17) (note that an intuitive approach was employed in the previous section). It may prove necessary to define as many rotating frames as there exist isotopes of different nature, in order to account for multi-impulsional *hetero-nuclear* experiments. For that purpose, we shall define a transform operator $\hat{T}$ by

$$\hat{T} = \exp\left(i\sum_j 2\pi\nu_{rj}\hat{I}_z^j t\right) \tag{2.31}$$

where ν_{rj} stands for the frequency of both the transmitter and the receiver, thus the reference frequency for *all* nuclei of the same isotopic nature. The inverse operator $\hat{T}^{-1}$ is deduced from (2.31) by changing the sign in the exponential argument. It may be noticed that $\hat{T}$ acts as an operator of rotation around the z axis, as indicated by formulae (2.27). $\hat{T}$ can be applied not only to $\hat{I}_x$, $\hat{I}_y$ and $\hat{I}_z$, but also to the total hamiltonian and to the density operator. In the same way as $\hat{T}\hat{I}_x\hat{T}^{-1}$ leads to $\hat{I}_{x'}$ where x' is an axis of the rotating frame, we can define $\hat{\sigma}'$ by

$$\hat{\sigma}' = \hat{T}^{-1}\hat{\sigma}\hat{T} \tag{2.32}$$

Our goal will be to derive the evolution equation of $\hat{\sigma}'$. This is accomplished first by differentiating (2.32) with respect to t:

$$\frac{\mathrm{d}\hat{\sigma}'}{\mathrm{d}t} = \left(-i\sum_j 2\pi\nu_{rj}\hat{I}_z^j\right)\hat{T}^{-1}\hat{\sigma}\hat{T} + \hat{T}^{-1}\frac{\mathrm{d}\hat{\sigma}}{\mathrm{d}t}\hat{T} + \hat{T}^{-1}\hat{\sigma}\hat{T}\left(i\sum_j 2\pi\nu_{rj}\hat{I}_z^j\right).$$

$\hat{\mathcal{H}}'$ can then be defined as $\hat{\mathcal{H}}' = \hat{T}^{-1}\hat{\mathcal{H}}\hat{T}$, where $\hat{\mathcal{H}}$ is the total hamiltonian governing the spin system, enabling one to recast the evolution equation (2.17) in the following way:

$$\hat{T}^{-1}\frac{\mathrm{d}\hat{\sigma}}{\mathrm{d}t}\hat{T} = 2i\pi(\hat{T}^{-1}\hat{\sigma}\hat{T}\hat{T}^{-1}\hat{\mathcal{H}}\hat{T} - \hat{T}^{-1}\hat{\mathcal{H}}\hat{T}\hat{T}^{-1}\hat{\sigma}\hat{T}) = 2i\pi[\hat{\sigma}', \hat{\mathcal{H}}']$$

Inserting this result in the expression of $\mathrm{d}\hat{\sigma}'/\mathrm{d}t$ provides actually the evolution equation of $\hat{\sigma}'$

$$\frac{\mathrm{d}\hat{\sigma}'}{\mathrm{d}t} = 2i\pi\left[\hat{\sigma}', \hat{\mathcal{H}}' + \sum_j \nu_{rj}\hat{I}_z^j\right]. \tag{2.33}$$

Since the interactions included in $\hat{\mathcal{H}}_0$ do not depend on the choice of the x and y axes (the only preferential direction is z), $\hat{\mathcal{H}}_0$ should be invariant under the action of the transform operator $\hat{T}$ and therefore should be left unchanged in $\hat{\mathcal{H}}'$. The same remark would apply for the hamiltonian including relaxation phenomena (Chapter 4), which anyway will be *disregarded here owing to the short duration of radio-frequency pulses*.

It remains to consider in $\hat{\mathcal{H}}'$ the interaction with the radio-frequency field. It can be noticed that the summation in the commutator (2.33) can be inserted in $\hat{\mathcal{H}}_0$ provided that ν_j is substituted by $\nu_j - \nu_{rj}$. *This amounts to considering the nuclear precession in the rotating frame*, as was already done in the previous section. Now, referring to the approach developed in Section 2.1.3, we shall retain only the component of the radio-frequency field which rotates in the same sense as nuclear precession. The hamiltonian describing the interaction of nuclear spins with that component is expressed as

$$\hat{\mathcal{H}}_{\mathrm{rf}} = -\sum_j\left(\frac{\gamma_j B_{1j}}{2\pi}\right)\left(\cos(2\pi\nu_{rj}t)\hat{I}_x^j - \sin(2\pi\nu_{rj}t)\hat{I}_y^j\right) \tag{2.34}$$

After applying the transformation $\hat{T}$, we obtain

$$\hat{\mathcal{H}}'_{\mathrm{rf}} = \hat{T}^{-1}\hat{\mathcal{H}}_{\mathrm{rf}}\hat{T} = -\sum_j \frac{\gamma_j B_{1j}}{2\pi}\hat{I}_x^i. \tag{2.35}$$

As expected, this transformation $\hat{T}$ (which allows switching from the laboratory frame to the rotating frame) removes the time dependence in the hamiltonian.

From all the above considerations, the evolution equations of $\hat{\sigma}'$, in the presence of a radio-frequency field and with the neglect of relaxation phenomena, can be written as

$$\frac{\mathrm{d}\hat{\sigma}'}{\mathrm{d}t} = 2i\pi\left[\hat{\sigma}', \hat{\mathcal{H}}_0 - \sum_j \nu_{1j}\hat{I}_x^j\right] \tag{2.36}$$

with $\nu_{1j} = \gamma_j B_{1j}/(2\pi)$ and where the Zeeman term in $\hat{\mathcal{H}}_0$ is supposed to involve precession frequencies expressed in the rotating frame. Moreover, if the radio-frequency field amplitude is assumed to be strong enough so that $\nu_{1j} \gg |\nu_j - \nu_{rj}|$ for any resonance ν_j present in the actual spectrum (this implies dealing with high power r.f. pulses), $\hat{\mathcal{H}}_0$ can be neglected in (2.36) which becomes

$$\frac{d\hat{\sigma}'}{dt} = 2i\pi\left[\hat{\sigma}', -\sum_j \nu_{1j}\hat{I}_x^j\right] \tag{2.37a}$$

which can be expressed in an even simpler way for the case of a homonuclear spin system:

$$\frac{d\hat{\sigma}'}{dt} = 2i\pi[\hat{\sigma}', -\nu_1\hat{F}_x] \tag{2.37b}$$

Denoting by τ_j the duration of the radio-frequency pulse which is applied to spin j and by α_j the quantity $2\pi\nu_{1j}\tau_j$ (also equal to $\gamma_i B_{1j}\tau_j$ and which is the flip angle introduced in Section 2.1.3), we can express the solution of differential equations (2.37a) and (2.37b) as

$$\hat{\sigma}'_+ = \exp\left(i\sum_j \alpha_j\hat{I}_x^j\right)\hat{\sigma}'_- \exp\left(-i\sum_j \alpha_j\hat{I}_x^j\right) \tag{2.38a}$$

$$\hat{\sigma}'_+ = \exp(i\alpha\hat{F}_x)\hat{\sigma}'_- \exp(-i\alpha\hat{F}_x) \tag{2.38b}$$

where $\hat{\sigma}'_+$ and $\hat{\sigma}'_-$ stand for the density operator before and after the r.f. pulse, respectively. It is nevertheless recommended to come back to $\hat{\sigma}$, $\hat{\sigma}'$ being just a convenient intermediate for handling the calculations. From (2.32), equation (2.38a) may be recast as

$$\hat{T}^{-1}\hat{\sigma}_+\hat{T} = \exp\left(i\sum_j \alpha_j\hat{I}_x^j\right)\hat{T}^{-1}\hat{\sigma}_-\hat{T}\exp\left(-i\sum_j \alpha_j\hat{I}_x^j\right)$$

Multiplying on the left by $\hat{T}$ and on the right by $\hat{T}^{-1}$ and noting that $\hat{T}\hat{I}_x\hat{T}^{-1} = \hat{I}_{x'}$ (x' belonging to the rotating frame), we arrive at

$$\hat{\sigma}_+ = \exp\left(i\sum_j \alpha_j\hat{I}_{x'}^j\right)\hat{\sigma}_- \exp\left(-i\sum_j \alpha_j\hat{I}_{x'}^j\right) \tag{2.39a}$$

Likewise, starting from (2.38b), we obtain

$$\hat{\sigma}_+ = \exp(i\alpha\hat{F}_{x'})\hat{\sigma}_- \exp(-i\alpha\hat{F}_{x'}). \tag{2.39b}$$

Since $\hat{\sigma}$ can be expanded on the product operator basis, equations (2.39a) and (2.39b) can be applied to any element of this basis. Let us recall that each element of this basis is representative of one of the possible states of the spin system. It is therefore essential to establish the relevant transformation rules. Because operators relative to two different spins commute and because $\hat{I}_{x'}^j$ acts only on $\hat{S}_j^{(r)}$ (see (2.21) for the general expression of the elements of the product operator basis $\hat{U}_r = N_r\prod_j\hat{S}_j^{(r)}$), using (2.39a) for $\hat{U}_r$ leads to

$$\hat{U}_{r+} = N_r\exp\left(i\sum_j \alpha_j\hat{I}_{x'}^j\right)\left(\prod_{j'=1}^n \hat{S}_{j'}^{(r)}\right)\exp\left(-i\sum_j \alpha_j\hat{I}_{x'}^j\right)$$

which can also be written:

$$\hat{U}_{r+} = N_r\prod_{j=1}^n \exp(i\alpha_j\hat{I}_{x'}^j)\hat{S}_j^{(r)}\exp(-i\alpha_j\hat{I}_{x'}^j). \tag{2.40}$$

An immediate consequence of (2.40) is that it is perfectly valid to *cascade the actions of r.f. pulses* or, in other words, to perform *successively* the rotations affecting the operators relevant to each spin. The examples given below illustrate the simplicity of the method and shed some light on the conditions under which the vectorial model is sound.

Radio-frequency pulses applied to operators of classes (I) and (II)

Operators of class (I) and (II) correspond to the most frequently encountered initial situations. It is therefore of some relevance to look at the effect of a simple *non-selective* pulse of flip angle α, recognizing that 'non-selective' means that the r.f. pulse acts in a similar way on both spins A and X. Without loss of generality, we can assume that the radio-frequency field is applied along the x' axis of the rotating frame. Because all spin manipulations take place within the rotating frame, the 'prime' will from now on be omitted. Thus, the notation used will be $(\alpha)_x$. As expected, results derived from the application of equations (2.39a), (2.39b) and (2.40) are in perfect agreement with the vectorial model

$$\hat{I}_x^{(\mathrm{A,X})} \xrightarrow{(\alpha)_x} \hat{I}_x^{(\mathrm{A,X})}$$
$$\hat{I}_y^{(\mathrm{A,X})} \xrightarrow{(\alpha)_x} \cos\alpha\,\hat{I}_y^{(\mathrm{A,X})} - \sin\alpha\,\hat{I}_z^{(\mathrm{A,X})}$$
$$\hat{I}_z^{(\mathrm{A,X})} \xrightarrow{(\alpha)_x} \cos\alpha\,\hat{I}_z^{(\mathrm{A,X})} + \sin\alpha\,\hat{I}_y^{(\mathrm{A,X})}$$

Radio-frequency pulses applied to operators of classes (I′), (III) and (IV)

Because one is dealing here with operators relating to two spins, the vectorial model may no longer apply; it must therefore be disregarded and the emphasis must be put on the structure of the product operators. For the sake of simplicity, we shall limit ourselves in the present discussion to $(\pi/2)_x$ of $(\pi)_x$ pulses. Let us first consider an A antiphase doublet along x, represented by $(2\hat{I}_x^{\mathrm{A}}\hat{I}_z^{\mathrm{X}})$. Under the application of a non-selective $(\pi/2)_x$ r.f. pulse, it is transformed into unobservable zero-quantum and two-quanta coherences, globally represented by $(2\hat{I}_x^{\mathrm{A}}\hat{I}_y^{\mathrm{X}})$; this is because a $\pi/2$ rotation around x leaves $\hat{I}_x^{\mathrm{A}}$ unchanged whereas it transforms $\hat{I}_z^{\mathrm{X}}$ into $\hat{I}_y^{\mathrm{X}}$.

Conversely, the A antiphase doublet along y, $(2\hat{I}_y^{\mathrm{A}}\hat{I}_z^{\mathrm{X}})$, is transformed into $(-2\hat{I}_z^{\mathrm{A}}\hat{I}_y^{\mathrm{X}})$, that is, into an observable coherence corresponding to an X antiphase doublet along y. There lies the principle of *coherence transfer* (Figure 2.16), widely exploited in most sequences aiming at establishing chemical shift correlations (one- or two-dimensional methods; see Chapter 5).

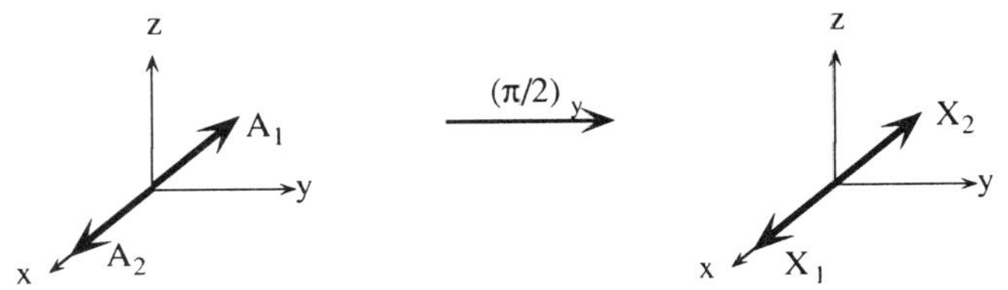

Figure 2.16 Principle of coherence transfer. An A antiphase doublet is transformed into an X antiphase doublet by application of a $(\pi/2)$ pulse

Finally, the effect on a $(\pi)_x$ r.f. pulse applied to the above configurations can be schematized as follows

$$2\hat{I}_x^A\hat{I}_z^X \xrightarrow{(\pi)_x} -2\hat{I}_x^A\hat{I}_z^X \text{ (exchange of A}_1\text{ and A}_2\text{)}$$
$$2\hat{I}_y^A\hat{I}_z^X \xrightarrow{(\pi)_x} 2\hat{I}_y^A\hat{I}_z^X \text{ (unchanged configurations).}$$

2.2.6 PRACTICAL RULES FOR MULTIPULSE NMR EXPERIMENTS

From the theoretical considerations and proofs given in the previous sections, a number of rules emerge which can help in analyzing pulse sequences, regardless of their complexity, without having to worry about the underlying theory. We propose here a survey of these rules, which are used in conjunction with some vectorial representations shown in the figures of the previous sections and which are not repeated here. The reader who does not wish to follow the above mathematical treatments may simply accept these schemes without rigorous justification. The present section is therefore self-consistent and the relevant figures will be quoted when necessary. The present guidelines concern a *system of two weakly coupled spin 1/2 nuclei, denoted by A and X*. This system constitutes the basis of the analysis of most sequences commonly employed; extension to systems involving more than two nuclei of spin 1/2 or to systems involving quadrupolar nuclei in an anisotropic medium (which show several distinct transitions) are just a matter of calculation. Concerning strongly coupled spin systems, it may be mentioned that the density matrix rather than the density operator is more appropriate because, with this latter approach, one could cope with mixing of the product operators defined in Figure 2.12. Nevertheless, the conclusions derived for a weakly coupled spin system remain generally valid to first approximation.

Equilibrium state

In any experiment, the initial state is the thermodynamic equilibrium represented by the longitudinal magnetizations weighted according to their respective gyromagnetic ratios

$$\gamma_A I_z^A + \gamma_X I_z^X$$

Actually, the above quantity represents a polarization rather than a magnetization because it stands, in the expression of magnetization, for the factor which arises from differences in energy level populations. This is to emphasize the fact that the polarization can be transferred from one nucleus to the other while the magnetic moment (which corresponds to another factor in the expression of magnetization) obviously cannot. Other states of the system will be formulated with the help of the quantities I_x^A, I_x^X, I_y^A and I_y^X (called operators) and with all possible products (called product operators) involving one of the components (I_x, I_y, I_z) of spin A and one of the components of spin X. For normalization purposes, a factor of 2 is placed in front of each of these products. Radio-frequency pulses which act in the course of the sequence may be non-selective (they act in a similar way on nuclei A and X) or selective in the sense that they will affect only A (or X) and will leave unchanged any quantity relative to X (or A).

Free precession periods

Provided that the state of the system, prior to a free precession interval (i.e. devoid of radio-frequency pulses), is represented by I_x^A, I_y^A, I_x^X, I_y^X, $2I_x^A I_z^X$, $2I_y^A I_z^X$, $2I_z^A I_x^X$ or $2I_z^A I_y^X$ (see Figure 2.12), it is valid, during this period to allow to precess the vectors A_1 and A_2 associated with the A doublet, as well as the vectors X_1 and X_2 associated with the X doublet, with their respective frequencies: $\nu_A + J/2$, $\nu_A - J/2$, $\nu_X + J/2$ and $\nu_X - J/2$ (see Figures 2.13 and 2.14).

If, at the beginning of the free precession period, the state of the system is represented by $2I_z^A I_z^X$, there is no further evolution.

In cases where zero-quantum or two-quanta coherences exist at the beginning of the free precession period, they are represented by products of the form $2I_x^A I_x^X$, $2I_x^A I_y^X$, $2I_y^A I_x^X$ or $2I_y^A I_y^X$ and, although *they are unobservable by the usual direct detection schemes*, they can be created by r.f. pulses applied to antiphase configurations (see below). Each term in the product evolves according to the precession frequency of the considered spin. Let us consider the example of $2I_x^A I_x^X$. Since I_x^A transforms into

$$I_x^A \cos(2\pi\nu_A t) - I_y^A \sin(2\pi\nu_A t)$$

$2I_x^A I_x^X$ becomes:

$$\begin{aligned} 2I_x^A I_x^X &\cos(2\pi\nu_A t)\cos(2\pi\nu_X t) + 2I_y^A I_y^X \sin(2\pi\nu_A t)\sin(2\pi\nu_X t) \\ &- 2I_x^A I_y^X \cos(2\pi\nu_A t)\sin(2\pi\nu_X t) - 2I_y^A I_x^X \sin(2\pi\nu_A t)\cos(2\pi\nu_X t) \end{aligned}$$

as visualized in Figure (2.15).

In any case, it is mandatory, at the end of the free precession period, to restore a representation in terms of products which is the only one to be valid when dealing with r.f. pulses (see below). An arbitrary configuration will therefore be decomposed according to the following rules:

- zero-quantum or twò-quanta coherences (whose evolution has just been treated) remain of course represented by the products $2I_x^A I_x^X$, $2I_x^A I_y^X$, $2I_y^A I_x^X$ or $2I_y^A I_y^X$;
- an in-phase A doublet along the x axis of the rotating frame is represented by I_x^A;
- an antiphase A doublet along the x axis of the rotating frame, which was, during the evolution period, schematized by the vectors A_1 and A_2, must be represented by $2I_x^A I_z^X$ if A_1 (the high frequency component) is along $-x$. Conversely, if A_2 (the low frequency component) is along $-x$, the product must be written $-2I_x^A I_z^X$. As the previous (theoretical) section may have been skipped, it could be of some interest to repeat here why this is so. Let us refer to Figure 2.10 which shows an antiphase doublet, and let us recall that an A doublet exists because of the J coupling with another spin X. The two lines in the A doublet are in fact associated with the two states of the spin X (parallel or antiparallel, which correspond actually to the two possible values $+1/2$ and $-1/2$ of I_z^X). We can therefore account for the opposite directions of A_1 and A_2 by multiplying I_x^A by the quantity I_z^X, which includes the sign property. Of

course, an in-phase doublet is such that both lines have the same sign and this state is properly described by I_x^A. These considerations are shown schematically in Figure 2.17. The factor of 2 is included for normalization purposes and for compensating the value $\pm 1/2$ of I_z^X. Finally, it can be noted that it is straightforward to switch to another axis of the rotating frame or to the X nucleus. For instance, an antiphase X doublet along the y axis with X_1 along $-y$ is represented by $2I_z^A I_y^X$.

Radio-frequency pulses

The transformation of a spin state under the application of a r.f. pulse is evaluated in a straightforward manner provided that the representation in terms of 'operators' (e.g. I_x^A) or of 'product-operators' (e.g. $2I_x^A I_z^X$) has been retained. Each operator is in fact subjected to a rotation by an angle equal to the flip angle which characterizes the r.f. pulse. For example, if a r.f. pulse of flip angle α is applied along the y axis of the rotating frame, I_x is transformed into $I_x \cos\alpha + I_z \sin\alpha$. It may be recalled that the axis along which a r.f. pulse is applied is dictated by the transmitter phase. Another example, whose result could not be predicted by a simple vectorial model, concerns the effect of a $(\pi/2)_x$ non selective pulse (i.e. a pulse applied along the x axis of the rotating frame and acting in the same manner on both spins A and X) upon an antiphase configuration defined by $2I_x^A I_z^X$. Relying on the rotation properties of such a non-selective pulse, we obtain $2I_x^A I_y^X$, that is a spin state involving one-quantum and two-quanta coherences which are not directly observable. Conversely, applying a non-selective $(\pi/2)_y$ pulse to the previous state ($2I_x^A I_y^X$) yields $2I_z^A I_y^X$; this corresponds to an X antiphase configuration along the y axis which is perfectly observable (X_2 lying along $+y$ and X_1 along $-y$). It may be noted that a related configuration would have been directly obtained by applying a non-selective $(\pi/2)_y$ pulse to $2I_x^A I_z^X$, yielding $-2I_z^A I_x^X$; it

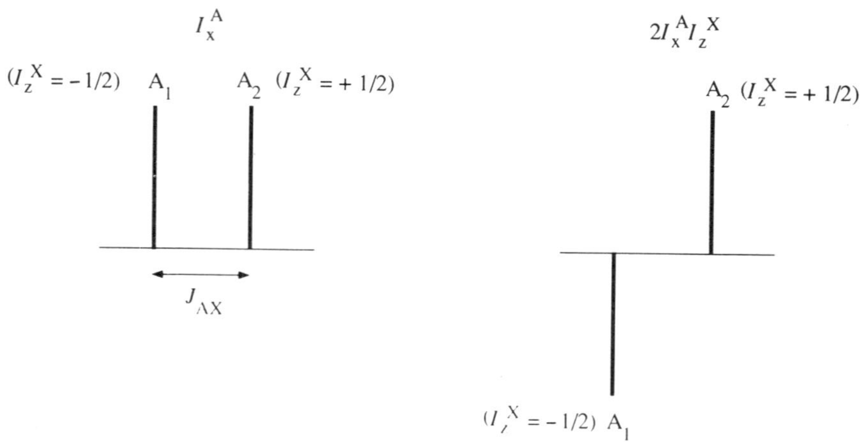

Figure 2.17 Left: the in-phase A doublet, A_1 and A_2 being associated with the two states of the spin X ($I_z = -1/2$ and $+1/2$, respectively). This state may be described simply by I_x^A since both lines have the same sign. Right: the corresponding antiphase doublet. To account for the different signs of lines A_1 and A_2, I_x^A is multiplied by I_z^X

can be recognized that, starting from an A antiphase configuration, one ends up with an X antiphase configuration. This is the *principle of coherence transfer*, widely used in many one- or two-dimensional experiments; it is visualized in Figure 2.18.

The method accommodates also selective pulses (i.e. pulses acting solely on a given spin). Let us consider again the spin state defined by $2I_x^A I_y^X$. A selective $(\pi/2)_x^A$ r.f. pulse (acting solely on A) obviously leaves this state unchanged, whereas a $(\pi/2)_y^A$ leads to $2I_z^A I_y^X$ (the X antiphase configuration obtained before with a $(\pi/2)$ non-selective pulse).

Finally, after having performed these various rotations on operators or on product operators, it is recommended to reconvert them according to the basic vectorial model (involving a vector associated with each transition) if they belong to classes (I), (II) and (III) (see Figure 2.12), in order to facilitate the analysis of the subsequent evolution period.

Particular attention must be paid to π pulses whose goal is to refocus 'chemical shifts' according to the spin echo strategy, which will be further considered and described in Chapters 4 and 5. Nevertheless, the following examples serve as a first acquaintance with the method. Let us first consider a single spin 1/2 nucleus to which the sequence of Figure 2.19 is applied. Since we are dealing with a single spin, the vectorial model applies and, whatever the frequency ν, nuclear magnetization is aligned with the y axis of the rotating frame by the end of the second interval of free precession. This sequence leads to the state in which the system was immediately after the first $(\pi/2)$ pulse and removes any chemical shift effect, including the one resulting from the spatial inhomogeneity of the static induction

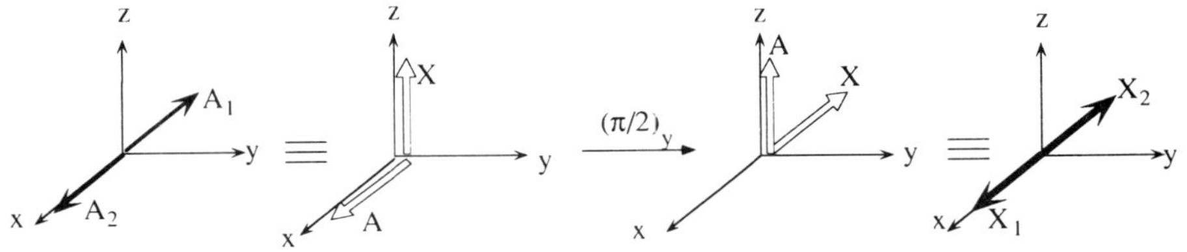

Figure 2.18 Detailed mechanisms of the coherence transfer shown in Figure 2.16

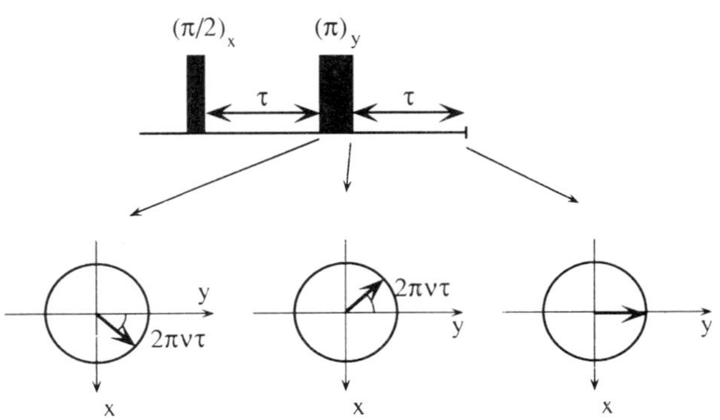

Figure 2.19 A spin echo sequence applied to a single spin 1/2

B_0. Nuclear magnetization is said to have been *refocused*; experimentally, this feature manifests itself by the reappearance of a signal named *echo*. It may be noted that, if phases of both pulses are identical (that is, a pulse sequence of the form $(\pi/2)_x - \tau - (\pi)_x - \tau$), refocusing occurs along $-y$. The actual result of a spin echo experiment depends strongly on the quality of the π pulse. A very efficient phase cycling, known under the acronym of EXORCYCLE (G. Bodenhausen *et al.*, *J. Magn. Reson.*, **27**, 511 (1977)) cancels essentially all imperfections of the refocusing pulse. It is based on the fact that, for ideal pulses, $(\pi)_y$ and $(\pi)_{-y}$ should lead to the same result, whereas $(\pi)_x$ and $(\pi)_{-x}$ should also lead to the same result but of opposite sign. From these observations, four experiments will generally be carried out and coadded with the following phases:

$$(\pi/2)_x - \tau - (\pi)_y - \tau - (\text{Acq})_+$$
$$(\pi/2)_x - \tau - (\pi)_{-y} - \tau - (\text{Acq})_+$$
$$(\pi/2)_x - \tau - (\pi)_x - \tau - (\text{Acq})_-$$
$$(\pi/2)_x - \tau - (\pi)_{-x} - \tau - (\text{Acq})_-$$

The concept of phase cycling is rather important in the practice of NMR. Many other schemes adapted to specific experiments will be indicated in the subsequent chapters.

In a general way, whenever one is interested in the evolution that occurs in the presence of the couplings alone, the spin echo sequence $\pi/2 - \tau - \pi - \tau$ will be used because it has the ability to remove any chemical shift effect. However, in the case of a coupled spin system, the actual effect of the central π pulse must be examined carefully. For example, let us consider the sequence $(\pi/2)_x^{\text{A}} - 1/4J - (\pi)_y - 1/4J$, where the initial $\pi/2$ pulse acts only on the A spin of the usual AX spin system. Immediately after the π pulse, the two A_1 and A_2 vectors are at right angles and symmetrically disposed with respect to the y axis since, due to the refocusing properties of the central π pulse, we can disregard any chemical shift effect (Figure 2.20).

This configuration can be decomposed into $(1/\sqrt{2})I_y^{\text{A}}$ and $(-1/\sqrt{2})(2I_x^{\text{A}}I_z^{\text{X}})$. If a $(\pi)_y$ pulse is applied to the A spin only, the sign of $(-1/\sqrt{2})(2I_x^{\text{A}}I_z^{\text{X}})$ is changed, A_1 and A_2 are exchanged and, subsequently to the second precession interval of duration $1/4J$, A_1 and A_2 end up together along the y axis, leading to a refocused in-phase doublet. Conversely, if in addition to $(\pi)_y^{\text{A}}$ we apply a $(\pi)_y^{\text{X}}$ pulse [or more simply if we apply a non-selective $(\pi)_y$ pulse], $(-1/\sqrt{2})(2I_x^{\text{A}}I_z^{\text{X}})$ is unchanged; A_1 and A_2 then keep precessing and end up with an antiphase configuration along $-x$ thus represented by $(-2I_x^{\text{A}}I_z^{\text{X}})$. If the duration of both intervals $(1/4J)$ is replaced by some arbitrary value denoted by $t_1/2$, it is easy to show that the echo formed at t_1 is *amplitude-modulated* according to $\cos(\pi J t_1)$. This modulation allows one to make a very accurate measurement of the J coupling and is the basis of a two-dimensional method known under the name of '*J-resolved*'. As already mentioned in Section (1.3.4), such a two-dimensional method involves two Fourier transformations, one applied to the time variable t_2 which corresponds to the physical acquisition of the NMR signal (here, half the echo starting at time t_1), the other to the time variable t_1 which has been defined above. The resulting map

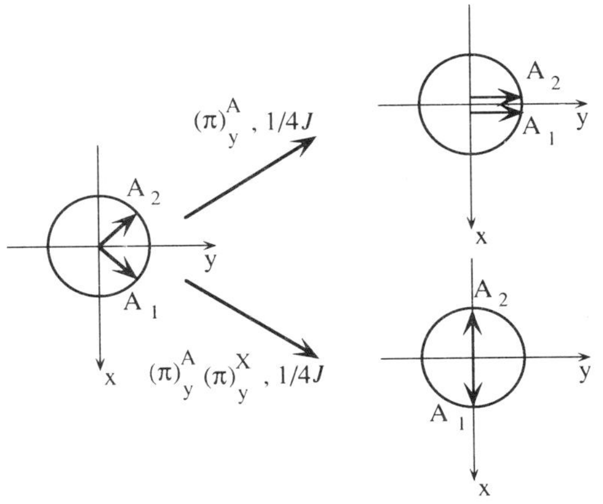

Figure 2.20 Effect of a selective π pulse (top) or of a non-selective π pulse (bottom) in an echo sequence applied to a two spin system with $\tau = 1/4J$

involves two frequency variables ν_1 and ν_2. The latter corresponds to the conventional NMR spectrum whereas the former contains the J coupling information.

Detection

The selection rules established in the first chapter imply that only one-quantum coherences can be physically detected. They are related to configurations of class (II) (see Figure 2.12) and the corresponding free induction decays (fid) yield, after Fourier transformation, the peak amplitude as well as their phases. For instance, if the x axis of the rotating frame is chosen as the receiver reference axis, I_x^A leads to a doublet of normal appearance, that is, a doublet viewed in absorption (Lorentzian peaks of positive amplitude), whereas I_y^A will be observed as a dispersive doublet (Lorentzian peak in the dispersive mode; see Chapter 3). It can be noted that the concept of absorption or dispersion is somewhat arbitrary, as it is related to the phase correction of the relevant spectra which permits, in a straightforward manner, change of an absorption-like spectrum for a dispersion-like spectrum (dephased by 90° with respect to the absorption spectrum). In any event, it should not be believed that a reference receiver adjusted for the x axis would not be able to detect signals of the I_y type; these would simply be considered as dispersive components, with of course the possibility of phase-correction in the final spectrum. As a further example, let us consider the spin state defined by $2I_x^A I_z^X$ where the receiver reference is again the x axis of the rotating frame. At the detection stage, this state will appear as an absorptive antiphase doublet with the line A_2 positive and the line A_1 negative. All these considerations indicate that, within the detection period, it is perfectly valid to refer to the model which associates a vector with each of the transitions involved provided that the phase of the receiver reference is accounted for; the latter specifies the axis of the rotating frame corresponding to absorptive signals.

Field gradient pulses

The aim of field gradient pulses in high-resolution NMR spectroscopy is to defocus or refocus nuclear magnetization. This is related to other specific applications such as the measurement of self-diffusion coefficients or imaging techniques which will be presented in Chapters 4 and 5 and which directly rest on the spatial labeling properties of field gradients. For now, we shall concentrate on the potentialities of applying gradient pulses with the objective of selecting a particular *spectral* feature. This may include the elimination of a given resonance (for instance a huge solvent line) or the selection of a given coherence type (the so-called selection of coherence pathways). In a general way, it will be shown that application of field gradient pulses can replace complicated phase cycling schemes (see Chapter 5) with, most of the time, improved performance. To make these statements more clear, let us consider first some very simple experiments involving static field gradient pulses. A single gradient pulse produces a spread in precession frequencies which, if the gradient is sufficiently strong and if the sample is homogeneous, cancels completely the resultant magnetization. This is because all possible precession angles occur within the sample with the same probability and lead to a net result equal to zero, as illustrated in Figure 2.21. In effect, the precession angle becomes location dependent; this is because the effective B_0 field can be expressed as $B_0 + g_0\mathrm{X}$, where X is the *spatial* coordinate corresponding to the gradient direction and g_0 the gradient strength ($g_0 = (\mathrm{d}B_0/\mathrm{dX})$) which is assumed to be uniform within the sample.

It may however be recognised that this process is reversible. Suppose we apply a second g_0 pulse of identical duration but of opposite polarity, this will *refocus* the spread in precession frequencies and restore transverse magnetization as it was after the $(\pi/2)_x$ pulse. In order to illustrate this feature further, let us consider the simple sequence of Figure 2.22 which involves a (π) selective inverting pulse

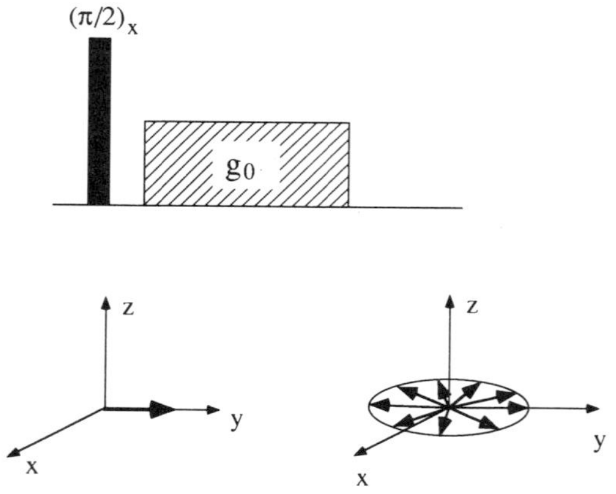

Figure 2.21 Defocusing properties of a static field gradient pulse symbolized by g_0. A standard $(\pi/2)_x$ pulse flips the nuclear magnetization toward the y axis. If a static field gradient pulse is applied for a sufficient time, it produces a spread (scattering) of precession frequencies so that, on average, the transverse components of nuclear magnetization totally cancel

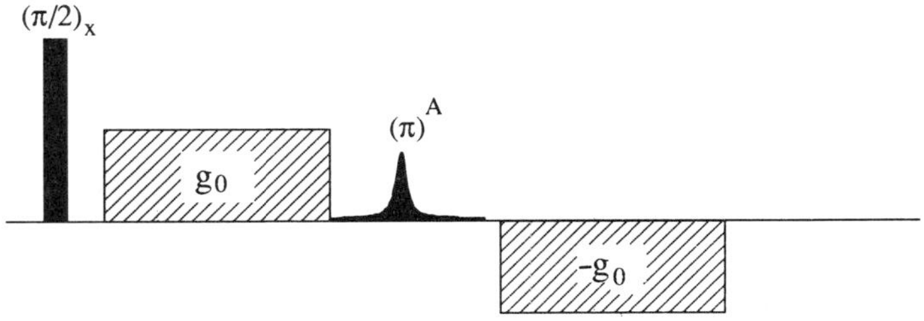

Figure 2.22 A schematic illustration of the possibility of eliminating the A resonance by using B_0 gradient pulses of opposite polarities along with a selective inverting pulse, $(\pi)^A$, acting only on the A resonance

applied to a given resonance (for instance, a solvent resonance). It can be immediately appreciated that resonances not affected by the π inverting pulse will be left unchanged after the two gradient (g_0;$-g_0$) pulses which act in a defocusing–refocusing manner provided that the usual precession can be neglected during the gradient pulses. Conversely, for the A resonance, an inversion occurs, so that A magnetization keeps defocusing under the second ($-g_0$) gradient pulse. As a whole, the spectrum, except the A resonance which disappears, is restored.

The next step is to worry about the fate of coherences under the application of gradient pulses. Because any coherence can be represented by an operator or a product operator, their behavior can be deduced from the transformation of the individual operators I_x, I_y and I_z. The evolution of these three operators under the application of a g_0 pulse can be visualized through the diagrams presented in Figure 2.23.

The hatched areas in Figure 2.23 are purposely limited so as to recognize whether the projection onto a given axis is $\cos\theta$ or $\sin\theta$. In effect, the nuclear magnetization is assumed to be randomly distributed. The dependence upon a sine or a cosine function is essential since we must perform, ultimately, a space average by assuming that all values of θ are equally probable. It is well known that $\langle\cos\theta\rangle = \langle\sin\theta\rangle = 0$ whereas $\langle\cos^2\theta\rangle = \langle\sin^2\theta\rangle = 1/2$. This means that, after the application of a g_0 gradient pulse, I_z remains of course unchanged while I_x, which gives rise to $I_x\langle\cos\theta\rangle - I_y\langle\sin\theta\rangle$, is zero by averaging over the whole sample;

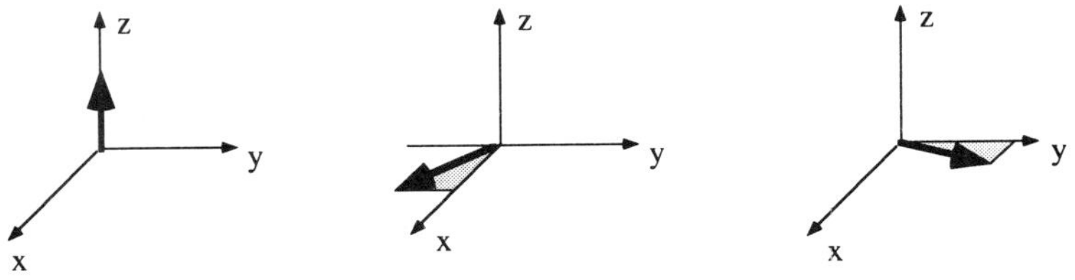

Figure 2.23 Under the application of a g_0 gradient pulse, I_z is unchanged but I_x and I_y are defocussed. The way in which I_z, I_x and I_y are respectively transformed is shown from left to right. The hatched areas mean that, for a given location within the sample, I_x transforms into $I_x\cos\theta - I_y\sin\theta$, whereas I_y transforms into $I_y\cos\theta + I_x\sin\theta$. By convention, the vicinity with respect to a given axis is associated with $\cos\theta$ whereas its remoteness is associated with $\sin\theta$

similar arguments hold for I_y. Let us look now at the fate of a coherence such as $2I_x^A I_y^X$. From Figure 2.23 and from the average relationships given above, it is a simple matter to look for the terms affected by a factor $\langle \cos^2\theta \rangle$ or $\langle \sin^2\theta \rangle$ which can lead to a non-zero result. Obviously, these are $2I_x^A I_y^X(\langle \cos^2\theta \rangle)$ but also $2I_y^A I_x^X(\langle \sin^2\theta \rangle)$, so that, under the application of a g_0 gradient pulse, we can write

$$(2I_x^A I_y^X) \xrightarrow{g_0} (2I_x^A I_y^X)/2 + (2I_y^A I_x^X)/2$$

It can be noted that the initial coherence is reduced by a factor of 1/2, whereas an effective coherence transfer takes place because the g_0 pulse induces the coherence $(2I_y^A I_x^X)$, affected again by a factor of 1/2. In more complex situations (see below and Chapter 5), it will be useful to know the average value of $\langle \cos^m\theta \sin^n\theta \rangle$. These quantities are given in Table 2.1 for an appropriate set of n and m values, recognizing that it is always possible to switch from $\cos^2\theta$ to $\sin^2\theta$ through the identity $\cos^2\theta = 1 - \sin^2\theta$.

As an example, the method will now be applied to one of the simplest pulse sequences used in correlation spectroscopy. This sequence, which involves two B_0 gradient pulses bracketing a $\pi/2$ r.f. pulse, is schematized in Figure 2.24, where are also shown the evolution of the I_x, I_y and I_z operators. In passing, it must be stressed that, when dealing with a sequence which includes gradient pulses, a situation at a given stage cannot be considered as definite if subsequent gradient pulses occur. Dramatic changes may be expected due to either defocusing or refocusing processes. Returning to the current example and by reference to Figure 2.24, it can be seen that I_x, I_y and I_z obey the following transformations

$$I_x \rightarrow \cos^2\theta I_x - \sin\theta\cos\theta I_y + \sin\theta I_z$$
$$I_y \rightarrow \sin\theta\cos\theta I_x - \sin^2\theta I_y - \cos\theta I_z$$
$$I_z \rightarrow \sin\theta I_x + \cos\theta I_y$$

Upon averaging over the whole sample, I_x is transformed into $I_x/2$, I_y into $-I_y/2$ and I_z simply cancels. In order to determine whether a given coherence survives and how it is transformed, it suffices to perform the relevant products and to look for the non-zero quantities of the type $\langle \cos^m\theta \sin^n\theta \rangle$ (as deduced from Table 2.1). For instance, $2I_x^A I_z^X$ (antiphase A doublet) leads to:

$$\langle \cos^2\theta \sin\theta \rangle (2I_x^A I_x^X) + \langle \cos^3\theta \rangle (2I_x^A I_y^X) - \langle \sin^2\theta \cos\theta \rangle (2I_y^A I_x^X)$$
$$- \langle \sin\theta \cos^2\theta \rangle (2I_y^A I_y^X) + \langle \sin^2\theta \rangle (2I_z^A I_x^X) + \langle \sin\theta\cos\theta \rangle (2I_z^A I_y^X)$$

Table 2.1 Values of $\langle \cos^m\theta \sin^n\theta \rangle$

m \ n	0	1	2	3	4	5	6	7
0	1	0	1/2	0	3/8	0	5/16	0
1	0	0	0	0	0	0	0	0
2	1/2	0	1/8	0	1/16	0	5/128	0
3	0	0	0	0	0	0	0	0
4	3/8	0	1/16	0	3/128	0	3/256	0

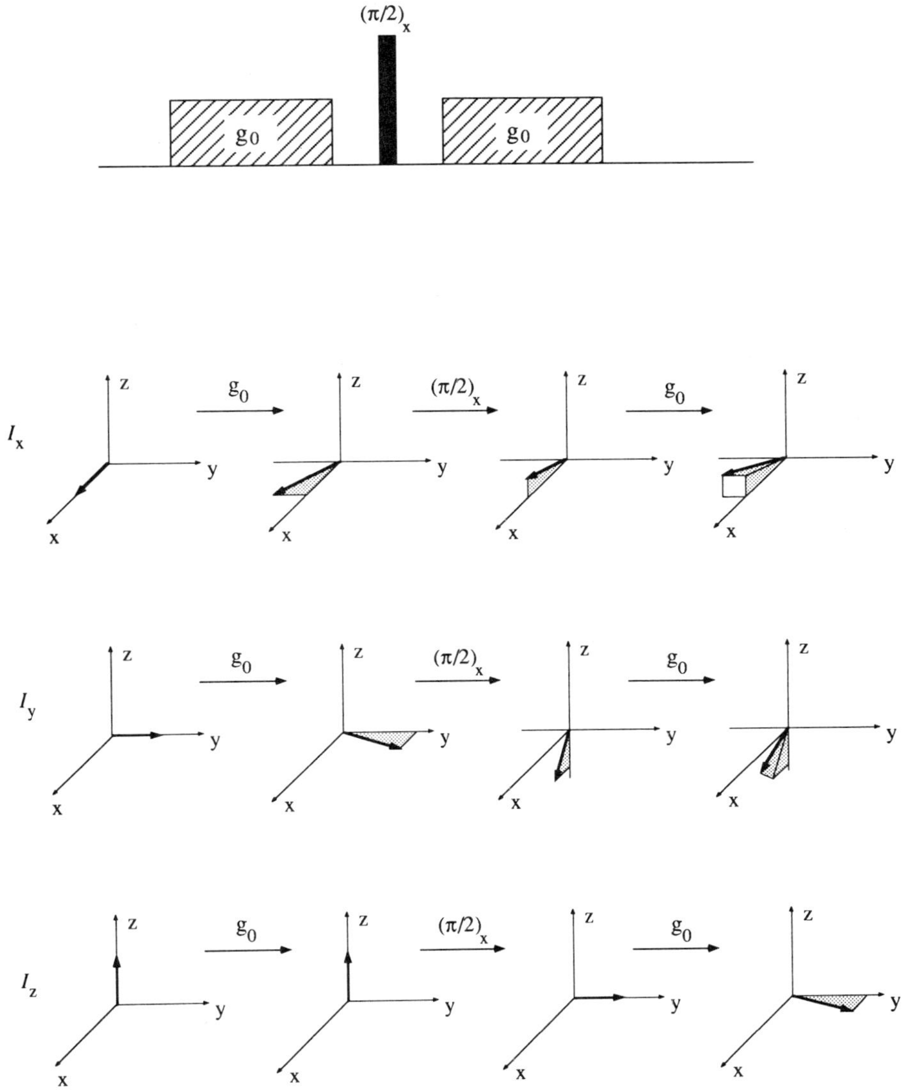

Figure 2.24 The evolution of I_x, I_y and I_z under the pulse sequence shown at the top of the figure, with the conventions of Figure 2.23. When the magnetization vector, defining the limit of a hatched area, does not lie within of the three principal planes (xy, xz or yz), this is indicated by a squared (or rectangular) section which allows one to determine the projection onto the x, y or z axes. For instance, the final situation of I_x (top right) is such that the projections onto x, y and z are respectively $\cos^2\theta$, $-(\sin\theta\cos\theta)$ and $\sin\theta$

With a little practice, it is unnecessary to write down all the terms listed above, and it is a simple matter to notice that

$$(2I_x^A I_z^X) \rightarrow (2I_z^A I_x^X)/2$$

since, among all the coefficients, only $\langle \sin^2\theta \rangle$ is non-zero.

A last point concerns another type of field gradients, which originates from the inhomogeneity of the radio-frequency field. Such gradients will be dubbed radio-frequency field gradients or simply B_1 gradients. They can be created very easily by a properly devised r.f. coil and present some interesting features (negligible rise and fall times when applied in the form of pulses, no interference with the normal operations of the spectrometer). They act on nutation rather than on precession but can be thought as producing a space-dependent rotation of nuclear magnetization, just like B_0 gradients. The only difference stems from the plane in which this rotation takes place. Whereas, for B_0 gradients, it is invariably the xy plane, it can be, for B_1 gradients, either the yz plane or the xz plane of the rotating frame according to their r.f. phase (x and y, respectively). Although there exists a formal equivalence between B_0 gradients and B_1 gradients (by combining B_1 gradient pulses with homogeneous, or standard, r.f. pulses; see Figure 2.25), the latter can often be employed in a much simpler way.

This can be illustrated by the application of a simple B_1 gradient pulse, $(g_1)_x$, to a two-spin state represented by $2I_y^A I_z^X$ (an A antiphase doublet along y). From the diagrams of Figures 2.26 and from the rules derived above, it is easily seen that

$$(2I_y^A I_z^X) \xrightarrow{(g_1)_x} (2I_y^A I_z^X)/2 - (2I_z^A I_y^X)/2$$

This process is seen to include a coherence transfer from A to X (second term of the above relationship) and can thus be employed for correlating spins which involve a J coupling.

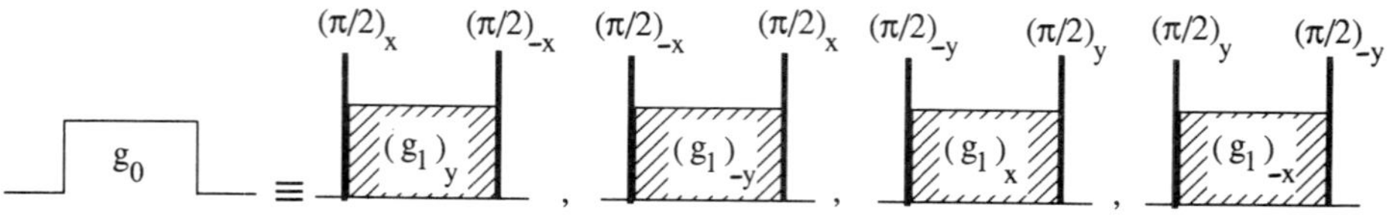

Figure 2.25 Equivalence between a B_0 gradient pulse (g_0) and B_1 gradient pulses (g_1) with relevant phases. ($\pi/2$) stand for normal hard pulses assumed to be produced by a perfectly homogeneous radio-frequency field

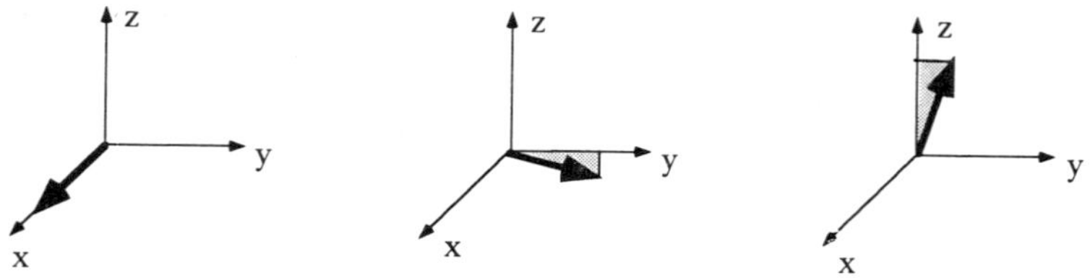

Figure 2.26 Under the application of a B_1 gradient pulse acting along the x axis of the rotating frame, I_x is unchanged while I_y and I_z are defocused. These transformations (from right to left) are schematized according to the conventions of Figure 2.23

Appendix: an Analytical Solution to the Bloch Equations

Assuming a radio-frequency field of amplitude B_1 applied along the x axis of the rotating frame and referring to equations (2.7) and (2.9), we can write the evolution of the three components of nuclear magnetization as

$$\begin{aligned}\frac{\mathrm{d}M_x}{\mathrm{d}t} &= -R_2 M_x + \Omega M_y \\ \frac{\mathrm{d}M_y}{\mathrm{d}t} &= -\Omega M_x - R_2 M_y + \omega_1 M_z \\ \frac{\mathrm{d}M_z}{\mathrm{d}t} &= -\omega_1 M_y + R_1(M_0 - M_z)\end{aligned} \tag{A1}$$

with $R_{1,2} = 1/T_{1,2}$, $\Omega = \gamma B_0 - \omega_r$ and $\omega_1 = \gamma B_1$.

The resolution of equations (A1) already published (G. A. Morris and P. B. Chilvers *J. Magn. Reson.*, **A107**, 236 (1994)) is based on the Laplace transform and some of its properties. The approach proposed below is somewhat more 'classical'. Whatever the employed method, it is necessary to search the roots of the characteristic equation associated with the evolution matrix which, according to equations (A1), can be written as

$$\begin{bmatrix} -R_2 & \Omega & 0 \\ -\Omega & -R_2 & \omega_1 \\ 0 & -\omega_1 & -R_1 \end{bmatrix}$$

The actual characteristic equation has the form

$$x^3 + (R_1 + 2R_2)x^2 + (R_2^2 + 2R_1R_2 + \Omega^2 + \omega_1^2)x + (R_1R_2^2 + R_2\omega_1^2 + R_1\Omega^2) = 0 \tag{A2}$$

By substituting x by $y = x + R_\mathrm{m}$, where $R_\mathrm{m} = (R_1 + 2R_2)/3$ is a sort of 'mean' relaxation rate, (A2) can be recast as

$$y^3 + py + q = 0 \tag{A3}$$

with

$$p = -R_\mathrm{d}^2/3 + (\Omega^2 + \omega_1^2)$$

and $q = -R_\mathrm{d}[(2/27)R_\mathrm{d}^2 + (2\Omega^2 - \omega_1^2)/3]$, where we have introduced the difference in relaxation rates $R_\mathrm{d} = R_2 - R_1$. We can look for a root of (A3) which has the form $y_0 = u^{1/3} + v^{1/3}$ with $uv = -(p/3)^3$ leading to the equation $u^2 + qu - (p/3)^3 = 0$. A sufficient condition for the roots of the latter equation to be real is that p is positive and we obtain:

$$u_\pm = -(q/2) \pm [(q/2)^2 + (p/3)^3]^{1/2}$$

Conversely, it may be instructive to wonder at which conditions a negative value of p would occur. This would involve a weak value of the radio-frequency field amplitude and a large difference between relaxation rates; in practice, $R_2 \gg R_1$, and $\gamma B_1 < R_2$ with Ω small enough. This would correspond physically to a r.f. field 'smaller than the linewidth', and mathematically to an overdamped situation. Whenever such a situation is encountered ($p < 0$), we may consider the r.f. field sufficiently weak so that *we can safely replace the r.f. pulse by a precession interval* [*according to equations (2.10)*]. This strategy will be retained and we are thus left with the above solution for u, which yields the root y_0 of equation (A3).

$$y_0 = \{-(q/2) + [(q/2)^2 + (p/3)^3]^{1/2}\}^{1/3} + \{-(q/2) - [(q/2)^2 + (p/3)^3]^{1/2}\}^{1/3} \tag{A4}$$

The complementary roots of equation (A3) follow from its factorization by $(y - y_0)$

$$(y - y_0)(y^2 + by + c) \equiv y^3 + py + q$$

Identifying the two sides of this equation leads to $b = y_0$ and $c = p + y_0^2$ and consequently to the resolution of:

$$y^2 + y_0 y + (p + y_0^2) = 0$$

This latter equation possesses two complex roots (again, by assuming that p is positive):

$$y = -(y_0/2) \pm (i/2)(4p + 3y_0^2)^{1/2}$$

Finally, the roots of the initial characteristic equation (A2) may be expressed as

$$\lambda_0 = -R_{\mathrm{m}} + y_0 \tag{A5}$$

$$\lambda_{1,2} = \lambda \pm ir \tag{A6}$$

with $\lambda = -R_{\mathrm{m}} - y_0/2$ and $r = (1/2)[4p + 3y_0^2]^{1/2}$.

The analytical solutions to the Bloch equations may therefore be written as

$$M_{x,y,z} = A_{x,y,z} \exp(\lambda_0 t) + \exp(\lambda t)(B_{x,y,z} \cos rt + C_{x,y,z} \sin rt) + D_{x,y,z} \tag{A7}$$

where the twelve coefficients A, B, C, D have still to be determined.

Before going through those calculations, it may be interesting to note that, in the special case where $T_1 = T_2$, the two damping terms $\exp(\lambda_0 t)$ and $\exp(\lambda t)$ reduce to $\exp(-R_m t)$, whereas the oscillatory contributions involve an angular velocity equal to $(\Omega^2 + \omega_1^2)^{1/2}$.

The coefficients A, B, C, D can be calculated from initial conditions ($t = 0$), boundary conditions ($t \to \infty$) or from relationships among them. Let us begin with the stationary state which is reached when $t \to \infty$ and which, mathematically, can be expressed by $(\mathrm{d}/\mathrm{d}t)M_{x,y,z} = 0$. These conditions, applied to equations (A3), lead to a set of three linear equations in D_x, D_y and D_z whose solution is

$$D_x = \frac{\omega_1 \Omega R_1}{\Omega^2 R_1 + \omega_1^2 R_2 + R_1 R_2^2} M_0 \tag{A8a}$$

$$D_y = \frac{\omega_1 R_1 R_2}{\Omega^2 R_1 + \omega_1^2 R_2 + R_1 R_2^2} M_0 \tag{A8b}$$

$$D_z = \frac{\Omega^2 R_1 + R_1 R_2^2}{\Omega^2 R_1 + \omega_1^2 R_2 + R_1 R_2^2} M_0 \tag{A8c}$$

Initial conditions, $M_{x,y,z}(0)$ and $((\mathrm{d}/\mathrm{d}t)M_{x,y,z})_{t=0}$, yield the expressions of coefficients $B_{x,y,z}$ and $C_{x,y,z}$ as a function of $D_{x,y,z}$ (see above) and $A_{x,y,z}$ (calculated later)

$$B_{x,y,z} = M_{x,y,z}(0) - A_{x,y,z} - D_{x,y,z} \tag{A9}$$

$$C_x = (1/r)[-(\lambda + R_2)M_x(0) + \Omega M_y(0) + (\lambda - \lambda_0)A_x + \lambda D_x] \tag{A10a}$$

$$C_y = (1/r)[-\Omega M_x(0) - (\lambda + R_2)M_y(0) + \omega_1 M_z(0) + (\lambda - \lambda_0)A_y + \lambda D_y] \tag{A10b}$$

$$C_z = (1/r)[R_1 M_0 - \omega_1 M_y(0) - (\lambda + R_1)M_z(0) + (\lambda - \lambda_0)A_z + \lambda D_z] \tag{A10c}$$

The calculation of $A_{x,y,z}$ stems from relationships existing between the coefficients introduced in equation (A7). By using the identity $R_2 D_x - \Omega D_y = 0$ and by integrating, with respect to time, the first of the Bloch equations, we can write

$$\int_0^\infty \frac{\mathrm{d}M_x}{\mathrm{d}t}\,\mathrm{d}t = -R_2\int_0^\infty (M_x - D_x)\,\mathrm{d}t + \Omega\int_0^\infty (M_y - D_y)\,\mathrm{d}t \tag{A11}$$

It turns out that the integration of $(M_x - D_x)$ is easily calculated from (A7). We obtain

$$\int_0^\infty (M_x - D_x)\,\mathrm{d}t = -(1/\lambda_0)A_x + [1/(\lambda^2 + r^2)](-\lambda B_x + rC_x) \tag{A12}$$

and similarly for $(M_y - D_y)$ and $(M_z - D_z)$, which can be inserted into (A11) for obtaining A_x. A_y and A_z follow from the same type of calculations and, finally, we arrive at

$$A_x = \frac{E_x + \Omega A_y}{R_2} \tag{A13a}$$

$$A_y = \frac{R_1 R_2 E_y - R_1 \Omega E_x + R_2 \omega_1 E_z}{R_1 \Omega^2 + R_1 R_2^2 + R_2 \omega_1^2} \tag{A13b}$$

$$A_z = \frac{E_z - \omega_1 A_y}{R_1} \tag{A13c}$$

with

$$\begin{aligned} E_x = \frac{\lambda_0}{(\lambda - \lambda_0)^2 + r^2}\{&(\lambda^2 + r^2 + 2\lambda R_2)D_x - 2\lambda\Omega D_y \\ &+ [\Omega^2 - r^2 - (R_2 + \lambda)^2]M_x(0) \\ &+ 2\Omega(R_2 + \lambda)M_y(0) - \omega_1 \Omega M_z(0)\} \end{aligned} \tag{A14a}$$

$$\begin{aligned} E_y = \frac{\lambda_0}{(\lambda - \lambda_0)^2 + r^2}\{&2\lambda\Omega D_x + (\lambda^2 + r^2 + 2\lambda R_2)D_y \\ &- 2\lambda\omega_1 D_z - 2\Omega(\lambda + R_2)M_x(0) \\ &+ [\omega_1^2 + \Omega^2 - r^2 - (\lambda + R_2)^2]M_y(0) \\ &+ \omega_1(R_1 + R_2 + 2\lambda)M_z(0) - \omega_1 R_1 M_0\} \end{aligned} \tag{A14b}$$

$$E_z = \frac{\lambda_0}{(\lambda - \lambda_0)^2 + r^2}\{2\lambda\omega_1 D_y + (\lambda^2 + r^2 + 2\lambda R_1)D_z - \omega_1 \Omega M_x(0)$$
$$- \omega_1(2\lambda + R_1 + R_2)M_y(0) + [\omega_1^2 - \Omega^2 - (\lambda + R_1)^2]M_z(0) + R_1^2 M_0\} \tag{A14c}$$

Bibliography

O. W. Sørensen, G. W. Eich, M. H. Levitt, G. Bodenhausen and R. R. Ernst: *Prog. NMR Spectrosc.*, **16**, 163 (1983)

R. R. Ernst, G. Bodenhausen and A. Wokaun: *Principles of Nuclear Magnetic Resonance in One and Two Dimensions*, Clarendon Press, Oxford, 1987, Chapter 2

T. C. Farrar: *An Introduction to Pulse NMR Spectroscopy*, The Farragut Press, Madison, 1989

M. Goldman: *Quantum Description of High Resolution NMR in Liquids*, Clarendon Press, Oxford, 1988

R. Freeman: *A Handbook of Nuclear Magnetic Resonance*, Longman, Harlow, 1988

C.P. Slichter: *Principles of Magnetic Resonance*, 3rd ed., Springer-Verlag, Berlin, Heidelberg, New York, 1989

T. C. Farrar and J. Harriman: *Density Matrix and its Application in Spectroscopy*, 2nd ed., The Farragut Press, 1995

S. W. Homans: *A Dictionary of Concepts in NMR*, Clarendon Press, Oxford, 1989

3 FOURIER TRANSFORM NMR AND DATA PROCESSING

3.1 Averaging (Accumulation) of Free Induction Decays

Although the current practice of NMR is intimately related to Fourier transformation, this mathematical operation serves actually for data processing at the ultimate stage of a pulse NMR experiment. As illustrated in Figure 3.1, it is almost mandatory to look for a method capable of overcoming the poor sensitivity

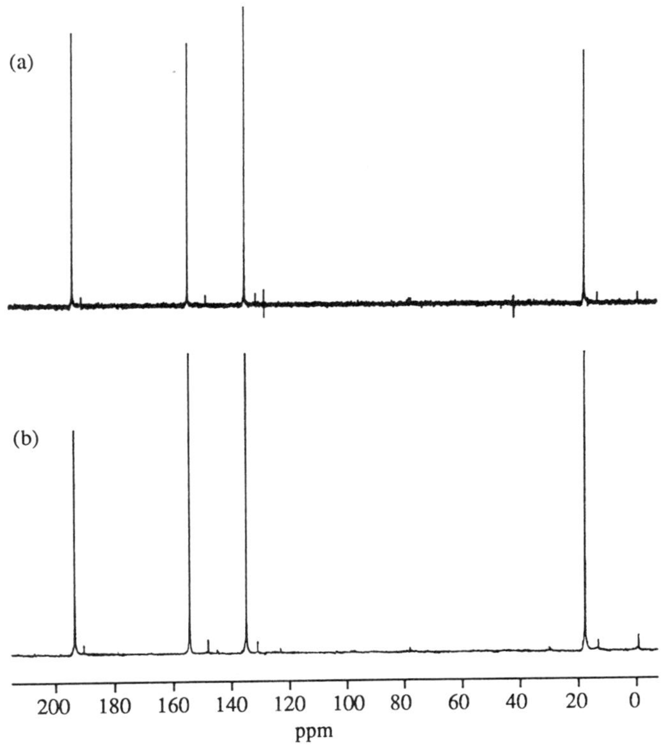

Figure 3.1 (a) Proton-decoupled carbon-13 spectrum of *trans*-crotonaldehyde as obtained after Fourier transformation of the fid resulting from the application of a single 90° pulse; (b) the same spectrum after accumulation of the free induction decays resulting from 256 consecutive experiments

inherent in some single NMR experiments. A possible approach consists of adding coherently the results of n identical, though independent, experiments; this is better performed in the pulse mode so that free induction decays (fids) are accumulated and subsequently subjected to Fourier transformation. Indeed, as shown below, upon accumulating n fids, the signal-to-noise ratio (S/N ratio, a quantity representative of the signal amplitude versus the noise amplitude) is improved by a factor of $\sqrt{n}$.

The S/N ratio of a single experiment can be evaluated by S_0/σ_0, S_0 being the amplitude of the NMR signal and σ_0 the standard deviation associated with noise; σ_0 may be estimated by the overall noise amplitude divided by 2.5 (Figure 3.2; the scaling factor of 2.5 originates from statistical considerations in relation to Gaussian random variables). If n experiments of identical nature are added, the signal increases and its amplitude becomes equal to nS_0. On the other hand, the noise pertaining to a given experiment can be regarded as the realization of a random variable. It is well known that the variance of the sum of n *independent* random variables is equal to the sum of the variances of these variables. This is precisely the case here, as the noise of a given experiment is independent of that of another experiment. Because the standard deviation is the square root of the variance, it is equal, for n coadded experiments, to $\sqrt{n}\,\sigma_0$ and the final S/N ratio is $\sqrt{n}\,S_0/\sigma_0$, and is thus improved by a factor of $\sqrt{n}$.

One of the interests of the NMR pulse method lies in the possibility of repeating experiments at a rate which is much faster (of the order of a second) than that of the earlier continuous-wave method which required several minutes to run a single spectrum. This feature, in addition to other advantages, prompted most NMR spectroscopists to switch to Fourier transform NMR in the early 1970s. In order to make clear the involvement of Fourier transformation in a pulse experiment, we shall outline the main results established in Chapter 2: a radio-frequency (r.f.) field whose frequency ν_r is close to the resonance frequency ν_0, applied for a short time interval τ (of the order of a few microseconds), flips the nuclear magnetization by an angle $\alpha = \gamma B_1 \tau$, where B_1 is the amplitude of the r.f. field and γ is the gyromagnetic ratio of the relevant nucleus. $\gamma B_1/2\pi$, which is expressed in Hz, must be much larger than the frequency difference $|\nu_r - \nu_0|$. Contrary to a widespread feeling, which rests on an incorrect hypothesis according to which spins respond linearly to an external perturbation, this latter condition ($\gamma B_1 \gg 2\pi|\nu_r - \nu_0|$) is not exactly related to the excitation spectrum of a pulse. Rather, one must refer to the concept of the rotating frame, where the r.f. field appears stationary. In that frame, one has to deal with an effective field whose longitudinal component,

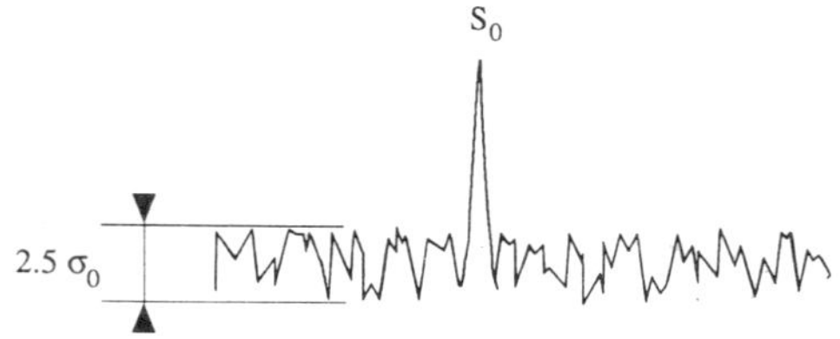

Figure 3.2 Estimation of the signal-to-noise ratio (S/N ratio): S_0/σ_0

expressed in frequency units, is equal to $\nu_r - \nu_0$, whereas the transverse component remains $\gamma B_1/2\pi$ and is therefore overwhelming provided that the above condition is fulfilled. Immediately after the r.f. pulse, the nuclear magnetization $\boldsymbol{M}$ will be subjected to a precessional motion around B_0 at its own frequency ν_0 and will therefore induce a signal (emf: electromotive force) in the receiver coil at the same frequency ν_0 (Figure 3.3). Because the emf is itself proportional to ν_0 and taking into account the expression of nuclear magnetization (see formula 2.2), we can see that, in a general way, the intensity of the *NMR signal is proportional to* $\gamma^3 B_0^2$. This allows us to quantify the S/N ratio improvement when going to higher field or when observing a nucleus of higher gyromagnetic ratio. The NMR response is

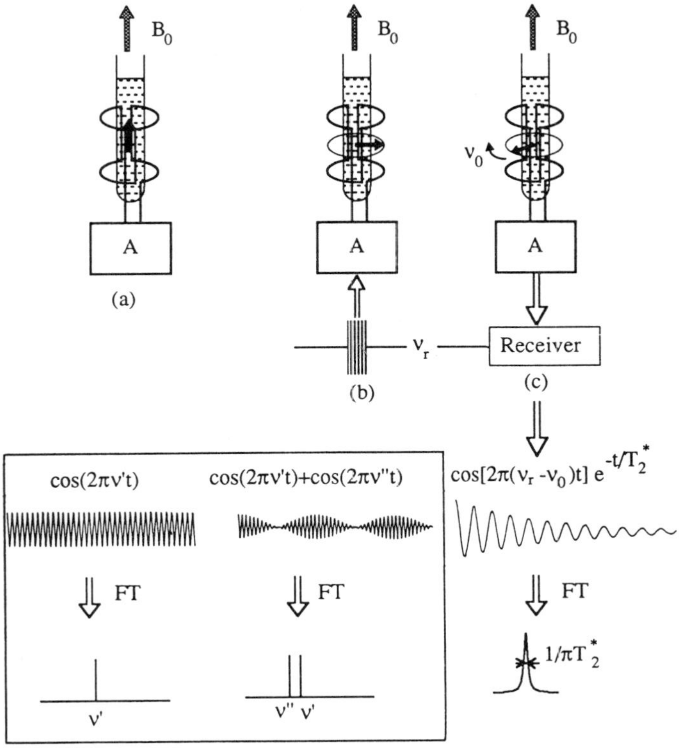

Figure 3.3 Principle of Fourier transform NMR. (a) At equilibrium, nuclear magnetization is along the static field B_0; (b) a radio-frequency pulse ν_r of appropriate duration, created by the coil surrounding the sample (A stands for tuning and matching circuits), flips the nuclear magnetization into a plane perpendicular to B_0; (c) subsequently, nuclear precession at frequency ν_0 takes place and induces within the coil a signal at the same frequency damped according to T_2^*. This signal is detected with respect to ν_r. Fourier transformation of this audio-frequency signal leads to the conventional spectrum: a Lorentzian line of width at half-height equal to $1/\pi T_2^*$. The mathematical operation of Fourier transformation plays the role of a frequency analyzer. Applied to one or several cosine functions (in the inset), it generates infinitely sharp peaks (Dirac peaks) at the actual frequencies of these cosine functions. Applied to a damped cosine function, it generates a peak whose width is related to the damping factor

generally detected with respect to the excitation frequency ν_r in such a way that one is left with a signal whose frequency is $\nu_r - \nu_0$ (in the audiofrequency domain). This time function is necessarily damped, the decay being generally exponential with a time constant denoted by T_2^*, which accounts for the effects of B_0 inhomogeneity in addition to the transverse relaxation time $T_2(T_2^* < T_2)$. In fact, the precessing component of magnetization, which lies in a plane perpendicular to B_0 (this component is also called transverse magnetization), tends to disappear by the combined effects of transverse relaxation and of B_0 inhomogeneity; this latter effect can be compared to destructive interferences. While the transverse magnetization disappears, longitudinal magnetization *independently* rebuilds and tends exponentially toward its equilibrium value according to a time constant T_1, also called the longitudinal relaxation time ($T_1 > T_2^*$).

Generally, the response of nuclear spins to a r.f. pulse consists of the superposition of damped sine (or cosine) functions, each of them being associated with a line in the spectrum. The corresponding fid is sometimes called an interferogram by reference to optical spectroscopy (Figure 3.4). A fid (or an interferogram) is a function of time from which the classical spectrum (a function of frequency) has to be reconstructed (see Section 2.1.4.). This is a very general problem possibly solved by the Fourier transformation, which has been known since the nineteenth century. However, to put it in practice for the exploitation of physical measurements, scientists had to wait until the mid-sixties (of this century!). At that time, there appeared (i) an efficient algorithm, the so-called fast Fourier transform, due to Cooley and Tukey, and (ii) mini-computers of reasonable cost capable of performing the relevant calculations in a sufficiently short time.

Using pulse techniques to improve the sensitivity of an NMR experiment nevertheless requires some precautions. Until now, it has been implicitly assumed that the spin system had recovered to thermal equilibrium before the application of any further pulse; this implies that the time T elapsed between two consecutive pulses should be of the order of $5T_1$ so that the quantity $\exp(-T/T_1)$ does not exceed 1%. Owing to the usual values of longitudinal relaxation times, the interest of the accumulation process would be questionable, unless one decides to choose a flip angle α smaller than 90°.

As a matter of fact, it is possible to determine the optimal value of α for a given repetition rate (the notations are defined in Figure 3.5) by assuming the existence of a stationary state. This amounts to postulating the same value, $M_z(0)$, for longitudinal magnetization after each r.f. pulse; $M_z(T)$ can then be calculated from $M_z(0)$ through the Bloch equations

$$M_z(T) = M_0 - [M_0 - M_z(0)]\exp(-T/T_1)$$

M_0 being the equilibrium magnetization. On the other hand, the r.f. pulse of flip angle α transforms $M_z(T)$ into $M_z(0) = M_z(T)\cos\alpha$. We can thus derive $M_z(T)$ from these two latter equations and calculate the amplitude of transverse magnetization, the quantity which will be actually detected:

$$M_z(T)\sin\alpha = M_0 \sin\alpha \frac{1 - \exp(-T/T_1)}{1 - \cos\alpha\exp(-T/T_1)}.$$

We are consequently looking for the angle α which maximizes $M_z(T)\sin\alpha$. By

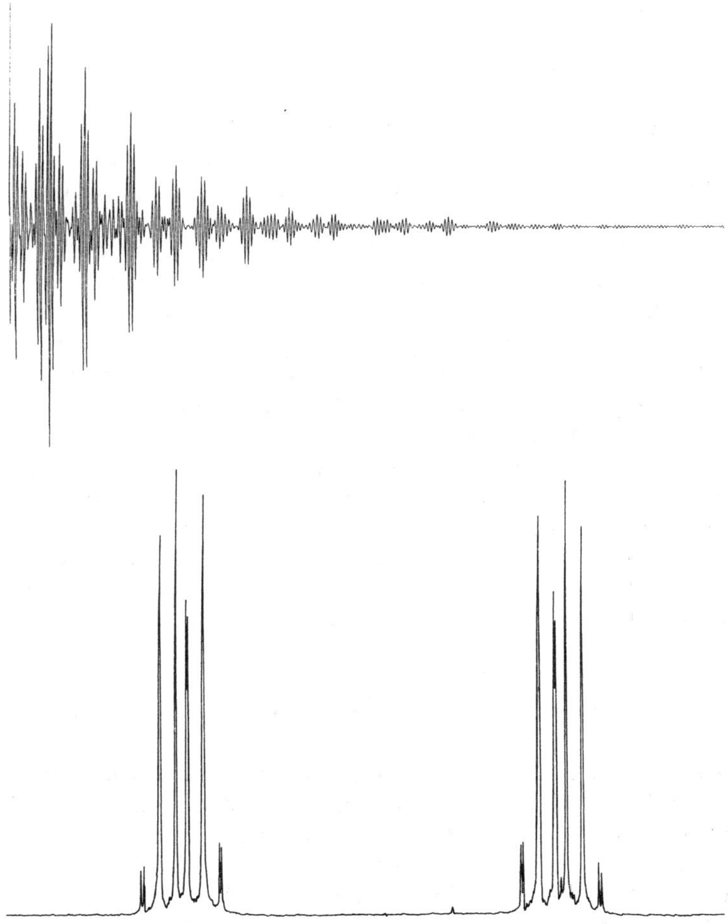

Figure 3.4 Interferogram composed of superimposed damped sine (or cosine) functions and its Fourier transform, which is just the conventional NMR spectrum (400 MHz ^1H spectrum of *ortho*-dichlorobenzene)

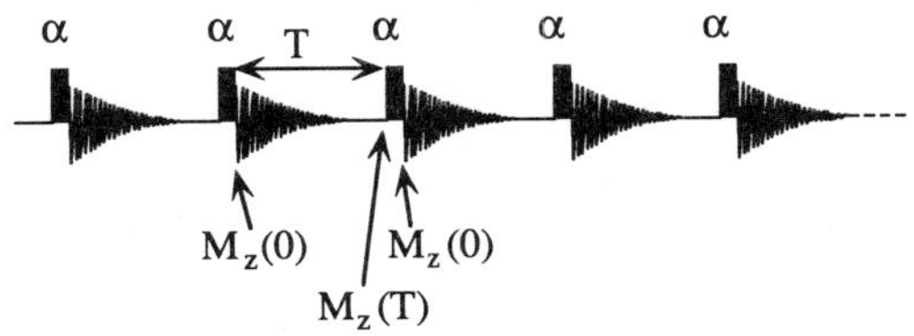

Figure 3.5 Successive pulses corresponding to a flip angle α and separated by an interval of duration T. M_z stands for the longitudinal component of nuclear magnetization

zeroing its derivative, it is a simple matter to obtain α_{opt}, also called the Ernst angle

$$\cos \alpha_{opt} = \exp(-T/T_1). \tag{3.1}$$

After some calculations not reported here, the S/N improvement obtained by pulse methods over continuous-wave methods can be estimated by the quantity $\sqrt{\nu_t/\Delta\nu}$, where ν_t is the spectral window under investigation and $\Delta\nu$ is a typical linewidth at half height.

This result would indicate that the interest in pulse methods (and the subsequent data treatment by Fourier transform) is limited to spectra involving sharp lines. In fact, these considerations are essentially of theoretical nature and one must take into account the practical aspects of pulse NMR. It is fairly easy to obtain an interferogram, which can eventually be subjected to data manipulations aimed at improving the quality of the final spectrum (see Section 3.6), whereas it is relatively difficult, if not impossible, to run a continuous-wave experiment, especially in the case of low sensitivity nuclei. In fact, pulse methods opened the way to the observation of numerous isotopes whose spectra proved to be very informative. There are other intrinsic advantages of the pulse method which made continuous-wave detection totally obsolete. To cite a few of them, we can mention the direct measurement of relaxation times and the various spin preparations leading to polarization or coherence transfers.

3.2 Fourier Transform and Two-Dimensional Spectroscopy

The advantage of Fourier transformation in nuclear magnetic resonance is not limited to sensitivity improvements. Indeed, as early as 1971, Jeener imagined an experiment involving two time variables t_1 and t_2; t_2 corresponds to the usual acquisition of the fid, possibly after the application of a read pulse, whereas, during the previous time interval t_1, the spin system has been subjected to perturbations, for instance a sequence of radio-frequency pulses, which confer to the observed signal a further sine (or cosine) modulation (in phase or amplitude) depending upon the variable t_1. The final signal, denoted as $s(t_1, t_2)$, will be processed by a double Fourier transform with respect to t_2 and t_1, provided that this latter time variable has been varied according to an appropriate increment (see below) and that the relevant interferograms have been conveniently stored. In practice, for each of them, a Fourier transformation with respect to t_2 will first be calculated; this leads to a series of spectra whose line positions are determined in the frequency domain ν_2. The amplitude or phase of those lines exhibiting a sine (or cosine) modulation according to the time variable t_1; a subsequent Fourier transform with respect to t_1 thus yields a surface $S(\nu_1, \nu_2)$ representative of a two-dimensional spectrum (Figure 3.6).

A peak appearing at frequencies ν'_1 and ν'_2 indicates a correlation of the two spins whose resonance frequencies are precisely ν'_1 and ν'_2. The nature of this correlation depends upon the events which occurred during the time interval t_1. Without entering the details of perturbations which can possibly be applied to the spin system (they will be actually considered in Chapters 4 and 5), we can briefly

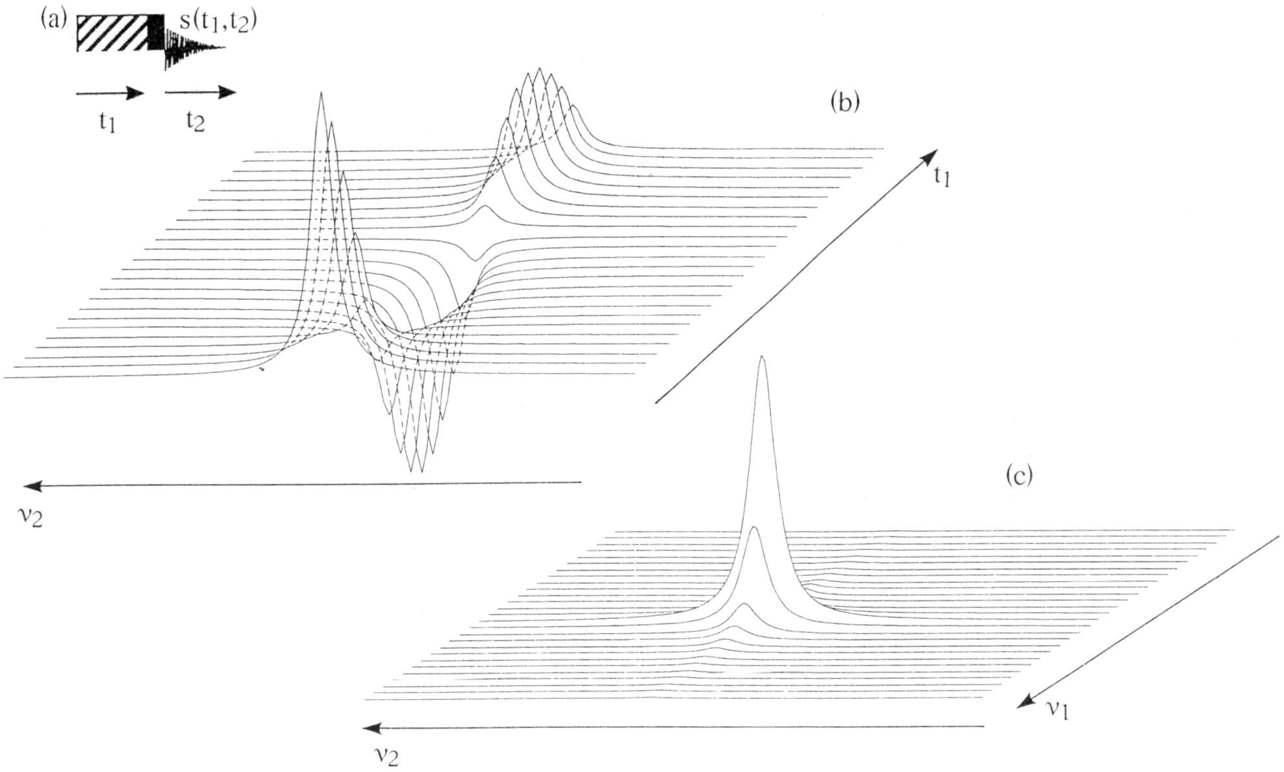

Figure 3.6 (a) A schematized hypothetical two-dimensional NMR experiment: events which occur during t_1, followed by a read pulse, confer to the signal s, physically acquired during t_2, an additional modulation depending on the time variable t_1; (b) Fourier transforms, with respect to t_2, of interferograms obtained for a series of t_1 values (in this example, the peak amplitude in a cosine function of the time variable t_1); (c) a Fourier transform with respect to t_1 leads to the two-dimensional spectrum $S(\nu_1, \nu_2)$

mention the main correlations regarding spectral, structural or dynamical parameters that are made accessible by two-dimensional methodologies. For the sake of simplicity, these correlations are described below by means of a simple amplitude modulation, although the actual experimental procedures lead often to a phase modulation.

Chemical shift correlation via J coupling (COSY)

This takes place between two nuclei resonating at frequencies ν_{01} and ν_{02} (Figure 3.7). The signal $\cos[2\pi(\nu_{r2} - \nu_{02})t_2]\exp(-t_2/T_2^*)$ acquired as a function of the variable t_2 (conventional fid) is modulated (multiplied) by a damped cosine function involving the other time variable t_1: $\cos[2\pi(\nu_{r1} - \nu_{01})t_1]\exp(-t_1/T_2^*)$. It will, however, be noticed that effective transverse relaxation times in both time domains are not necessarily identical.

Correlation between a spin–spin coupling and chemical shift

The t_1 modulation is based on an interaction between spins, for instance the indirect coupling (see p. 93: *J*-resolved spectroscopy); it can then be expressed as $\cos(\pi J t_1)\exp(-t_1/T_2^*)$.

Chemical shift correlation via an incoherent coupling (NOESY)

This can arise, for instance, from the direct (dipole–dipole) coupling in relation to molecular reorientation which leads to relaxation phenomena known as nuclear Overhauser effect (NOE). Chemical exchange yields the same type of correlation.

Correlation between chemical shift and spatial localization

When this takes place via the spin density, the t_1 function, of the type $\cos(2\pi\gamma g X t_1)\exp(t_1/T_2^*)$, comes from the application of a field gradient g along the spatial X direction and gives rise to localized spectroscopy techniques. An encoding with respect to the other spatial directions can be envisaged provided that two more time variables associated with these spatial directions are put in to use. Moreover, imaging procedures rest most of the time on a single chemical species (water, as far as biomedical applications are considered). As a consequence, the chemical shift dimension can be omitted and affected to a space dimension.

In this respect, magnetic resonance imaging (MRI) is just a special case of 2D NMR, corresponding to the X and Y spatial dimensions (with a given slice in the third spatial direction, Z, defined through selective excitation procedures).

For each of these four examples, the amplitude factor of the t_1 modulation (omitted in the expressions given above) is regained by the second Fourier transform, since the intensity of the cross peak (at frequencies $\nu'_1 = \nu_{r1} - \nu_{01}$ and $\nu'_2 = \nu_{r2} - \nu_{02}$) is directly proportional to the interaction responsible for the observed correlation. This feature is of prime importance for the two latter cases: interatomic distances can be estimated through correlations originating from the

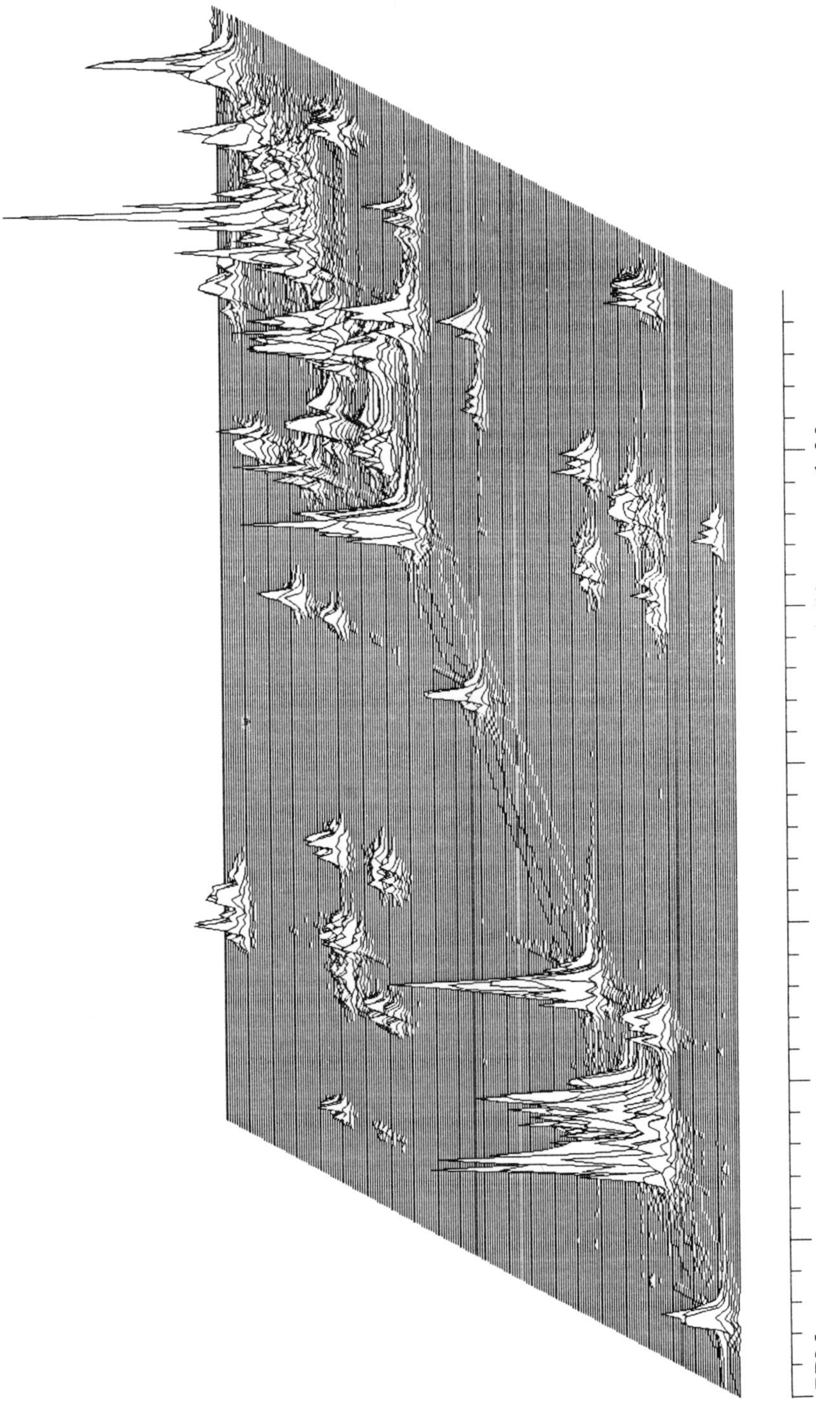

Figure 3.7 Chemical shift correlations. The conventional spectrum is along the diagonal. Other peaks (cross peaks) are at cross positions with respect to two diagonal peaks. Cross peaks indicate a correlation which arises from the existence of a *J* coupling

nuclear Overhauser effect, whereas the spatial distribution of molecules can be probed through localized NMR experiments.

3.3 Properties of the Fourier Transformation

In mathematical terms, the Fourier transform $F(\nu)$ of a time function $f(t)$ is expressed by the following integral:

$$F(\nu) = \int_{-\infty}^{\infty} f(t) \exp(-2i\pi\nu t)\, dt \tag{3.2}$$

with a reciprocal property:

$$f(t) = \int_{-\infty}^{\infty} F(\nu) \exp(2i\pi\nu t)\, d\nu \tag{3.3}$$

Equation (3.3) represents the inverse Fourier transform, which differs from the direct Fourier transform only by the sign of the exponential argument.

Equation (3.3) can be derived easily from (3.2) by considering the integral of $F(\nu)\exp(2i\pi\nu t')$ with respect to ν. This integral can be written as

$$\int_{-\infty}^{\infty}\int_{-\infty}^{\infty} f(t)\exp[2i\pi\nu(t'-t)]\, d\nu\, dt = \int_{-\infty}^{\infty} f(t)\left\{\int_{-\infty}^{\infty} \exp[2i\pi\nu(t'-t)]\, d\nu\right\} dt.$$

It turns out that $\int_{-\infty}^{\infty}\exp[2i\pi\nu(t'-t)]\, d\nu$ is a realization of the so-called 'Dirac distribution' $\delta(t-t')$, which is defined by (see the next section)

$$\int_{-\infty}^{\infty} \delta(t-t')f(t)\, dt = f(t');$$

this relationship completes the proof of equation (3.3). A similar proof leads to the Parseval theorem, which states that the norm of a function is equal to the norm of its Fourier transform:

$$\int_{-\infty}^{\infty} F^*(\nu)F(\nu)\, d\nu = \int_{-\infty}^{\infty} f^*(t)f(t)\, dt \tag{3.4}$$

where the asterisk denotes the complex conjugate. This theorem means that the global information is restored by a Fourier transformation.

3.3.1 FOURIER TRANSFORM OF A DIRAC PEAK

A Dirac distribution can be denoted by $\delta(x-a)$ and is defined by the fundamental relationship

$$\int_{-\infty}^{\infty} \delta(x-a)f(x)\, dx = f(a). \tag{3.5}$$

The following properties are a consequence of (3.5)

$$\delta(x-a) = \begin{cases} \infty & \text{for } x = a \\ 0 & \text{for } x \neq a \end{cases}$$

and

$$\int_{-\infty}^{\infty} \delta(x)\, dx = 1$$

and provide the usual picture of a Dirac distribution: a peak at abscissa a of infinite height, infinitely sharp and of area equal to unity.

From (3.5), it is straightforward to calculate the Fourier transform of a Dirac peak; this yields a cosine function for the real part and a sine function for the imaginary part:

$$f(t) = \int_{-\infty}^{\infty} \delta(\nu - \nu')\exp(2i\pi\nu' t)\,\mathrm{d}\nu = \cos(2\pi\nu' t) + i\sin(2\pi\nu' t). \tag{3.6}$$

Relying on the inverse property of Fourier transformation, we can arrive at the statements of Figure 3.3: the Fourier transform of $\exp(2i\pi\nu' t)$ is a Dirac peak at frequency ν'. Because the Fourier transformation is primarily an integral, it is a linear operation. Therefore, the Fourier transform of a composite signal containing the frequencies ν', ν'', ν''' ... exhibits Dirac peaks at frequencies ν', ν'', ν''' .... These considerations constitute a mathematical justification for the role of frequency analyzer played by the Fourier transformation.

3.3.2 FOURIER TRANSFORMATION OF A RECTANGULAR FUNCTION

The calculation is again straightforward. For a rectangular function of amplitude equal to unity and extending from 0 to T, we obtain:

$$\int_0^T \exp(-2i\pi\nu t)\,\mathrm{d}t = T\frac{\sin(2\pi\nu T)}{2\pi\nu T} - i\frac{\sin^2\pi\nu T}{\pi\nu} \tag{3.7}$$

The real part can be seen to be the sinc function $(\sin x/x)$ multiplied by T (Figure 3.8).

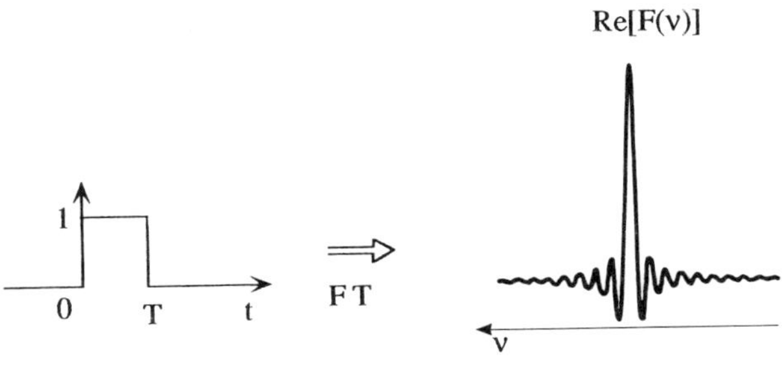

Figure 3.8 The real part of the Fourier transform of a rectangular function. Zeros occur at $\nu = \pm 1/2T$; $\pm 1/T$; $\pm 3/2T$ etc

3.3.3 FOURIER TRANSFORMATION OF A GAUSSIAN FUNCTION

With $f(t) = \exp(-t^2/2a^2)$, one obtains:

$$F(\nu) = \int_{-\infty}^{\infty} \exp(-t^2/2a^2)\exp(-2i\pi\nu t)\,\mathrm{d}t = a\sqrt{2\pi}\exp(-2\pi^2a^2\nu^2) \tag{3.8}$$

This indicates that the Fourier transform of a Gaussian function is another Gaussian function. This is an interesting feature which can be advantageous when devising windows to improve the visual quality of spectra (see below). The result given in (3.8) can be derived from the calculation of $dF/d\nu$ followed by an integration by parts in order to change the above integral into a differential equation.

3.3.4 FOURIER TRANSFORMATION OF A PRODUCT

Let $F(\nu)$ and $G(\nu)$ be the two Fourier transforms of the functions $f(t)$ and $g(t')$. Their product can be written as:

$$F(\nu)G(\nu) = \int_{-\infty}^{+\infty}\int_{-\infty}^{+\infty} f(t)g(t')\exp[-2i\pi\nu(t + t')]\,dt\,dt'$$

or by substituting t' by $\tau = t + t'$:

$$F(\nu)G(\nu) = \int_{-\infty}^{+\infty}\left[\int_{-\infty}^{+\infty} f(t)g(\tau - t)\,dt\right]\exp(-2i\pi\nu\tau)\,d\tau.$$

The quantity between the square brackets is called the convolution product and is generally represented by $f{*}g$:

$$f{*}g(\tau) = \int_{-\infty}^{+\infty} f(t)g(\tau - t)\,dt. \tag{3.9}$$

The product of two Fourier transforms is therefore given by the Fourier transform of their convolution product. Moreover, by reference to the inverse Fourier transform, we can state that the Fourier transform of a product of functions is equal to the convolution product of their transforms:

$$f{*}g(t) = \int_{-\infty}^{+\infty} F(\nu)G(\nu)\exp(2i\pi\nu t)\,d\nu \tag{3.10}$$

or

$$F{*}G(\nu) = \int_{-\infty}^{\infty} f(t)g(t)\exp(-2i\pi\nu t)\,dt. \tag{3.11}$$

This property, although somewhat formal at first sight, is rather important for the practice of Fourier transform NMR. As an example, let us consider the truncated interferogram at Figure 3.9 (top). This amounts to multiplying the total interferogram by a rectangular function (whose Fourier transform is shown in Figure 3.8). The convolution product of a Lorentzian function (see below) by a sinc function results in a sort of mixing of the representative curves (Figure 3.9, bottom) leading to distorted Lorentzian peaks flanked by damped oscillations.

3.4 Fourier Transformation of a Damped Sine (or Cosine) Function. (One-Dimensional NMR)

3.4.1 QUADRATURE DETECTION

As already mentioned in Section 2.1.4, the response of the spin system to a radio-frequency pulse can be written as:

$$S(t) = \exp(-2i\pi\nu' t)\exp(-t/T_2^*) \tag{3.12}$$

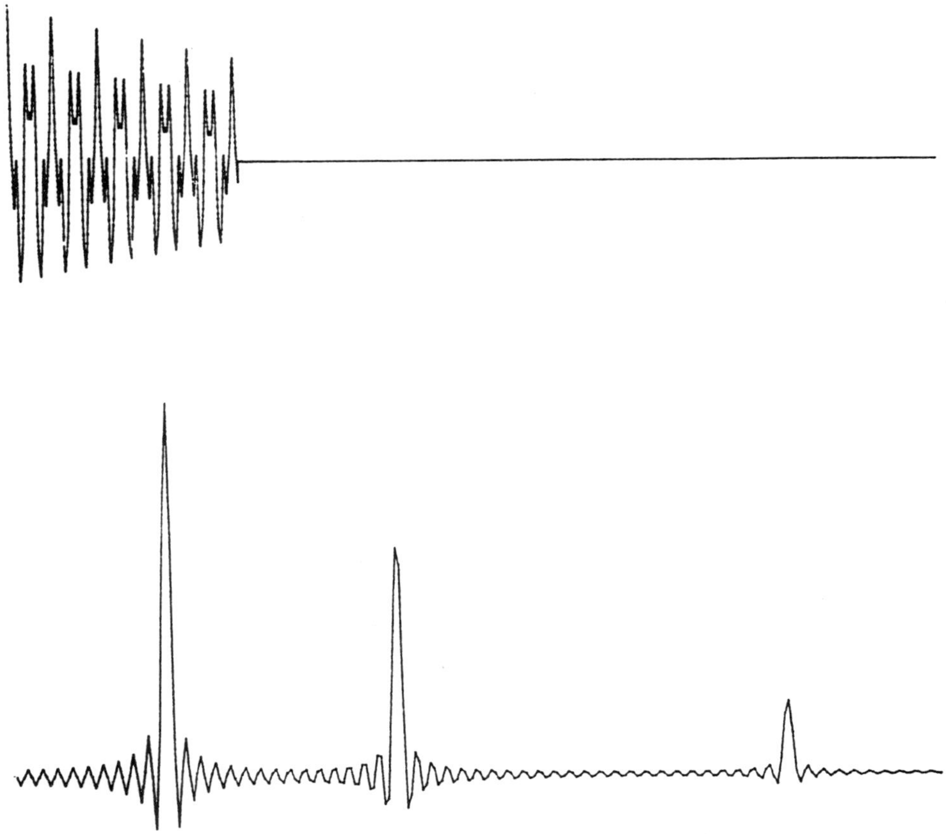

Figure 3.9 Truncated interferogram and the corresponding Fourier transform

For simplicity, the signal amplitude has been set to unity. It may appear peculiar that (3.12) involves a complex number. In fact, this quantity represents the experimental reality for most spectrometers presently in use, owing to their quadrature detection capability. This detection mode will now be described. The NMR signal of frequency ν_0 coming out of the probe is preamplified and then demodulated with respect to a reference frequency which is precisely the excitation frequency ν_r. This demodulation is performed by means of a mixer whose output is composed of two signals at frequencies $\nu_r - \nu_0$ and $\nu_r + \nu_0$. We are only interested in the low-frequency signal $\nu_r - \nu_0$; the high-frequency signal is anyway eliminated by the receiver which, because of its bandwidth, can only accommodate low-frequency (audio-frequency) signals. The result of this demodulation scheme is therefore the function $\cos[2\pi(\nu_r - \nu_0)t]$. Now, suppose that *simultaneously* a similar demodulation is performed with a reference dephased by 90° (Figure 3.10). This will result in a low-frequency signal of the form $\sin[2\pi(\nu_r - \nu_0)t]$.

By combining the signals arising from the two channels schematized in Figure 3.10, denoting by ν' the frequency difference $\nu_r - \nu_0$ and introducing the damping

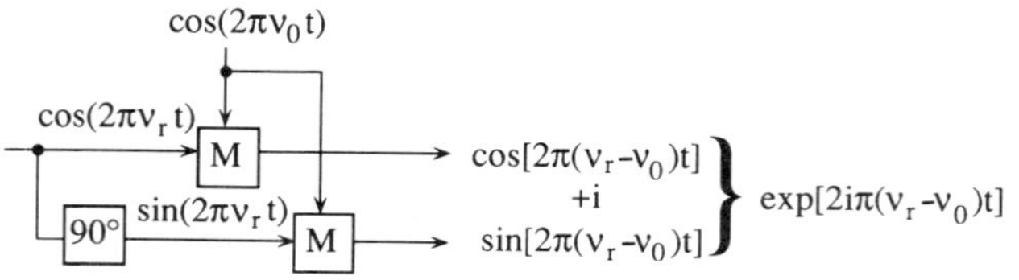

Figure 3.10 Principle of the quadrature detection mode

factor $\exp(-t/T_2^*)$, we retrieve the quantity given in (3.12), which can then be subjected to a *complex* Fourier transform to yield:

$$\begin{aligned}\int_0^{\infty} \exp(2i\pi\nu' t)\exp(-t/T_2^*)\exp(-2i\pi\nu t)\,dt \\ = \frac{T_2^*}{1+4\pi^2T_2^{*2}(\nu'-\nu)^2} + i\frac{T_2^{*2}2\pi(\nu'-\nu)}{1+4\pi^2T_2^{*2}(\nu'-\nu)^2} \\ = A(\nu'-\nu) + iD(\nu'-\nu). \end{aligned} \qquad (3.13)$$

As already pointed out, the real part (A) is the classical absorption spectrum (Lorentzian line at frequency $\nu = \nu'$ of width at half height equal to $1/\pi T_2^*$), whereas the imaginary part (D) is the dispersion spectrum (Figure 3.11).

Examination of equation (3.13) and of Figure 3.11 would suggest that the imaginary part D does not provide any additional information with respect to the real part A. In practice, the consideration of both A and D appears mandatory in order to perform the so-called phase corrections. The latter are required because the experimental detection procedure induces almost invariably a phase angle φ which adds to $2\pi\nu' t$ in the exponential argument. This phase angle generally involves a frequency-independent contribution φ_0, due to the demodulation reference (zero-order phase correction); a second contribution, directly proportional to

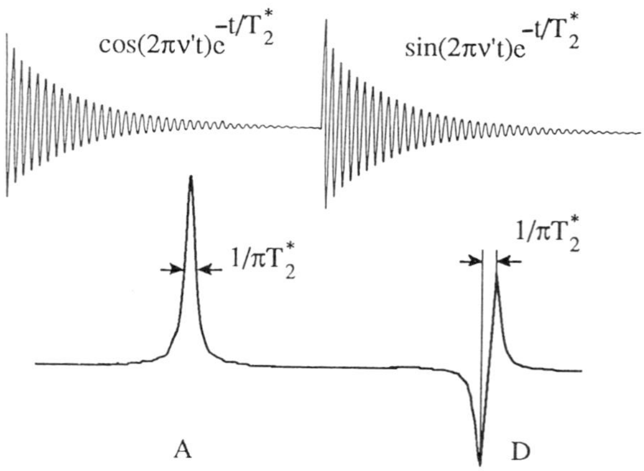

Figure 3.11 Fourier transform of $\exp(2i\pi\nu' t)\exp(-t/T_2^*)$. A: real part (absorption spectrum); D: imaginary part (dispersion spectrum)

the frequency ν' and thus expressed as $\nu'\varphi_1$, must also be accounted for (first-order phase correction). The latter has two origins: (i) the low-pass filter positioned in front of the analog-to-digital converter (see below), (ii) a dead-time d inserted between the end of the radio-frequency pulse and the beginning of data acquisition to avoid the spurious oscillations which follow the r.f. pulse; acquisition is thus started with a retardation d entailing a phase modification equal to $2\pi\nu' d$. Altogether, zero order and first order phase changes are accounted for by the inclusion of a factor $\exp(i\varphi)$ in the integral of (3.13), with the angle φ equal to $\varphi_0 + \nu'\varphi_1$. As a consequence, the real part of the Fourier transform is a mixture of absorption and dispersion:

$$\mathrm{Re} = \cos\varphi \mathrm{A}(\nu' - \nu) - \sin\varphi \mathrm{D}(\nu' - \nu).$$

Likewise, for the imaginary part, one has:

$$\mathrm{Im} = \cos\varphi \mathrm{D}(\nu' - \nu) + \sin\varphi \mathrm{A}(\nu' - \nu).$$

The pure absorption spectrum is regained by combining the real and imaginary parts:

$$\mathrm{A}(\nu' - \nu) = \cos\varphi \mathrm{Re} + \sin\varphi \mathrm{Im}.$$

This operation (Figure 3.12) is generally carried out by trial and error.

Another solution to the phase problem is to calculate the magnitude spectrum M, defined as:

$$\mathrm{M} = (\mathrm{Re}^2 + \mathrm{Im}^2)^{1/2} = (\mathrm{A}^2 + \mathrm{D}^2)^{1/2} = \frac{T_2^*}{\sqrt{1 + 4\pi^2 T_2^{*2}(\nu' - \nu)^2}}. \tag{3.14}$$

M becomes independent φ; however, the line is no longer Lorentzian and is somewhat broader (with a width at half height of $\sqrt{3}/\pi T_2^*$) with more important wings (Figure 3.13).

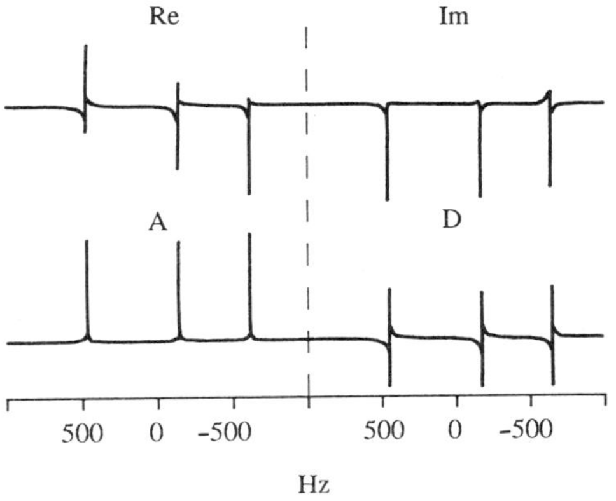

Figure 3.12 Raw spectrum (Re, Im) for which a frequency-dependent dephasing can be noted. (A, D): phase corrected spectrum

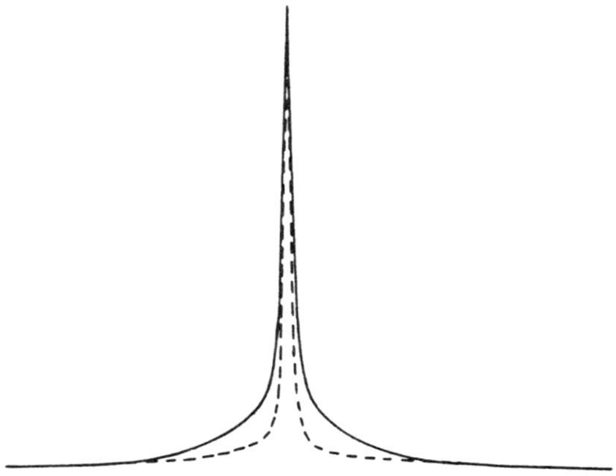

Figure 3.13 Lorentzian line (dashed trace) and the corresponding magnitude spectrum (full trace)

3.4.2 SINGLE CHANNEL DETECTION

In that case, omitting the amplitude (set to unity) and a possible phase angle, the detected signal has the form $\cos(2\pi\nu' t)\exp(-t/T_2^*)$. We then proceed with a *real Fourier transform* which, owing to the identity $\cos(2\pi\nu' t) = (1/2)[\exp(2i\pi\nu' t) + \exp(-2i\pi\nu' t)]$, leads to $(1/2)[\mathrm{A}(\nu-\nu') + \mathrm{A}(\nu+\nu')]$ for the real part and to $(1/2)[\mathrm{D}(\nu-\nu') + \mathrm{D}(\nu+\nu')]$ for the imaginary part. As an immediate consequence, we note that the sign of ν' cannot be determined. Thus, it will be useless to display the negative part in the frequency domain which is redundant regarding the positive part. Moreover, the lack of sign information dictates that the reference frequency ν_r must be set at one extremity of the spectral width SW. Conversely, quadrature detection enables one to position the reference frequency at the center of the spectral window, with a twofold advantage:

- the amplitude of the r.f. field $\gamma B_1/2\pi$ can be limited so as to fulfill the condition greater than SW/2 and not greater than SW, as for single channel detection;
- the S/N ratio is increased by a factor of $\sqrt{2}$, as explained later.

3.4.3 SEQUENTIAL QUADRATURE DETECTION

Simultaneously sampling the output of the two channels of Figure 3.10 may entail certain technical problems which can be circumvented by turning to the sequential mode known under the acronym of TPPI (time proportional phase incrementation). In order properly to understand this procedure, we must introduce the concept of 'Nyquist frequency' (denoted by ν_{max} in the following) which will be justified mathematically later (Section 3.7.1). Measuring a time-dependent signal implies the definition of a sampling interval Δt. A first consequence of the digitization of the time variable is that the spectral width examined without ambiguity lies in the range $[0 - \nu_{max}]$ (with $\nu_{max} = 1/2\Delta t$) if single channel detection is employed and $[-\nu_{max}, +\nu_{max}]$ for quadrature detection. The TPPI procedure dictates the use of a sampling rate half that indicated above: $\Delta t' = 1/4\nu_{max}$. Moreover, the phase of the detection reference must be incremented by $\pi/2$ at each sampling step by using *alternately* one or other of the two channels of Figure 3.10 and applying to the detected signal the appropriate sign (given in Table 3.1).

By taking into account the phase of the detection reference (one of the two quadrature channels) and the sign given to the detected signal, we can recognize that the signal detected at $t = k\Delta t'$ is affected by a phase angle equal to $\varphi = k\pi/2 = 2\pi\nu_{max}t$. In other words, the actual signal (disregarding the amplitude and damping factors) has the form $\cos[2\pi(\nu_r - \nu_0)t + \varphi] = \cos[2\pi(\nu_r + \nu_{max} - \nu_0)t]$. This amounts to shifting the reference frequency by the quantity ν_{max} (Figure 3.14). If we apply a *real* Fourier transform (as if we were dealing with single channel detection), we then explore without ambiguity the frequency zone ranging from $\nu_r + \nu_{max} - \nu_0 = 0$ to $\nu_r + \nu_{max} - \nu_0 = 1/2\Delta t' = 2\nu_{max}$, thus ranging from $\nu_r - \nu_0 = -\nu_{max}$ to $\nu_r - \nu_0 = \nu_{max}$. The properties of quadrature detection are consequently regained without the possible difficulties associated with the simultaneous sampling of the two quadrature channels.

Table 3.1 Conditions to be used for sequential quadrature detection. The integer k defines the acquisition process and $t = k\Delta t'$. 0 and $\pi/2$ refer to the two quadrature channels of Figure 3.10

k	0	1	2	3	4	5	6	7
	Detection channel							
	0	$\pi/2$	0	$\pi/2$	0	$\pi/2$	0	$\pi/2$
Sign applied to the detected signal	+	+	−	−	+	+	−	−

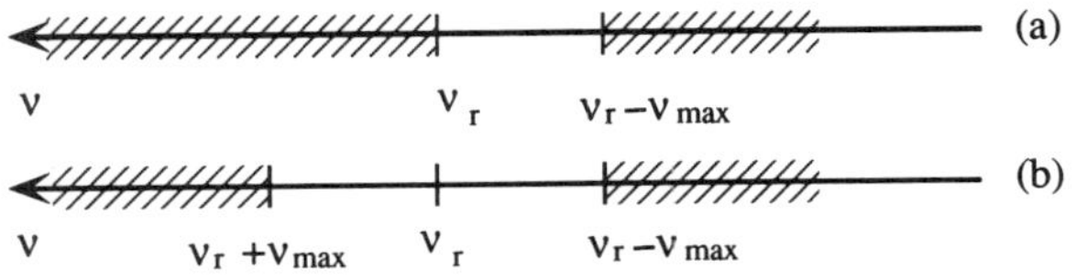

Figure 3.14 Frequency zones examined unambiguously (without folding back; see Section 3.7.1) for (a) single channel detection; (b) sequential quadrature detection

3.5 Double Fourier Transformation of the Product of Two Sine (or Cosine) Functions (Two-Dimensional NMR)

We shall denote by t_2 the time domain dimension corresponding to the physical acquisition of free induction decays. Since quadrature detection is assumed, the function of the time variable t_2 is of the form $\exp(2i\pi\nu'_2t_2)\exp(-t_2/T_2^*)$. According to the experimental procedure effective in the t_1 dimension, we shall deal either with a 'phase modulation' $\exp(2i\pi\nu'_1t_1)\exp(-t_1/T_2^*)$, which is the analog of quadrature detection, or with an amplitude modulation $\cos(2\pi\nu'_1 t_1)\exp(-t_2/T_2^*)$, which is analog of single channel detection.

3.5.1 PHASE MODULATION

Phase modulation is generally achieved by combining interferograms which are obtained through an appropriate cycling of the transmitter phase. This can also be obtained in a more genuine fashion by application of field gradients (see Chapter 5) or, more simply, in the case of echo modulation by J couplings. The relevant signal (omitting the amplitude factor) has the following analytical expression:

$$s(t_1, t_2) = \exp(\pm 2i\pi\nu'_1 t_1)\exp(2i\pi\nu'_2 t_2)\exp(-t_1/T_2^*)\exp(-t_2/T_2^*) \qquad (3.15)$$

where it can be recalled that ν'_1 and ν'_2 are resonance frequencies.

Whenever experimental conditions are so chosen that the arguments of the exponents in t_1 and t_2 are of opposite sign, one is dealing with a 'type n detection', which leads to physically detected signals (in the t_2 dimension) involving an echo, as the inhomogeneity of the static magnetic field is compensated for when the condition $t_2/t_1 = \nu'_1/\nu'_2$ is fulfilled (the quantity $\cos[2\pi(\nu'_1 t_1 - \nu'_2 t_2)]$ in (3.15) becomes equal to unity); see Figure 3.15. The overall aspect of the relevant signals suggests the use of adapted windows for improving the spectral resolution (such as a sine bell; see below). For the other sign combination ('type p detection'), $\mathbf{B}_0$ inhomogeneity effects are cumulated. Signals obtained under such conditions are dubbed 'anti-echoes'.

In both cases, the *complex* Fourier transform with respect to t_1 of the series of spectra, themselves obtained from *complex* Fourier transforms with respect to t_2, can be expressed according to absorption (A) and dispersion (D) spectra in both dimensions ($A_1 = A(\nu'_1 - \nu_1)$, $D_1 = D(\nu'_1 - \nu_1)$, $A_2 = A(\nu'_2 - \nu_2)$, $D_2 = D(\nu'_2 - \nu_2)$; ν_1 and ν_2 are the frequency variables):

$$S(\nu_1, \nu_2) = A_1A_2 + D_1D_2 + i(A_1D_2 - D_1A_2). \qquad (3.16)$$

The simultaneous occurrence of absorption and dispersion spectra in both real and imaginary parts of the final two-dimensional map precludes the obtention of a pure absorption spectrum, but rather leads to spectra affected by the 'phase twist' phenomenon (Figure 3.16), which, unfortunately, cannot be corrected by combinations of real and imaginary parts.

One remedy is the amplitude spectrum $\sqrt{A_1^2 + D_1^2}\sqrt{A_2^2 + D_2^2}$, with the drawbacks of poor resolution in both dimensions (ν_1 and ν_2) and increased wings. When such two-dimensional spectra are displayed in the 'contour plot' mode, this entails cross-shaped peaks and more overlap with nearby peaks (Figure 3.17). Nevertheless, phase modulation possesses some advantages which are reminiscent of

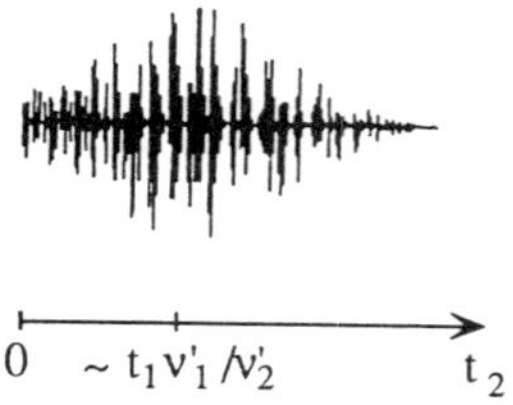

Figure 3.15 An interferogram of type n

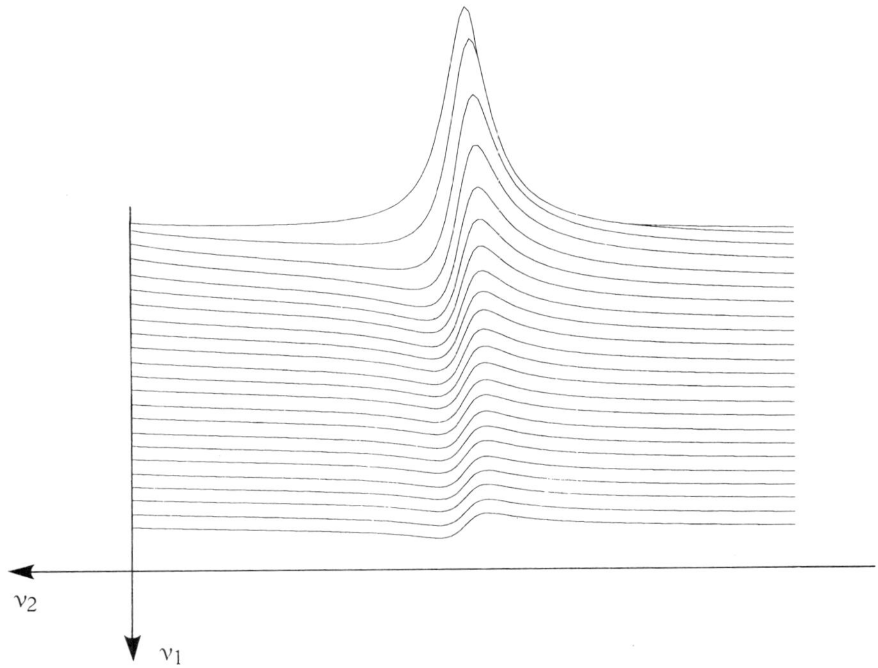

Figure 3.16 'Phase twist' resulting from a double *complex* Fourier transform

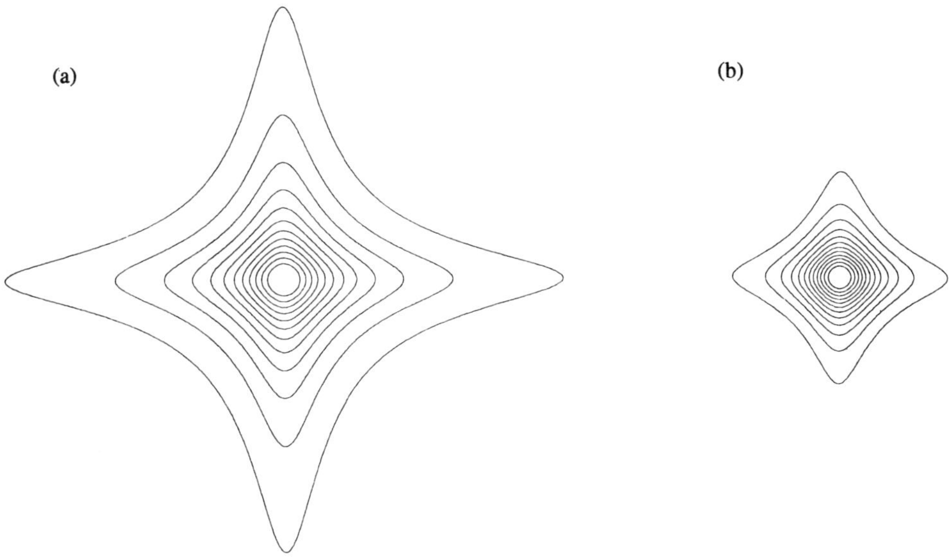

Figure 3.17 'Contour plot' of a two-dimensional peak: (a) amplitude spectrum; (b) pure absorption spectrum. The same scale has been employed in each case

quadrature detection in one-dimensional spectroscopy: the sign of ν'_1 is perfectly determined and consequently the spectral window in ν_1 can be reduced by a factor of 2 the data storage; the reference frequency for the time domain t_1 can be positioned somewhere near the middle of the spectrum just as for the t_2 domain (the carrier frequency is thus identical for both time domains).

3.5.2 AMPLITUDE MODULATION

Amplitude modulation means that, in the t_1 domain, one is simply dealing with a cosine (or sine) function, so that the signal is of the form:

$$s(t_1, t_2) = \cos(2\pi\nu'_1 t_1)\exp(2i\pi\nu'_2 t_2)\exp(-t_1/T_2^*)\exp(-t/T_2^*) \qquad (3.17)$$

Obviously, such a function does not allow for sign discrimination in the ν_1 dimension but can lead to a pure absorption spectrum in both dimensions provided that the second Fourier transform (with respect to t_1) is applied solely to the real part of the ν_2 domain. However, the latter must have been phase-corrected according to the conventional procedures of one-dimensional NMR so that it is represented by the absorption spectrum $A(\nu_2 - \nu'_2) = A_2$. Hence, after the first *complex* Fourier transform (with respect to t_2), one has:

$$s'(t_1, \nu_2) = \cos(2\pi\nu'_1 t_1)\exp(-t_1/T_2^*)(A_2 + iD_2) \qquad (3.18)$$

Only the real part will be retained and yields $A_2(\mathrm{Re}_1 + i\,\mathrm{Im}_1)$ after a *real* Fourier transform with respect to t_1. A phase correction procedure can then be applied to the ν_1 dimension to arrive at the final two-dimensional spectrum, whose real part can be written as:

$$S(\nu_1, \nu_2) = A_2 A_1 = A(\nu'_2 - \nu_2)A(\nu'_1 - \nu_1). \qquad (3.19)$$

This surface is indeed made of pure absorption spectra in both dimensions (Figure 3.17) with their inherent advantages concerning spectral resolution and the possibility to separate resonances which would have otherwise overlap in an amplitude spectrum.

3.5.3 PURE ABSORPTION SPECTRA DERIVED FROM PHASE MODULATION IN THE t_1 DOMAIN

Fortunately, it is feasible to benefit from the resolution advantages of amplitude modulation even if the experiment is carried out in the phase modulation mode. However, this implies special procedures for data acquisition.

A first method (D.J. States *et al.*, *J. Magn. Reson.*, **48**, 286 (1982)) consists of treating the t_1 domain in a separate way for the real part, on the one hand, and for the imaginary part, on the other hand. The quantity $\exp(2i\pi\nu'_1 t_1)\exp(2i\pi\nu'_2 t_2)$ will be split into

$$\begin{aligned} s_c(t_1, t_2) &= \cos(2\pi\nu'_1 t_1)\exp(2i\pi\nu'_2 t_2) \\ s_s(t_1, t_2) &= \sin(2\pi\nu'_1 t_1)\exp(2i\pi\nu'_2 t_2) \end{aligned} \qquad (3.20)$$

In practice, two different memory blocks are needed for completing this operation. Complex Fourier transformation of s_c (with respect to t_2) leads to $\cos(2\pi\nu'_1 t_1)$ $(A_2 + iD_2)$, whereas Fourier transformation of s_s leads to $\sin(2\pi\nu'_1 t_1)$ $(A_2 + iD_2)$.

Retaining only the real parts of these two Fourier transforms yields $A_2[\cos(2\pi\nu'_1 t_1) + i\sin(2\pi\nu'_1 t_1)]$, whose complex Fourier transform (with respect to t_1) is seen to be equal to $A_2(A_1 + iD_1)$. The real part of this latter expression is identical to (3.19), i.e. to the product of two pure absorption spectra in both dimensions. In the above, possible phase corrections have been omitted; it is a simple matter to reintroduce these corrections eventually in the two-dimensional map.

A second method (D. Marion and K. Wüthrich, *Biochem. Biophys. Res. Commun.*, **113**, 967 (1983)), which is widely employed, stems from the application of the TPPI procedure (quadrature sequential detection) to the time domain t_1. Experimentally, this amounts to incrementing by $\pi/2$ the phase of the r.f. pulse responsible for the t_1 modulation in concert with the t_1 increment, that is, at each new experiment. The t_1 increment must be equal to $1/4\nu_{1max}$ (see Section 3.4.3), so that amplitude modulation still obtains, with $A_2\cos(2\pi\nu'_1 t_1)$ replaced by

$$A_2\cos[2\pi(\nu'_1 + \nu_{1max})t_1]. \qquad (3.21)$$

This provides the possibility, by means of a real Fourier transform with respect to t_1, to explore the spectral window ranging from $-\nu_{1max}$ to $+\nu_{1max}$, again without changing the carrier frequency, and to obtain as a final result a pure absorption spectrum represented by the product A_1A_2.

A variant of these methods, named States-TPPI (D. Marion *et al.*, *J. Magn. Reson.*, **85**, 393 (1989)), has been devised so that axial peaks (which arise from longitudinal magnetization at the end of the t_1 period and which consequently are not modulated as a function of t_1) do not appear at the frequency $\nu_1 = 0$, i.e. at the center of the spectrum, but rather at the edge of the spectrum, i.e. at the Nyquist frequency. This is accomplished through the method of States et al. where, for every t_1 increment, the phase of both the preparation pulse(s) (which govern the t_1 evolution) and the receiver are shifted by 180°. This is equivalent to the normal procedure for signals modulated in t_1. However, since the phase of the preparation pulse(s) has no effect on the axial peaks, these peaks appear as being modulated according to the Nyquist frequency owing to the sign change of the receiver (this can be understood on the basis of the approach used to explain the TPPI procedure).

3.6 Data Manipulation in the Time Domain

Before considering various post-processing treatments of the fid (interferogram), we must emphasize phase cycle procedures whose goal is to attenuate or even remove the effects of some instrumental artifacts and which therefore avoid numerical corrections. As already mentioned, such procedures imply that a predetermined number of interferograms has to be accumulated and consists of the modification, from one scan to the other, of the phase of both the transmitter and the receiver. As a first example, let us consider the two following experiments:

$$(\pi/2)_x(\mathrm{Acq})_x$$

$$(\pi/2)_{-x}(\mathrm{Acq})_{-x}$$

The NMR signals are effectively added in a coherent way since the receiver (Acq) phase follows exactly the transmitter phase. In contrast a d.c. component arising from the receiver will be removed by the subtraction process (d.c. is a symbol widely used in electrical engineering, meaning 'direct current'; it stands here for a time-independent signal). This is evidently closely related to the States-TPPI procedure discussed in the previous section. In the same type of idea, artifacts associated with quadrature detection, resulting from an imbalance between the different channels, will be attenuated by 'rotating' in a coordinated fashion the transmitter and the receiver phases according to the cycle x, y, $-x$, $-y$:

$$(\pi/2)_x(\text{Acq})_x$$
$$(\pi/2)_y(\text{Acq})_y$$
$$(\pi/2)_{-x}(\text{Acq})_{-x}$$
$$(\pi/2)_{-y}(\text{Acq})_{-y}.$$

This phase cycling, quasi-systematically used, is known under the acronym of CYCLOPS. It may be useful to recall the meaning of the receiver phase in the context of quadrature detection. It has of course nothing to do with the possible phase corrections performed on the final spectrum. Rather, it concerns the assignment of the two quadrature detection channels and also the sign applied to the detected signal. The relevant conventions are given in Table 3.2.

The numerical treatments themselves, which are carried out before Fourier transformation, consist in multiplying the interferogram $f(t)$ by a time dependent function $h(t)$, sometimes called a window or apodization function, in order to improve either the spectral resolution or the apparent S/N ratio. Some windows are claimed to improve both these features, but most of the time they are mutually exclusive (an improvement of the S/N ratio generally entails a degradation of the spectral resolution). An important point of concern is the conservation of the signal intensities (peak areas), especially if the NMR spectrum must be used for analytical purposes. The peak intensity is obtained through integration over the frequency domain:

$$\mathrm{A} = \int_{-\infty}^{\infty}\left[\int_0^{\infty} f(t)\exp(-2i\pi\nu t)\,\mathrm{d}t\right]\mathrm{d}\nu = \int_0^{\infty} f(t)\left[\int_{-\infty}^{\infty}\exp(-2i\pi\nu t)\,\mathrm{d}\nu\right]\mathrm{d}t = f(0). \tag{3.22}$$

This latter result ($A = f(0)$) stems from the fact that $\int_{-\infty}^{\infty}\exp(-2i\pi\nu t)\,\mathrm{d}\nu$ is the Dirac function $\delta(t)$ (see Section 3.3), and means that the intensity of an NMR

Table 3.2 Definition of the receiver phase according to the assignment of the two quadrature detection channels. Re and Im stand for the real and the imaginary parts of the fid, respectively

Receiver phase	x	y	$-x$	$-y$
Re	(channel 1)	(channel 2)	−(channel 1)	−(channel 2)
I_m	(channel 2)	−(channel 1)	−(channel 2)	(channel 1)

signal is representative of the corresponding nuclear magnetization, itself proportional to the interferogram amplitude (at time $t = 0$). When the function $f(t)$ is multiplied by an apodization function, (3.22) becomes:

$$A = \int_{-\infty}^{\infty}\int_{0}^{\infty} f(t)h(t)\exp(-2i\pi\nu t)\,d\nu = f(0)h(0). \qquad (3.23)$$

Hence, a sufficient condition for preserving the peak areas after application of a window $h(t)$ is that $h(0) = 1$.

3.6.1 EXPONENTIAL MULTIPLICATION

This is the most widely used window. It leads to an improvement of the S/N ratio at the expense of line broadening. The way in which exponential multiplication works can be understood intuitively by noticing that noise manifests itself towards the end of the interferogram when the NMR signal has decayed to a nearly zero value. Multiplying $f(t)$ by an exponential function whose argument is negative, $\exp(-t/\tau)$, contributes to reduce noise in this terminal part of the interferogram. This noise reduction is regained in the frequency domain after Fourier transformation (Figure 3.18). However, lines broaden because the effective damping factor $1/T_2^*$ is replaced by $1/T_2^* + 1/\tau$ (the linewidth at half height, originally equal to $1/\pi T_2^*$, becomes $(T_2^* + \tau)/(\pi T_2^* \tau)$). It can be noticed, by reference to (3.23), that exponential multiplication does not alter signal intensities.

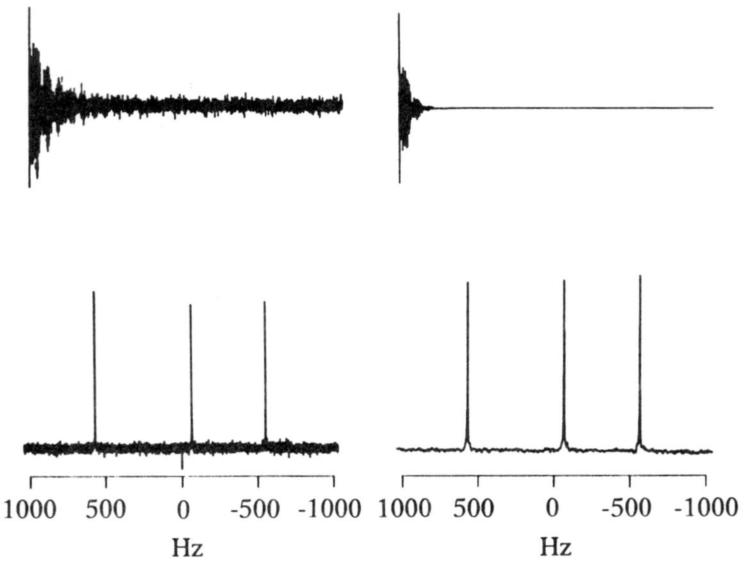

Figure 3.18 Left: raw data in the time domain and the corresponding spectrum obtained after Fourier transformation. Right: the same fid multiplied by a decaying exponential and the corresponding spectrum which exhibits an improved S/N ratio but broader lines

In order to evaluate mathematically the modifications due to exponential multiplication, we can realize that noise is in fact made up by a superimposition of sine (or cosine) functions. Denoting by N the amplitude of such a sine function (considered to be representative), we can realize that it yields, in the frequency domain, a signal whose height is equal to NT, where T represents the acquisition time. This result is easily derived from the convolution product of a sine function by a rectangular function of length equal to T. As a consequence, the S/N ratio without any data manipulation can be evaluated according to:

$$(r_{S/N})_0 = ST_2^*/NT. \tag{3.24}$$

where S is the signal amplitude [the numerator of (3.24) originates from formula (3.13), which provides the amplitude of an absorption signal]. Equation (3.24) suggests that the S/N ratio improves as T decreases. The acquisition time should therefore be kept to a minimum within a limit corresponding to the absence of truncation effects (in principle, $T \approx 5T_2^*$ for avoiding signal distortions in the frequency domain. See Figure 3.9). This feature is illustrated by the two spectra of Figure 3.19.

Returning to exponential multiplication, we know that the time constant τ must be chosen smaller than T_2^* to prevent any signal distortion. The exponential multiplication affects the sine function representative of noise which will, in the frequency domain, lead to a signal of amplitude $N\tau$, whereas ST_2^* must be replaced by $ST_2^*\tau/(T_2^* + \tau)$. Hence, after the exponential multiplication, the S/N ratio can be expressed as:

$$(r_{S/N})_\tau = \frac{ST_2^*}{N(T_2^* + \tau)} \tag{3.25}$$

Comparison of (3.24) and (3.25) indicates that, for $T \approx 5T_2^*$, the improvement of the signal-to-noise ratio cannot be greater than a factor of 5, which would anyway be obtained for $\tau \to 0$, that is, for infinite broadening. In order to keep the broadening factor within reasonable limits, we can adopt the compromise $\tau \approx T_2^*$ (matched filter). In those conditions, the S/N ratio is improved by a factor of 2.5 with respect to the

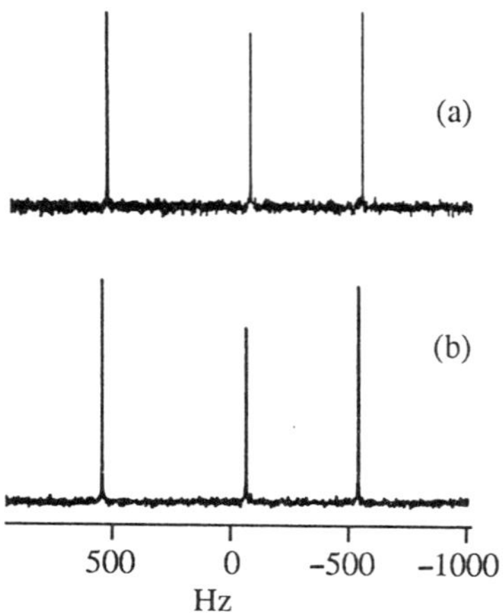

Figure 3.19 Interferograms corresponding to spectra (a) and (b) have been acquired in 2 s and 1 s, respectively. For peaks of identical height, noise is seen to be twice as large in (a) as in (b)

raw interferogram. However, if an acquisition time of the order of $2T_2^*$ is retained (this choice often prevails in carbon-13 spectroscopy because distortions are merely undetectable) and for the 'matched filter' conditions ($\tau \approx T_2^*$), formulae (3.24) and (3.25) show that no improvement in the S/N ratio can be expected. Under such conditions, this apodization function can be considered as illusory.

3.6.2 WINDOWS DESIGNED FOR IMPROVING SPECTRAL RESOLUTION OR SPECTRUM LEGIBILITY

After the operation of Fourier transformation, some baseline distortion may occur, which would suggest the occurrence of broad lines. The latter may correspond to real resonances due to protons of solid materials existing within the probe, or to spurious oscillations which take place after the excitation pulse. In both cases, their attenuation in the time domain is much faster than that of those NMR signals that one wishes to observe. For avoiding unwanted baseline distortions, a convenient method consists in reducing the amplitude of the very first part of the interferogram. On the other hand, resolution may be insufficient for separating or visualizing nearby resonances. This difficulty can be overcome by multiplying the central part of the interferogram by a factor greater than one; this amounts to increasing the spectral resolution artificially (Figure 3.20). It turns out that most of the windows which have been proposed entail a degradation of the S/N ratio, and also peak distortions similar to those which result from the truncation of the interferogram.

The *sine bell* window is widely employed because it fulfills the criteria of

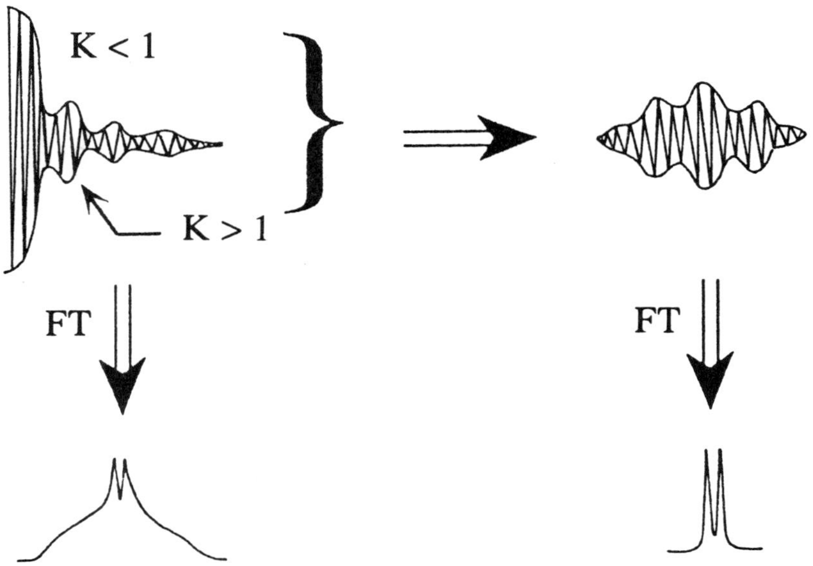

Figure 3.20 Multiplicative factors K for schematically improving the appearance of spectra. $K < 1$ applied to the beginning of the interferogram reduces baseline distortions. $K > 1$ applied to the central part artificially increases the spectral resolution

Figure 3.20; furthermore, it is very easy to use and leads to legible spectra. Data manipulation consists in multiplying the interferogram by the function

$$h(t) = \sin(\pi t/T) \tag{3.26}$$

where T is the total acquisition time. This operation amounts to multiplying the interferogram by the first half-period of a sine function (hence the term sine bell). Instead of a Lorentzian line, one obtains a peak flanked by sidelobes whose analytical expression is given by:

$$F(\nu) = T_2^*(\pi T_2^*/T) \times \frac{1 + (\pi T_2^*/T)^2 - 4\pi^2 T_2^{*2}(\nu' - \nu)^2}{[1 + (\pi T_2^*/T)^2 + 4\pi^2 T_2^{*2}(\nu' - \nu)(\nu' - \nu + 1/T)][1 + (\pi T_2^*/T)^2 + 4\pi^2 T_2^{*2}(\nu' - \nu)(\nu' - \nu - 1/T)]} \tag{3.27}$$

In the latter expression, the signal amplitude has been set to unity for simplicity and it has been assumed that $\exp(-T/T_2^*)$ is negligibly small.

The signal area must necessarily be zero because $h(0) = 0$; in the actual spectrum, this is reflected by lobes with negative parts on each side of the peak. The maximum of $F(\nu)$ (for $\nu = \nu'$) which corresponds to the signal height in the frequency domain is equal to:

$$S = T_2^* \frac{(\pi T_2^*/T)}{1 + (\pi T_2^*/T)^2} \tag{3.28}$$

and differs from the signal height of a Lorentzian line by the factor $(\pi T_2^*/T)/[1 + (\pi T_2^*/T)^2]$, whereas noise is affected by the factor (T/π) (which replaces the factor T in the absence of any apodization function). Therefore, the S/N ratio is reduced (with respect to a conventional treatment) by $\pi^2(T_2^*/T)/[1 + \pi^2(T_2^*/T)^2]$. An acquisition time equal to $5T_2^*$ entails a reduction of the S/N ratio by a factor of 5, which drops to 2 for $T = 2T_2^*$. Now, in spite of poorer sensitivity and sidelobes, this window is advantageous regarding spectral resolution, which may be improved because the linewidth at half height calculated from (3.27) is equal to:

$$\Delta\nu_{1/2} = \frac{1}{\pi T_2^*}\sqrt{\{[1 + (\pi T_2^*/T)^2]^2 + 4\}^{1/2} - 2} \tag{3.29}$$

The relevant factor with respect to a Lorentzian line is given by the square root in (3.29). Typical values are 0.66 and 1.42 for $T = 5T_2^*$ and $T = 2T_2^*$, respectively. The other interesting feature of this window is of course the elimination or attenuation of baseline distorsions shown in Figure 3.21. There exist variants of the sine bell window which involve a phase angle in the sine function.

Another widely used window is the Gaussian function, possibly shifted. Its goal is again an improvement of the spectral resolution with a slight sensitivity loss. In that case, it can be recalled that the original Lorentzian line is convoluted by a Gaussian function. Popular windows used in two-dimensional spectroscopy include a sine squared function with a dephasing of 45° and combination of exponential and Gaussian multiplications with appropriate arguments.

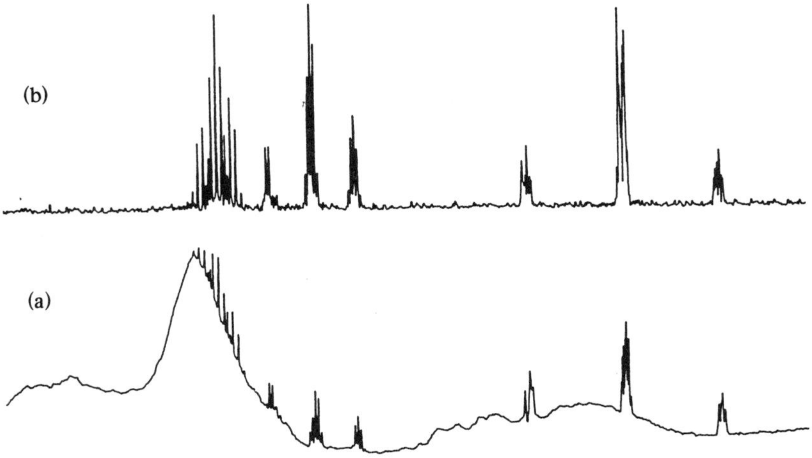

Figure 3.21 (a) Raw spectrum; (b) its Fourier transform after an exponential multiplication followed by the application of a sine bell. The exponential multiplication tends to reduce the sensitivity loss due to the sine bell apodization function

3.7 Discrete Fourier Transformation

3.7.1 GENERAL CONSIDERATIONS

The heart of a NMR Fourier transform spectrometer is a computer, which is run in the real-time mode, as it must interact with the transmit–receive system at well defined moments; it drives the various functions of the instrument, it controls pulse sequences, and it acquires data for storage and addition in memory. Immediately after a read pulse, the computer must be in a state such that it is ready to convert the fid signal (which emerges as an analog signal, i.e. a voltage) into numerical form (binary numbers). This conversion is carried out by a device called an analog-to-digital converter (ADC, presently available as integrated circuits) with constant intervals Δt between two consecutive data points; Δt is called the sampling rate or dwell time. At the outset, the computer memories contains a series of binary numbers X_0, X_1, X_2 . . ., X_k . . ., X_{n-1}, which represent the amplitude of the interferogram at time 0, Δt, $2\Delta t$, . . . $k\Delta t$, . . ., $(n-1)\Delta t$. X_k may stand for a complex number $X_k = u_k + iv_k$, where u_k and v_k arise from the two channels of the quadrature detection system; in that case, it occupies two memory locations.

Because the data have been discretized, appropriate equations must be set up. The numerical (discrete) form of Fourier transformation can be expressed as

$$A_r = \sum_{k=0}^{n-1} X_k \exp(-2i\pi rk/n) \tag{3.30}$$

which provides the amplitude of the spectrum (in the frequency domain) at a frequency equal to $r/n\Delta t$ for $r = -n/2, -n/2+1, \ldots, n/2-1, n/2$. Equation (3.30) and the range of possible r values will now be justified. Starting from the continuous form of Fourier transformation

$$F(\nu) = \int_0^{\infty} f(t) \exp(-2i\pi\nu t)\,dt,$$

we can arrive at the discrete form by substituting t by $k\Delta t$ and by introducing a quantity denoted by ν_{max} (the subscript will be clarified later)

$$\nu_{max} = 1/2\Delta t \tag{3.31}$$

so that $\exp(-2i\pi\nu t)$ can be written as

$$\exp(-2i\pi\nu t) = \exp(-2i\pi\nu k\Delta t) = \exp(-2i\pi k\nu/2\nu_{max}) \tag{3.32}$$

Now because $\exp(-2ikm\pi) = 1$, k and m being integers, it is easy to realize that

$$F(\nu + 2\nu_{max}) = F(\nu - 2\nu_{max}) = F(\nu);$$

$$F(\nu + 4\nu_{max}) = F(\nu - 4\nu_{max}) = F(\nu);$$

$$F(\nu + 6\nu_{max}) = F(\nu - 6\nu_{max}) = F(\nu); \text{etc.}$$

Therefore, for a signal at frequency ν, we obtain a series of replications (or alias, hence the term *aliasing*) at frequencies $\nu + m\nu_{max}$ with $m = \pm 2, \pm 4, \pm 6 \ldots$ It can be noted that this feature is a direct consequence of the discretization of information. As another consequence, the unique interval where no replication occurs (i.e. such that a signal truly belonging to this interval does not generate any replication within this interval) is therefore $[-\nu_{max}, \nu_{max}]$. This interval must be described by n data points, since the original n data points of the time domain will simply be transformed so as to yield the frequency domain. For this reason, the integer r, which defines the frequency scale, must vary between $-n/2$ and $+n/2$. We arrive naturally at (3.30) by identifying r/n with $\nu/2\nu_{max}$; this indeed corresponds to $[-\nu_{max}, +\nu_{max}]$ for the frequency range. The integral of the continuous form is evidently replaced by a discrete summation, whereas $dt = (T/n)\,dk$ reduces to a scaling factor T/n, which has been omitted in (3.30).

Let us consider now a signal outside the spectral window $[-\nu_{max}, \nu_{max}]$. Because of the aliasing phenomenon, it will actually appear in this latter window. We shall rather say that it is *folded back* in that window. Its frequency is consequently erroneous. This is again an immediate consequence of the discretization of information and of the choice of an inappropriate sampling rate Δt. Aliasing and, equivalently, folding back are shown schematically in Figure 3.22.

The frequency $\nu_{max} = 1/2\Delta t$ is called the Nyquist frequency, and the ability to explore the spectral window $[-\nu_{max}, \nu_{max}]$ is related to the Shannon theorem which states that, for obtaining a reliable result from a Fourier transform, a sinusoid must be defined by at least two points per period (Figure 3.23). If one is dealing with a complex Fourier transform, $2n$ data points have to be considered in the time domain and in the frequency domain (n data points for the real part and n data points for the imaginary part). In the frequency domain, the digital resolution is given by the frequency interval between two consecutive points: $\Delta\nu = 2\nu_{max}/n = 1/n\Delta t = 1/T$, since only the real part, after the phase correction operations, is of concern. To improve the digital resolution without increasing the acquisition time, one may resort to an approach which consists of adding zeroes in zones of length T or $3T$ or $7T$ etc. This procedure is known under the name of *zero-filling*, and is perfectly valid so long as the interferogram has decayed to a quasi-zero value by the end of the standard acquisition time T.

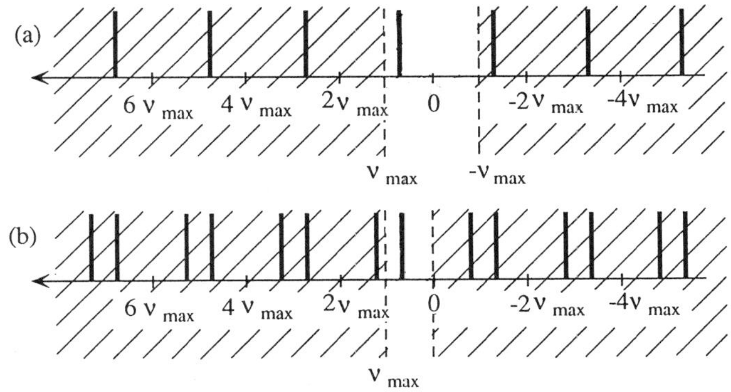

Figure 3.22 Aliasing (replications) or folding back, due to the discretization of information, shown schematically without accounting for possible phase modifications: (a) complex Fourier transform; (b) real Fourier transform with additional folding back arising from the symmetry with respect to the zero frequency. For preventing erroneous frequency measurements, no signal must be located in the dashed zones

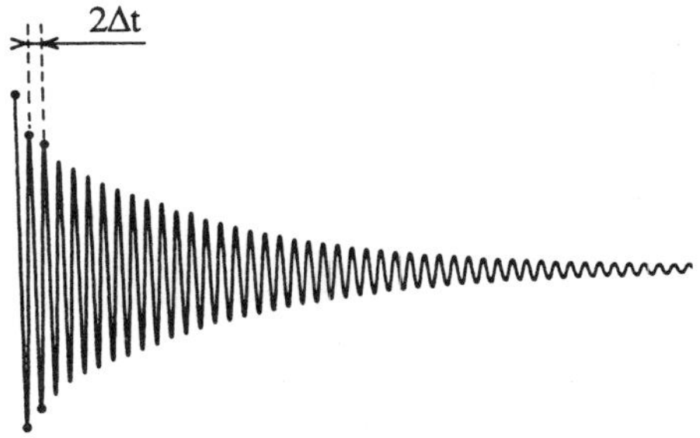

Figure 3.23 Pictorial proof of the Shannon theorem. To avoid any misinterpretation of oscillatory functions, at least two data points per period must be acquired

We turn now to another problem directly related to the discretization of the information, and which is a consequence of the folding back phenomenon. All signals, whatever their origin, whose specific frequency is outside $[-\nu_{max}, \nu_{max}]$ will be folded back within this spectral window (the spectral window of interest). This is especially true for noise which will tend to degrade the S/N ratio of the spectrum which lies in this spectral window. There fortunately exists a simple remedy which stems from the insertion, prior to the analog-to-digital conversion, of a low-pass filter with a cutoff frequency equal to ν_{max}. In the same type of idea, it has been shown that 'oversampling' techniques (a sampling rate much shorter than necessary) are very convenient in the sense that they improve (artificially) the resolution of the ADC. To be efficient, these techniques nevertheless require *digital* filtering of the spectral zone of interest.

The above discussion concerns complex Fourier transformation or, in other words, signals acquired according to the quadrature detection mode. The single channel detection mode, with as corollary a real Fourier transform, presents some special features which will now be discussed. It has been shown before that a signal at frequency ν is seen by a real Fourier transform (single channel detection) at its actual frequency but also at $-\nu$. The aliasing phenomenon, associated with the discretization of information, induces its own replications which must be combined with the symmetry about the zero frequency, so that one has now the following equalities

$$F(\nu) = F(\nu + 2\nu_{max}) = F(\nu - 2\nu_{max}) = F(\nu + 4\nu_{max}) = F(\nu - 4\nu_{max})$$

and so on, as in the case of the complex Fourier transform, which must be appended by

$$F(+\nu) = F(-\nu) = F(-\nu + 2\nu_{max}) = F(-\nu - 2\nu_{max}) = F(-\nu + 4\nu_{max})$$
$$= F(-\nu - 4\nu_{max}) \text{ etc.}$$

as shown in Figure 3.22(b). As mentioned before, the usable spectral window is $[0, \nu_{max}]$. Since the acquisition implies only n data points (in contrast to the quadrature detection, which requires $2n$ data points), the final digital resolution is again equal to $(\nu_{max})/(n/2)$ because of the necessity to have at hand real and imaginary parts for phase correction purposes. The final digital resolution is thus the same as for quadrature detection for a spectral window which is half the size. Another consequence is the folding back of noise situated in the zone $[0, -\nu_{max}]$ (the low-pass filter with a cutoff frequency of ν_{max} is evidently unable to sense the sign of the considered signals). This entails a further degradation of the S/N ratio by a factor equal to $\sqrt{2}$ (it is easy to see that the variance relating to noise is multiplied by 2). All these features should favor quadrature detection in any situation.

3.7.2 THE FAST FOURIER TRANSFORM (FFT) ALGORITHM

This algorithm, proposed in 1965 by Cooley and Tukey, has made the Fourier transform very popular, since the mathematical operations for several thousands of data points are performed in a time of the order of a second *without any simplification or approximation* (for an exhaustive discussion, see W. T. Cochran *et al.*, *Proc. IEEE*, **55**, 1664 (1967)). The only requirement is that the number of data points to be transformed is a multiple of 2, that is, $n = 2^u$ or $u = \log_2 n$ ($n = 4, 8, 16, 32, 64, 128, 256, 512, 1024, 2048 \ldots$), the idea being to reduce the number of multiplications (relatively demanding in computer time) by successive partitions of the time domain. Formula (3.30) will be recast in the following form as needed by the forthcoming calculations

$$\mathrm{A}_r = \sum_{k=0}^{n-1} W^{rk}\mathrm{X}_k, \text{ with } W = \exp(-2i\pi/n). \tag{3.33}$$

It may be recalled that X_k is the complex amplitude of the interferogram at a time $t = kT/n$ (where T is the total acquisition time) and that A_r is the spectrum amplitude in the frequency domain; r is an integer between $-n/2$ and $(n/2) - 1$. In fact, the zone $[-n/2, -1]$ is replicated by the zone $[n/2, n-1]$ and, for simplicity, A_r will be calculated for $r = 0, 1, 2, \ldots, n-1$. At the outset, the second half of the frequency domain will be displaced at the left of the first half so that the spectrum is displayed in the usual manner. We turn now to the algorithm itself. If n can be divided by 2, the X_k can be arranged in two classes: those of even subscript which will be labeled by the subscript $k' = 0, 1, 2, \ldots, n/2 - 1$ (with $k = 2k'$) and those of odd subscript labeled by $k'' = 0, 1, 2, \ldots, n/2 - 1$ (with $k = 2k'' + 1$). This yields two new Fourier transforms, each of which operates on $n/2$ data points:

$$\begin{aligned} B_r &= \sum_{k'=0}^{n/2-1} (W')^{rk'} X_{2k'} \\ C_r &= \sum_{k''=0}^{n/2-1} (W')^{rk''} X_{2k''+1} \end{aligned} \tag{3.34}$$

with $r = 0, 1, \ldots, n/2 - 1$ and $W' = \exp[-2i\pi/(n/2)] = W^2$. By proper subscript manipulation, one can easily express the A_r (related to the Fourier transform of n points) as a function of the B_r and C_r (related to the Fourier transform of $n/2$ points):

$$A_r = B_r + W^r C_r$$

$$A_{r+n/2} = B_r + W^{r+n/2} C_r$$

for $r = 0, 1, \ldots, n/2 - 1$ and where $W^{r+n/2}$ can be substituted by $-W^r$. The complexity of the problem has thus benefited from a two-fold reduction. This process can be repeated provided that n is of the form 2^u: the separation into two classes ($X_{k'}$ and $X_{k''}$) will keep going until there remains only one data point which is then identical to its Fourier transform. The direct application of (3.30) would necessitate n^2 multiplications whereas, with the Cooley–Tukey method, this number drops to $n \log_2 n$. For the example of the transformation of 4096 points, the number of multiplications goes from 16.7 million down to 49 152.

3.8 Methods other than Fourier Transformation for Analyzing or Processing Free Induction Decays

The conventional spectrum is actually retrieved after processing the time domain signals by Fourier transformation, provided that the interferogram is not corrupted by parasitic signals (only random noise is properly treated) and that it has been fully sampled. Indeed, the elimination of too many points at the beginning of the interferogram entails an important first order phase correction which results in an undulating baseline (P. Plateau *et al.*, *J. Magn. Reson.*, **54**, 46 (1983)) precluding the observation of relatively broad lines; on the other hand, truncation of the interferogram (lack of significant data points at the extremity) results in sidelobes affecting each peak (see Figure 3.9).

A remedy to these problems may be the data manipulations described in Section 3.6. However, it has been noticed that these data manipulations generally entail some alteration of one or several spectral characteristics. The idea is therefore to devise mathematical procedures able to return, from an incomplete data set possibly affected with unwanted signals, the spectral characteristics of each line (frequency, linewidth, amplitude and phase). Such procedures are not transformations and will not necessarily succeed, especially if they are iterative in nature. Anyhow, they will require computational times which are orders of magnitude larger than those of Fourier transformation. Many algorithms have been proposed. The simplest are based on the least-squares method and consist of refining for each line the four spectral parameters indicated above, with the obvious prerequisite that initial values should be available. The method can be applied directly to the time domain data (see for instance: F. Montigny *et al.*, *Chem. Phys. Lett.*, **170**, 175 (1990)). Conversely, there exist methods which do not require *a priori* knowledge. Two of them, linear prediction and maximum entropy (and their numerous variants), seem to have found some acceptance and their basic principles will now be briefly discussed.

3.8.1 LINEAR PREDICTION

This method (first application to NMR data: H. Barkhuijsen *et al*., *J. Magn. Reson.*, **61**, 465 (1985)) is based on the assumption that a data point X_k in the time domain (the subscript k is equivalent to the time variable; see previous sections) can be expressed linearly according to the data points already sampled

$$X_k = \sum_{m=1}^{M} a_m X_{k-m} \qquad (3.35)$$

where the integer M is smaller than k and must be appropriately chosen (see below). Equation (3.35) amounts to stating that it is possible to predict a data point from the data points of smaller rank (*forward prediction*). It can be mentioned that a similar equation exists for data points of larger rank (*backward prediction*). In fact, equation (3.35) is strictly valid for a NMR signal composed of M damped sinusoids, or rather (if quadrature detection is assumed) of signals whose time-dependent part is of the form

$$z_{m'} = \exp[(-b_{m'} + 2\pi i \nu_{m'})\Delta t] \qquad (3.36)$$

where $b_{m'} = 1/T^*_{2m'}$ is the damping factor of the $m' -$ th signal (with $T^*_{2m'}$ its effective transverse relaxation time), $\nu_{m'}$ is its precession frequency in the rotating frame and Δt is the sampling interval. Inserting the amplitude $c_{m'}$ of this signal yields a theoretical expression for X_k

$$X_k = \sum_{m'=1}^{M} c_{m'}(z_{m'})^k \qquad (3.37)$$

Likewise the actual expression of the data point of index $(k - m)$ can be written as

$$X_{k-m} = \sum_{m'=1}^{M} c_{m'}(z_{m'})^{k-m}$$

The clue consists of multiplying the previous equation by a coefficient $(-a_m)$ which will possess some special properties (see below) and to carry out a summation over m from 0 to M. This yields:

$$-\sum_{m=0}^{M} a_m X_{k-m} = -\sum_{m'=1}^{M} c_{m'}(z_{m'})^{k-M} \sum_{m=0}^{M} a_m (z_m)^{M-m} \qquad (3.38)$$

The a_m coefficients can be so chosen that they fulfill the following equality

$$\sum_{m=0}^{M} a_m (z_m)^{M-m} = 0 \text{ with } a_0 = -1 \qquad (3.39)$$

This means that the quantities $z_{m'}$ are the roots of the above polynomial expression (whose degree is M, making it valid to consider the M $z_{m'}$ as possible roots) and that the right hand side member of (3.38) is zero. The left hand side can then be written as

$$X_k - \sum_{m=1}^{M} a_m X_{k-m} = 0$$

which is identical to (3.35). Once this property is demonstrated, one can envisage using linear algebra methods to determine the coefficients a_m for an appropriate set of equations such as (3.35). Because experimental data are involved, it is recommended to select an overdimensioned M value for taking into account (i) more signals than initially expected, and (ii) noise contributions. This stage of the calculation may be performed by purely matricial methods based possibly on the concept of singular values. The next step is the insertion of the a_m coefficients into equation (3.39) in order to calculate the roots of this equation, which provide the precession frequencies and the damping factors.

Finally, the signal amplitudes represented by the quantities $c_{m'}$ can be deduced from equation (3.37) by standard least-squares procedures. This step completes the method which therefore provides, at least in principle, the whole set of spectral characteristics for each signal (amplitude, frequency, damping factor and phase). In particular, the 'deconvolution' of signals overlapping in the frequency domain is automatically achieved. Likewise, quantitation from NMR data is readily obtained, as the signal amplitudes in the time domain are just the peak area in the frequency domain. They can be evaluated without worrying about artifacts which lead to distortion in the frequency domain after Fourier transform operations. As mentioned above, a value of M much greater than the number of expected NMR signals occurring in the time domain accounts for contributions from noise. The latter are easily recognized through low c_m coefficients.

A more restricted application of the method may be quite valuable in practice. It does not require a complete analysis in terms of spectral characteristics but rather makes use of the coefficients extracted from equation (3.35), provided that artifact-free data points have been used. In turn, the prediction coefficients permit the user to reconstruct missing parts of the interferogram or data points corrupted by spurious signals. A Fourier transformation can therefore be applied to improved data, and yields a spectrum free from distortions.

The user must however be aware of the limitations inherent in approaches based on linear predictions, which are recalled below.

- The linear prediction equation (3.35) is strictly valid only in the case of the superimposition of damped sinusoids;
- Data should be acquired according to a constant sampling rate;
- It is impossible to insert into those calculations prior knowledge of some spectral parameters.

3.8.2 MAXIMUM ENTROPY

This method (first application to NMR data: S. F. Gull and G. J. Danniel, *Nature*, **272**, 686 (1978)) may be considered as more general than linear prediction in the sense it does not require any model (such as damped sinusoids) and can accommodate prior knowledge (for instance, well defined chemical shifts which will not have to be calculated or refined). Moreover, it allows the user to interact with the ongoing calculations. The first foundation of the method is the classical χ^2 test, which can be expressed in this context as

$$\chi^2 = \sum_{k=0}^{n-1} (\mathrm{TF}(F_k) - \mathrm{X}_k)^2/\sigma^2 \tag{3.40}$$

where X_k are the time domain experimental data (according to the usual notations for the free induction decay) and F_k are the relevant spectrum amplitudes in the frequency domain. $\mathrm{TF}(F_k)$ represents the transfer function between the frequency and the time domains; it is generally the inverse Fourier transform. σ is the standard deviation of noise, supposed to be perfectly white, and n is the number of data points. Standard statistical theories state that when χ^2 becomes equal to n, the set of F_k constitutes the best estimation of the spectrum which can fit the experimental data. Unfortunately, because of the presence of noise or artifacts, a great variety of spectra approximately fulfilling the condition $\chi^2 = n$ can be found. Hence an additional constraint is needed. The method stems from a search for the maximum of a quantity called entropy, by reference to statistical thermodynamics. This quantity may be of the form

$$S = -\sum_{k=0}^{n-1} P_k \ln (P_k) \tag{3.41}$$

where

$$P_k = F_k \Big/ \sum_{k=0}^{n-1} F_k.$$

Equation (3.41) arises from considerations based on information theory which are beyond the scope of this book. Intuitively, we can simply assume that S reflects the amount of information present in the signal; its maximization should therefore lead to the optimum spectrum. The combination of the two constraints represented by (3.40) and (3.41) can be treated by the Lagrange multipliers method. It can be noted that prior knowledge (frequencies, amplitudes, linewidths...) can be easily inserted in the F_k. Various algorithms are presently available. Some final remarks can be made about this method.

- Unwanted effects in the Fourier transform spectrum, due to truncation or artifacts, should in principle disappear.
- The S/N ratio and also the spectral resolution should be improved, since the maximum information is extracted from the experimental data.
- The method does not, however, provide a full analysis of the NMR signals (frequencies, amplitudes, linewidths and phases) but rather improved results with respect to a simple Fourier transform.

Bibliography

R. R. Ernst: Sensitivity enhancement in magnetic resonance, *Adv. Magn. Reson.*, **2**, 1 (1966)

T. C. Farrar and E. D. Becker: *Pulse and Fourier Transform NMR*, Academic Press, New York, 1971

R. N. Bracewell: *The Fourier Transform and its Applications*, 2nd ed. McGraw-Hill, New York, 1978

F. Roddier: *Distributions et Transformation de Fourier*, McGraw-Hill, Paris, 1985

D. Shaw: *Fourier Transform NMR Spectroscopy*, 2nd ed., Elsevier, Amsterdam, 1984

R. R. Ernst, G. Bodenhausen and A. Wokaun: *Principles of Nuclear Magnetic Resonance in One and Two Dimensions*, Clarendon, Oxford, 1987, Chapters 4 and 6

D. S. Stephenson: Linear prediction and maximum entropy methods in NMR spectroscopy, *Progr. NMR Spectrosc.*, **40**, 516 (1988)

R. de Beer and D. van Ormondt: Analysis of NMR data using time domain fitting procedures, *NMR: Basic Princ. Progr.*, **26**, 201 (1992)

4 DYNAMIC PHENOMENA IN NMR

In this chapter, we shall be dealing with the influence of molecular motions (rotation, translation, exchange between sites and so on) upon the dynamic behavior of an ensemble of nuclear spins. Up to now, we have been concerned mainly with spin dynamics arising from precession or application of radio-frequency pulses, considering that molecular motions, which modulate the interactions undergone by nuclear spins, could be treated in a phenomenological way. In this respect, we were led to introduce time constants which manifest themselves by an exponential evolution of nuclear magnetization: T_1, the longitudinal relaxation time, and T_2, the transverse relaxation time. We shall attempt in this chapter to express these relaxation times according to the various interactions to which nuclear spins are subjected. In a second stage, more attention will be paid to molecular motions which actually make these interactions time-dependent.

In turn, a correct interpretation of relaxation parameters should provide us with information about molecular motions. These include (i) overall motions, translational or rotational, governed essentially by the molecular volume and the viscosity of the medium in which are embedded the molecules under investigation, (ii) local motions such as internal rotations around C–C bonds, or segmental motions in molecules or aggregates of appreciable size, and (iii) exchange between two different molecules or between two distinct sites within the same molecule. Moreover, we shall see that under certain circumstances the two relaxation times T_1 and T_2 are not sufficient and that a complete treatment requires additional relaxation parameters such as cross-relaxation or cross-correlation rates. Before considering these relatively subtle points, we shall first be interested in experimental methods which yield the classical relaxation parameters with accuracy and reliability.

4.1 Experimental Determination of Relaxation Parameters

4.1.1 LONGITUDINAL RELAXATION

Any experimental method aiming at the measurement of a dynamic parameter invariably starts by a perturbation which moves the system out of its equilibrium state; this is followed by an evolution period whose duration is generally denoted by τ, which is such that the dynamical parameters of interest manifest themselves. The ultimate stage consists of reading the actual state of the system. Generally, this measurement is performed for different durations of the evolution period so as to evaluate properly the dynamic parameters. Regarding longitudinal relaxation, it seems obvious that the first two stages (initial perturbation and evolution) should involve only longitudinal magnetization. However, the read stage necessarily implies transverse components which are the only ones giving rise to a detectable signal. In other words, longitudinal magnetization, after having evolved, must be converted into *single quantum coherences*. The efficiency of the whole process will be optimal, thus warranting the greatest accuracy for the determination of the considered relaxation parameter, provided that the initial perturbation is maximum. For longitudinal relaxation, this corresponds to a *complete inversion* by means of a 180° r.f. pulse, in such a way that magnetization *remains longitudinal* even during the evolution period. Subsequent to the evolution period, the ongoing value of the longitudinal magnetization is read with a 90° pulse which converts it into transverse magnetization, directly detectable. This is the well-known *inversion recovery* experiment which is schematized in Figure 4.1.

In a first approach, we shall limit ourselves to the hypothesis where nuclear magnetization conforms to the Bloch equations. Concerning the longitudinal component, one has:

$$\frac{\mathrm{d}M_z}{\mathrm{d}t} = -\frac{M_z - M_0}{T_1} \tag{4.1}$$

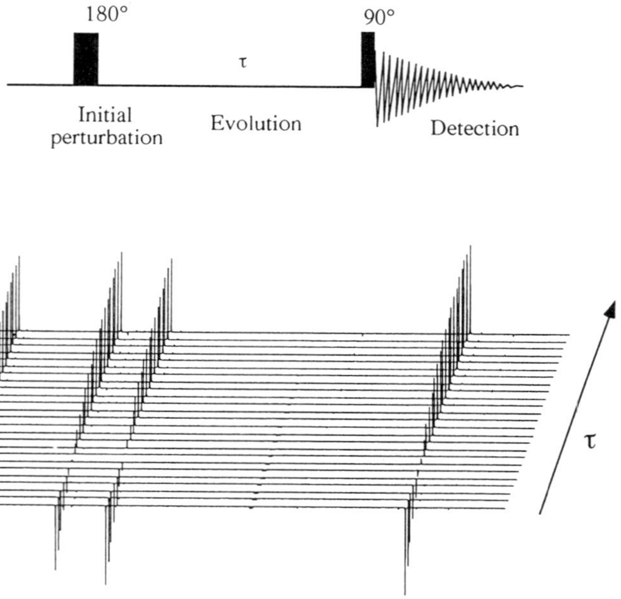

Figure 4.1 Scheme of an inversion recovery experiment yielding the longitudinal relaxation time. Free induction decays resulting from the 90° read pulse are subjected to a Fourier transform and lead to a series of spectra corresponding to the different τ values (evolution period). Spectra are generally displayed with a shift between two consecutive values of τ. The analysis of the amplitude evolution of each peak from $-M_0$ to M_0 provides an accurate evaluation of T_1. The example concerns carbon-13 T_1 of *trans*-crotonaldehyde whose values, are from left to right: 20.5 s, 19.8 s, 23.3 s and 19.3 s

This first order differential equation can easily be solved to yield

$$M_z(\tau) = M_0 + [M_z(0) - M_0]\exp(-\tau/T_1) \tag{4.2}$$

We can verify that the right hand side of (4.2) is equal to $M_z(0)$ for $\tau = 0$ and to M_0 for $\tau \to \infty$. For the ideal case of a perfect inverting (180°) pulse, $M_z(0) = -M_0$ and (4.2) reduces to

$$M_z(\tau) = M_0[1 - 2\exp(-\tau/T_1)] \tag{4.3}$$

or, in a logarithmic form

$$\ln\left[\frac{M_0 - M_z(\tau)}{2M_0}\right] = -\frac{\tau}{T_1}. \tag{4.4}$$

This latter form is especially convenient for a quick determination of T_1 which is the inverse of the slope of the linear representation of $\ln\{[(M_0 - M_z(\tau)]/2M_0\}$ vs. τ (Figure 4.2). M_0 can be measured by means of a single read pulse or for a time τ of the order of $5T_1$ (which insures a return to equilibrium of more than 99% of the magnetization).

It turns out that, in the case of an imperfect inverting pulse, the factor of 2 in equation (4.3) must be substituted by an unknown factor K (<2); it is thus recommended to turn to a non-linear fit (based on the criterion of least-squares) of $M_z(\tau) = M_0[1 - \mathrm{K}\exp(-\tau/T_1)]$, where the three quantities M_0, K and T_1 have to be refined, starting for example from values deduced from (4.4).

The quality of inversion can be greatly improved by relying on the concept of *composite pulses* (see for instance M.H. Levitt *et al.*, *Adv. Magn. Reson.*, **11**, 48 (1983)) which consists of replacing a single pulse by a *cluster*, whose goal is to correct in a self-consistent way and, at least, to first order the various pulse imperfections. In the present context, an imperfection of the read pulse does not matter; however, a composite pulse can be used to improve the efficiency of the initial inverting pulse, either if its duration has been misadjusted or if, because of spatial inhomogeneity of the radio-frequency field, the flip angle is not exactly 180° in all regions of the sample. Let us suppose that the inverting pulse is applied along the x axis of the rotating frame; self compensation can be seen to be effective if we resort to the cluster $(\pi/2)_x(\pi)_y(\pi/2)_x$, where $(\pi/2)_x$ may be considered as a pulse of flip angle $90° - \varepsilon$ (whose duration is underestimated by the experimenter or due to B_1 inhomogeneity). We shall nevertheless assume that the $(\pi)_y$ pulse is devoid of

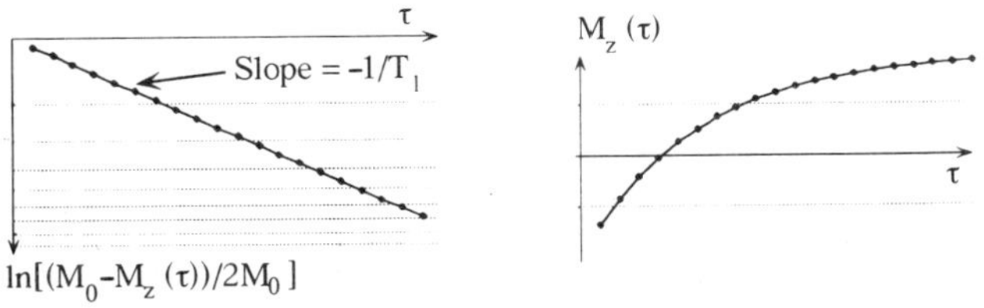

Figure 4.2 Left: semi-logarithmic plot leading to the longitudinal relaxation time. Right: non-linear fit of the data corresponding to the same experiment

imperfections; this constitutes the 'first order approximation'. From the magnetization motion sketched in Figure 4.3, we can recognize that the imperfection represented by ε has been removed. For this type of composite pulse, imperfections of the central $(\pi)_y$ pulse have not been considered, nor have possible off-resonance effects (by reference to Section 2.1.3, it can be recalled that magnetization rotates around B_{eff} which, for a pulse applied along the x axis of the rotating frame, is tilted about x in the xz plane). It has been empirically demonstrated that both problems are attenuated by choosing a $(4\pi/3)_y$ pulse in place of the $(\pi)_y$ pulse, so that an efficient composite inverting pulse can be schematized as

$$(\pi/2)_x(4\pi/3)_y(\pi/2)_x$$

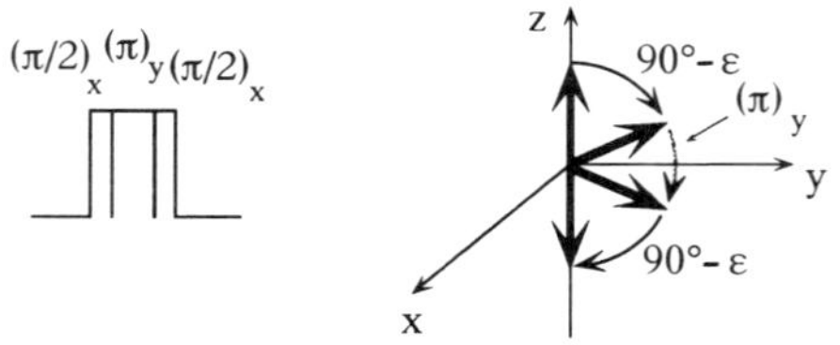

Figure 4.3 Compensation of pulse imperfections by means of an inverting 'composite pulse'

Finally, it may be stressed that the T_1 measurement, as above described, is meaningful only with the condition that the system returns to equilibrium between two experiments with different τ values, or between two consecutive scans if accumulation is necessary for improving the signal-to-noise ratio (this requires a waiting time of the order of $5T_1$). Variants of the inversion-recovery method have been developed for which this latter condition is lifted, thus avoiding prohibitively long measuring times (D. Canet *et al.*, *J. Magn. Reson.*, **18**, 199 (1975)).

4.1.2 TRANSVERSE RELAXATION; TRANSLATIONAL DIFFUSION; RELAXATION IN THE ROTATING FRAME

As stated before, the initial perturbation should be maximal with respect to the equilibrium state. Since we are dealing here with transverse magnetization, this maximal perturbation is obviously a 90° pulse. However, we have noticed in numerous instances (see previous chapters) that signals collected after a simple read pulse decay exponentially according to a time constant T_2^* which differs from the genuine T_2 by a contribution due to inhomogeneity of the static field B_0:

$$\frac{1}{T_2^*} = \frac{1}{T_2} + (B_0 \text{ inhomogeneity}) \tag{4.5}$$

Experimental methods must therefore be devised for removing this contribution.

Hahn sequence

It has been noted in Section 2.2.6 that the 'spin echo' technique affords a way to get rid of B_0 inhomogeneity, which manifests itself as an additional precession similar to that of a chemical shift. A 180° pulse applied after an interval τ has the virtue of refocusing any precession after another interval τ; this amounts to cancelling any chemical shift effect (Figure 4.4). This pulse sequence, $(\pi/2)_x-\tau-(\pi)_y-\tau$–acquisition, is the basic Hahn sequence, which yields in principle the true T_2 because any precession effect is removed, leaving a tranverse magnetization attenuated according the transverse relaxation time. By reference to the Bloch equations relating to transverse magnetization,

$$\frac{\mathrm{d}}{\mathrm{d}t}M_{x,y} = -\frac{M_{x,y}}{T_2} \tag{4.6}$$

we can acknowledge that the Fourier transform of the half-echo leads to a signal of amplitude $M_0 \exp(-2\tau/T_2)$. For a set of τ values, it thus appears possible to extract an accurate value of T_2 for each line in the spectrum. Unfortunately, this analysis does not take into account translational diffusion phenomena, which will be considered in the next section.

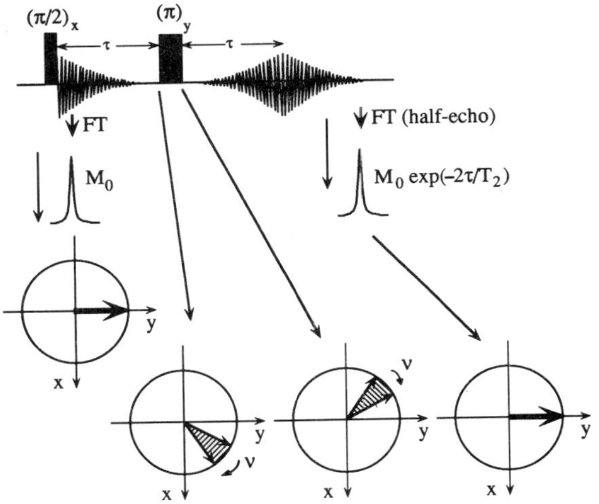

Figure 4.4 Basic Hahn sequence for the measurement of the transverse relaxation time T_2. Any precession motion characterized by the frequency ν in the rotating frame is refocused. This precession may arise either from chemical shift or from B_0 inhomogeneity (dashed area)

Effects of translational diffusion in the presence of B_0 inhomogeneities: the PGSE sequence

We shall assume that the field B_0 is not perfectly homogeneous; for simplicity and without loss of generality, we shall make the hypothesis that it varies linearly across the whole sample in the X direction of the laboratory frame so that the field sensed by a molecule at abscissa X is of the form

$$B(\mathrm{X}) = B_0 + g_0\mathrm{X} \tag{4.7}$$

Here g_0 stands for a *uniform gradient* of the static field. Let us recall that such a gradient can be purposely created with the aim of producing NMR images (Sections 1.3.1 and 5.4) or measuring the translational diffusion coefficient (see below). This gradient can also result from a nonexistent or incomplete compensation of the genuine inhomogeneity of B_0 (imperfect *shimming*; see Section 1.3.1). In any event, the precession frequency depends on the *location* of the considered molecule via the *spatial* dependence of B_0.

We have just seen that any precession effect is refocused by an echo sequence whose goal is precisely to get rid of B_0 inhomogeneity. However, this feature is impaired if, during the refocusing process, molecules translate from a location X to a location X′ for which the precession frequency is different from that at location X. This should result in a signal attenuation or an additional 'defocusing' which arises from self-diffusion phenomena. The latter are characterized by a coefficient D which is involved in a classical equation of the form:

$$\partial\psi/\partial t = D\partial^2\psi/\partial X^2 \tag{4.8}$$

where ψ is a property subjected to self-diffusion effects.

In order to be more explicit with regard to magnetic resonance, let us turn to the complex transverse magnetization, already used in Section 2.1.4

$$M_t(X, t) = M_x(X, t) + iM_y(X, t) \tag{4.9}$$

recalling that x and y refer to the rotating frame whereas X is a spatial coordinate (laboratory frame). It is convenient to separate the effects attributable to the static field gradient from those arising from precession in a perfectly homogeneous B_0 and from transverse relaxation; this is accomplished by factorizing $M_t(X, t)$ in the following way:

$$M_t(X, t) = \psi \exp[-(2i\pi\nu_0 + 1/T_2)t] \tag{4.10}$$

where ν_0 represents the precession frequency in a field B_0.

The quantity ψ must also contain the modification of precession frequency due to the gradient g_0, so that (4.8) must be modified as

$$\partial\psi/\partial t = -i\gamma g_0 X\psi + D\partial^2\psi/\partial X^2 \tag{4.11}$$

According to a standard mathematical method for solving this type of equation, we look for a solution of the form

$$\psi(t) = A(t)\exp(-i\gamma g_0 Xt)$$

where $A(t)$ is independent of the spatial variable X. Inserting the above expression for $\psi(t)$ in (4.11), we can derive the following solution

$$\psi(t) = A(0)\exp(-i\gamma g_0 Xt)\exp(-D\gamma^2 g_0^2 t^3/3). \tag{4.12}$$

Applying this result to both halves of the echo sequence, and recalling that M_t stands for the complex transverse magnetization $M_t = M_x + iM_y$, we obtain:

$$M_t(2\tau) = M_0\exp(-2\tau/T_2)\exp[-(D\gamma^2 g_0^2/3)(2\tau^3)] \tag{4.13}$$

It can be recognized that the refocusing process (compensation of all precession effects) is accounted for in (4.13). It should also be stressed that translational diffusion in the presence of a gradient generates a decay depending on τ^3, whereas transverse relaxation produces a decay depending on τ. In principle, this affords the possibility of determining both the relaxation time T_2 and the self-diffusion coefficient D, provided that the gradient strength (g_0) is known. However, the method suffers from a major drawback due to the continuous application of a gradient (which must be sufficiently strong for making the experiment sensitive to D), precluding the obtention of a high resolution spectrum after the Fourier transformation of the half echo and thus the determination of the diffusion coefficient of several species present in the sample. To circumvent this problem, the pulsed gradient spin echo method (PGSE, E. O. Stejskel and J. E. Tanner, *J. Chem. Phys.*, **42**, 288 (1965)) has been devised. The sequence is sketched in Figure 4.5; it affords the possibility of acquiring the half echo at a moment where no gradient is on. Moreover, the interval Δ enhances the effects of translational diffusion; a complete analysis of the sequence based on the previous theory would yield:

$$M_t(2\tau) = M_0 \exp(-2\tau/T_2) \exp[-D\gamma^2 g_0^2 \delta^2(\Delta + 2\delta/3)]. \tag{4.14a}$$

In practice, τ and Δ are fixed and properly chosen so that $M_t(2\delta)$ is sensitive to D. Generally $\Delta \gg \delta$ and we can write

$$M_t(\delta) \approx M_0 \exp(-2\tau/T_2) \exp(-D\gamma^2 g_0^2 \Delta\delta^2). \tag{4.14b}$$

δ can be varied from one experiment to the other, so that $\ln(M_t)$ vs. δ^2 is a straight line whose slope is equal to $-D(\gamma^2 g_0^2 \Delta)$, which thus yields an accurate value of D.

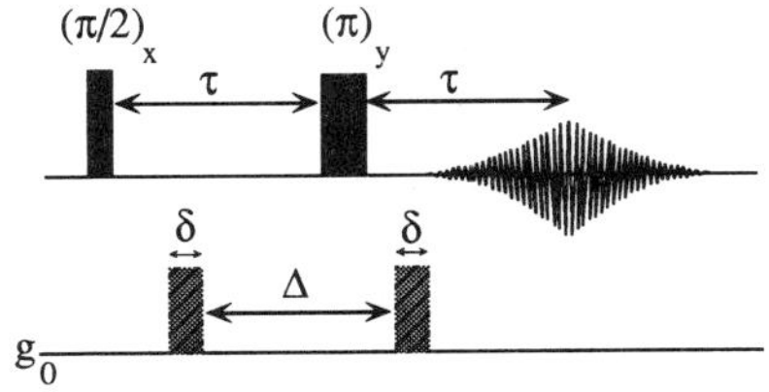

Figure 4.5 The PGSE sequence: spin echo sequence involving two B_0 gradient (denoted by g_0) pulses of duration δ. The two gradient pulses must be located on each side of the $(\pi)_y$ refocalization pulse without the need to be symmetrically disposed with respect to this r.f. pulse

Self-diffusion measurements by radio-frequency field gradients (B_1 gradients)

A B_1 gradient, delivered for instance by a single-turn coil, can be also used for measuring self-diffusion coefficients. The sequence is even simpler than with B_0 gradients; moreover, the signal is attenuated according to T_1, whereas, in the case of the classical PGSE experiment and because transverse magnetization is involved, this attenuation is governed by T_2. This latter feature may be quite advantageous for systems which possess short transverse relaxation times while longitudinal relaxation times remain at a relatively high value. The simplest sequence (D. Canet *et al.*, *J. Magn. Reson.*, **81**, 1 (1989); R. Dupeyre *et al.*, *J. Magn. Reson.*, **95**, 581 (1991)) is depicted in Figure 4.6.

In order to analyze this experiment, let us assume that the B_1 gradient acts along the x axis of the rotating frame and let us denote by θ the flip angle produced by a gradient pulse of duration δ for molecules located at the abscissa X.

$$\theta = \gamma g_1 X\delta = 2\pi q X \tag{4.15}$$

with

$$q = \gamma g_1 \delta/2\pi$$

If at the outset magnetization is at thermal equilibrium (M_0), the three magnetization components immediately after the first gradient pulse can be expressed, in the rotating frame, as

$$M_x = 0$$
$$M_y = M_0 \sin\theta$$
$$M_z = M_0 \cos\theta$$

We shall first disregard diffusion effects during Δ so that we are left with precession and relaxation. Denoting the precession angle by ψ and defining E_1 and E_2 by $\exp(-\Delta/T_1)$ and $\exp(-\Delta/T_2^*)$, respectively, we obtain at the end of the interval Δ

$$M_x = M_0 E_2 \sin\theta \sin\psi$$
$$M_y = M_0 E_2 \sin\theta \cos\psi$$
$$M_z = M_0(1 - E_1) + M_0 E_1 \cos\theta$$

The second gradient pulse entails a further nutation of the y and z components by an angle denoted by θ', which will be made equal to θ by the end of the calculation. For the phase $+x$ of this gradient pulse, one has

$$M_x = M_0 E_2 \sin\theta \sin\psi$$
$$M_y = (M_0 E_2 \sin\theta \sin\psi)\cos\theta' + [M_0(1 - E_1)]\sin\theta' + (M_0 E_1 \cos\theta)\sin\theta' \tag{4.16}$$
$$M_z = -(M_0 E_2 \sin\theta \cos\psi)\sin\theta' + [M_0(1 - E_1)]\cos\theta' + (M_0 E_1 \cos\theta)\cos\theta'$$

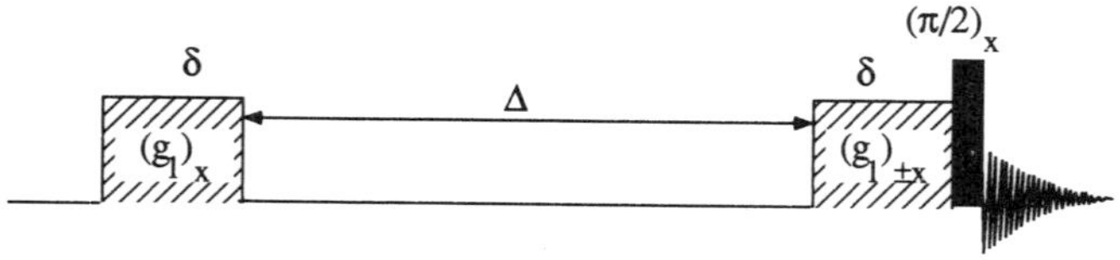

Figure 4.6 A pulse sequence for measuring self-diffusion coefficients by means of B_1 gradient pulses (g_1) of duration δ. The phase of the second gradient pulse is cycled in alternate scans. The read pulse $(\pi/2)_x$ is supposed to correspond to a homogeneous B_1 field

Now, we must perform an average over the whole sample, that is, an average over the angle θ, with (see Table 2.1)

$$\langle \cos\theta \rangle = \langle \sin\theta \rangle = \langle \cos 2\theta \rangle = \langle \sin 2\theta \rangle = 0$$

$$\langle \cos^2\theta \rangle = \langle \sin^2\theta \rangle = 1/2$$

These relationships imply that the gradient pulses are of appropriate strength and duration so that they produce a total defocusing of magnetization. Inserting these averages in (4.16), we obtain

$$\begin{aligned} \langle M_x \rangle &= 0 \\ \langle M_y \rangle &= 0 \\ \langle M_z \rangle &= -(M_0/2)E_2 \cos\psi + (M_0/2)E_1 \end{aligned}$$

Similar calculations for the phase $-x$ of the second gradient pulse lead to

$$\begin{aligned} \langle M_x \rangle &= 0 \\ \langle M_y \rangle &= 0 \\ \langle M_z \rangle &= (M_0/2)E_2 \cos\psi + (M_0/2)E_1 \end{aligned}$$

Thus, after two scans, only the z component of magnetization, weighted by E_1, is retained and can of course be measured by the subsequent read pulse. We have now to account for the diffusion phenomenon. Neglecting its effects during the gradient pulses, we shall view diffusion as changing the nutation angle θ' into $\theta' + \varphi$. The additional angle φ is related to the translational motion which occurs during the time interval Δ (G. S. Karczmar *et al.*, *Magn. Reson. Med.*, **7**, 111 (1988)). Inserting φ into equation (4.16) leads to

$$\begin{aligned} M_x &= M_0 E_2 \sin\theta \sin\psi \\ M_y &= (M_0 E_2 \sin\theta \cos\psi)\cos(\theta' + \varphi) + [M_0(1 - E_1)]\sin\theta' \\ &\quad + (M_0 E_1 \cos\theta)\sin(\theta' + \varphi) \\ M_z &= -(M_0 E_2 \sin\theta \cos\psi)\sin(\theta' + \varphi) + [M_0(1 - E_1)]\cos\theta' \\ &\quad + (M_0 E_1 \cos\theta)\cos(\theta' + \varphi) \end{aligned}$$

It must be emphasized that the term $M_0(1 - E_1)\cos\theta'$ in M_z has its origin in the reconstruction of magnetization by T_1 processes, starting after the first gradient pulse. Hence, this magnetization is obviously not spatially labeled by this pulse and cannot be affected by translational diffusion. In addition to the space average over θ (with $\theta' = \theta$; as already done above), we must now perform a time average for the sine and cosine functions which involve the angle φ (this second average is denoted below by an overbar). Taking into account the phase cycling, the final result can be expressed as

$$\begin{aligned} \langle \bar{M}_x \rangle &= 0 \\ \langle \bar{M}_y \rangle &= -M_0 E_2 \cos\psi \overline{\sin\varphi} \\ \langle \bar{M}_z \rangle &= M_0 E_1 \overline{\cos\varphi} \end{aligned} \qquad (4.17)$$

These equations have been derived under the reasonable assumption that the two averages (space and time) can be separated. We turn now to the evaluation of $\overline{\sin\varphi}$ and $\overline{\cos\varphi}$, which depend on the diffusion along the gradient axis. The angle φ can be written as $2\pi qr$, where r is the displacement during Δ. It is obvious that the probabilities for a displacement in one direction and for the same displacement in the opposite direction are identical. Therefore $\overline{\sin\varphi} = 0$ and the only non-zero magnetization component is $\langle \overline{M}_z \rangle$, thus we monitor the evolution of the sole longitudinal magnetization through $\overline{\cos\varphi}$. The evaluation of *that* latter quantity rests on the distribution function of r, which is Gaussian, $1/\sqrt{4\pi D\Delta}\exp(-r^2/4D\Delta)$, and can be seen to be consistent with the diffusion equation (4.8). Hence

$$\overline{\cos\varphi} = \frac{1}{\sqrt{4\pi D\Delta}}\int_{-\infty}^{+\infty} \cos(2\pi qr)\exp\left(-\frac{r^2}{4D\Delta}\right)\mathrm{d}r = \exp(-4\pi^2 q^2 D\Delta)$$

which yields, for the observed signal

$$S(\delta) = M_0 \exp(-\Delta/T_1)\exp(-D\gamma^2 g_1^2 \Delta\delta^2) \tag{4.18}$$

Equation (4.18) bears a close resemblance to (4.14b), except for the attenuation due to relaxation, which involves T_1 and not T_2. The method seems to be exceptionally robust and devoid of the instrumental problems associated with B_0 gradients. Its major drawback is the difficulty to generate large B_1 gradients.

Improvement of T_2 determination by the Carr–Purcell–Meiboom–Gill (CPMG) method

Acknowledging that the B_0 field can never be made perfectly homogeneous, the idea of Carr and Purcell was to minimize the effects of diffusion in a T_2 experiment. This goal is achieved by the train of 180° pulses of Figure 4.7 and stems from the fact that the differential equation (4.11) must be solved for each interval following a 180° pulse since, at that point, the magnetization sign is suddenly modified; consequently, new boundary conditions prevail.

Returning to the analysis of the Hahn sequence and extending it to the present situation, we obtain the amplitude of the nth echo:

$$M_t(2n\tau) = M_0 \exp(-2n\tau/T_2)\exp[-(D\gamma^2 g_0^2/3)(2n\tau^3)]. \tag{4.19}$$

We note that, for an evolution whose duration is identical to that of a Hahn sequence, the argument of the exponential relating to diffusion has been divided by

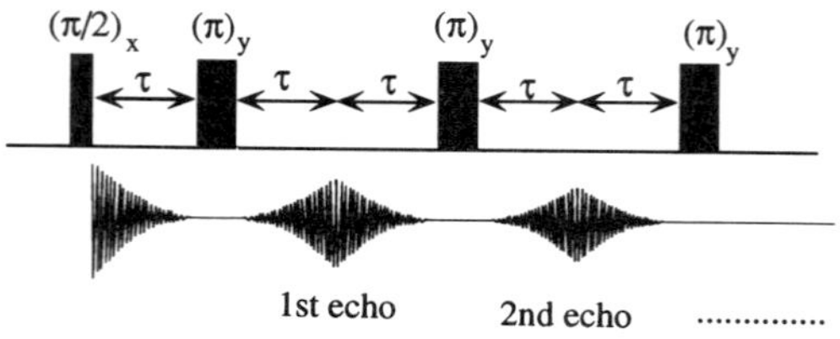

Figure 4.7 CPMG pulse sequence. An echo is formed halfway between two consecutive π pulses. The echo amplitude (or the Fourier transform of the half-echo) provides an evaluation of T_2 less affected by diffusion than in the simple Hahn sequence. The phase change of π pulses with respect to the initial $\pi/2$ pulse cancels the effect of their imperfections

n^2. Therefore, the remedy which can make translational diffusion effects negligible consists simply of increasing n, which amounts to bringing the π pulses closer to each other.

The trick introduced by Meiboom and Gill is to *dephase* all π pulses in the Carr–Purcell train by an angle of 90° with respect to the initial $\pi/2$ pulse. As shown in Figure 4.8, without this phase change, imperfections of the π pulses are

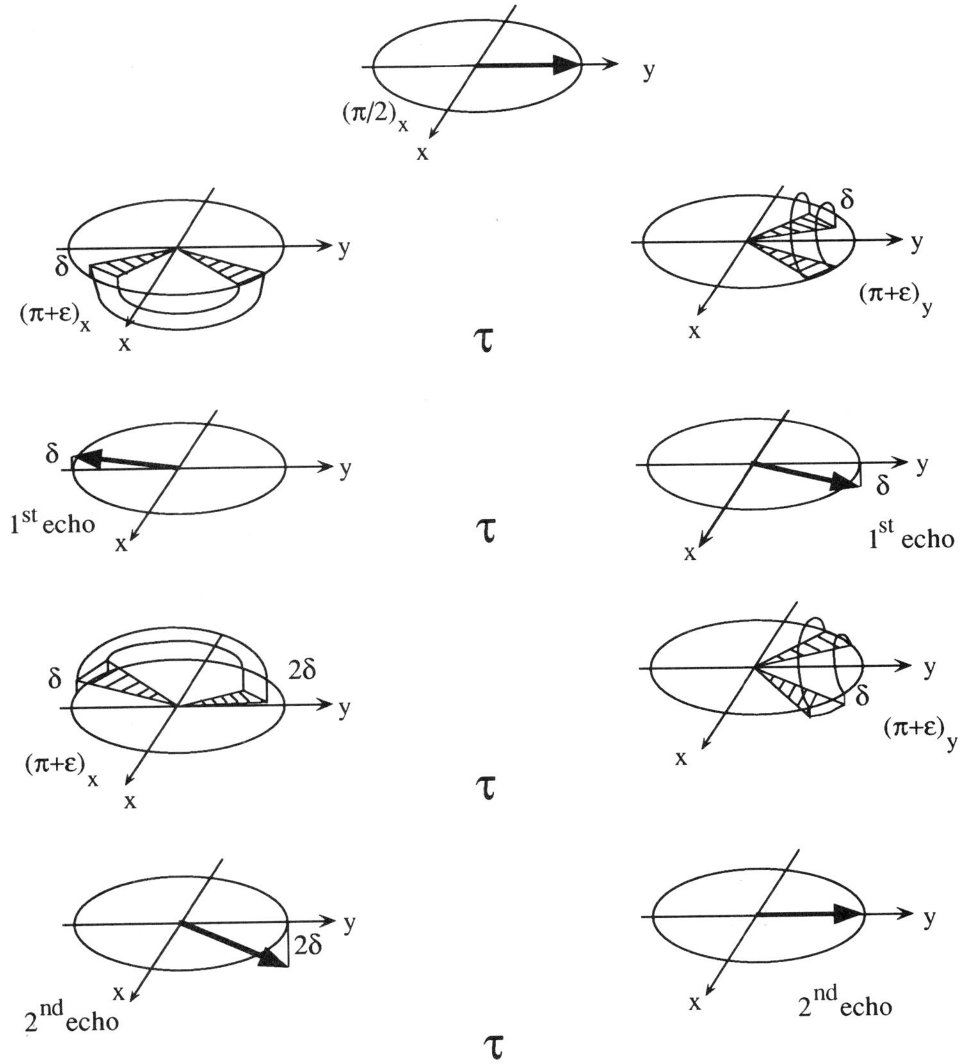

Figure 4.8 π pulse imperfections, denoted by $(\pi + \varepsilon)$, are responsible for a flip angle larger by δ. Left: errors cumulate in the simple Carr–Purcell sequence when π pulses have the same phase as the initial $\pi/2$ pulse. Right: self compensation at each even numbered echo by simply shifting the phase of all π pulses by 90° (CPMG sequence)

cumulative, whereas with the 90° phase change a self-compensation occurs for all echoes of even number.

The CPMG experiment can be handled in two ways, as follows. If the spectrum involves only one resonance (or if linewidths do not allow for the separation of several resonances), a *single experiment* can be run with acquisition of the amplitude of each echo along the pulse train (for sensitivity enhancement, accumulations can be carried out). This experiment is especially valuable for determining the relative proportions of two species which differ by their transverse relaxation time, for instance the two types of water (free and bound) in various materials (wood is a good example) or water and lipids in foodstuffs (Figure 4.9). For this type of measurement, a 'low resolution' spectrometer (without any shim system) proves to be quite sufficient. If the spectrum involves several well resolved resonances, one can proceed by Fourier transformation of the half echo (possibly with signal accumulation). As many experiments as necessary are performed by varying n so that the successive values of $2n\tau$ [see formula (4.19)] lead to T_2 with the required accuracy (Figure 4.10).

In a general way, the spectroscopist must be aware of the difficulties associated with the determination of transverse relaxation times. In homonuclear coupled spin systems, the magnetization decay is affected by a 'J modulation', which can be easily understood on the basis of Section 2.2.6 and Figure 2.20. In the case of a heteronuclear system with observation of the rare nucleus (e.g. carbon-13), it is strongly recommended to avoid proton decoupling during the evolution period. This is because decoupling cannot be infinitely perfect and thus induces an extra line broadening which behaves like a transverse relaxation mechanism; it therefore leads to apparent T_2 values much shorter than expected. Of course, proton decoupling can be switched on for signal acquisition in order to deal with a simpler spectrum. Let us finally mention that a time of about $5T_1$ must elapse between two consecutive experiments in order to allow for a complete return of the system to thermal equilibrium.

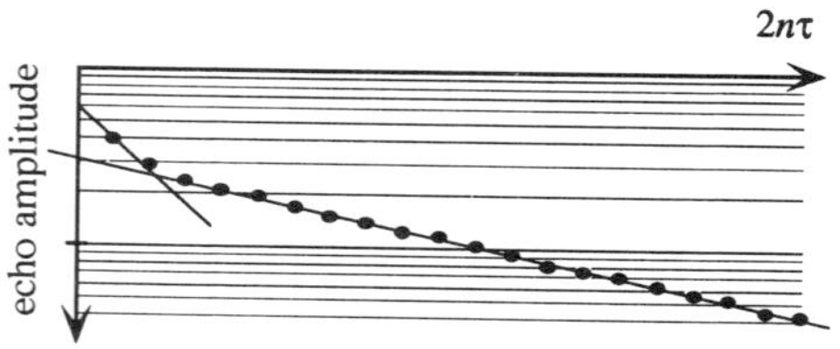

Figure 4.9 Biexponential evolution (in fact semi-logarithmic plot) of the amplitude of echoes resulting from a CPMG pulse train in the case of a sample which involves two species differing in their transverse relaxation time

Measurement of the transverse relaxation time by the spin-lock technique ($T_{1\rho}$, relaxation time in the rotating frame)

Most of the drawbacks mentioned above can be circumvented by a simple experiment, though somewhat instrumentally demanding. The sequence is depicted

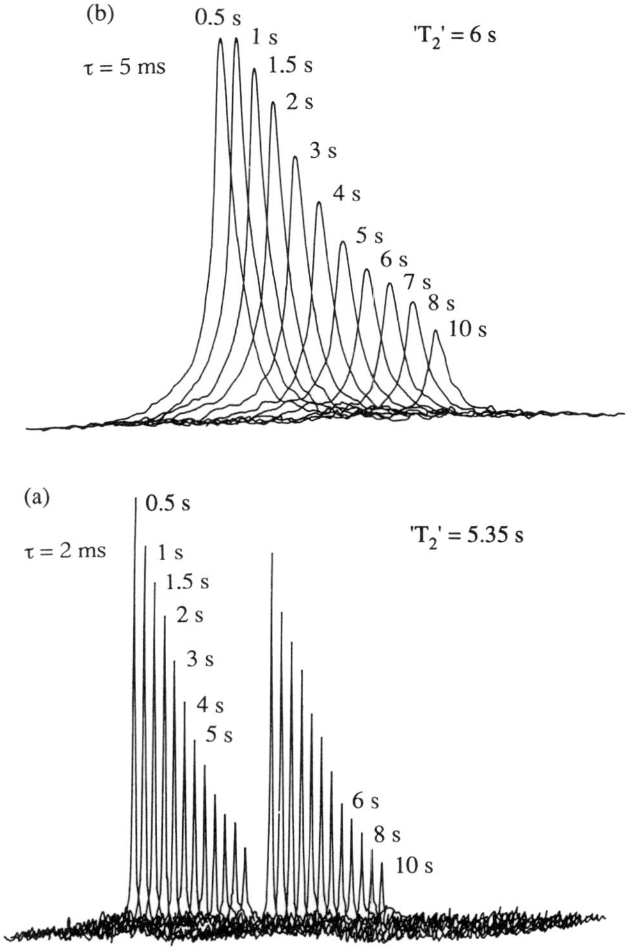

Figure 4.10 (a) Measurement of the transverse relaxation time in the carbon-13 doublet of formic acid (no proton decoupling) according to the sequence of Figure 4.7 with $\tau = 2$ ms and for successive values of n (the value of $2n\tau$ is given above each spectrum. (b) Same as (a) except that proton decoupling was applied only during signal acquisition and that $\tau = 5$ ms

in Figure 4.11. First, a standard $(\pi/2)_x$ flips the nuclear magnetization toward the y axis of the rotating frame; thereafter a radio-frequency field B_1 is applied along that direction for a duration τ. This r.f. field must be sufficiently strong to avoid off-resonance effects (see Section 2.1.3) but not too strong so as to prevent probe deterioration. During the τ interval, magnetization should nutate around B_1; since the two are collinear, magnetization is stationary along the y axis of the rotating frame. It is said to be locked, hence the terminology of 'spin-lock' associated with this experiment.

Any modification of the magnetization thus arises from relaxation phenomena.

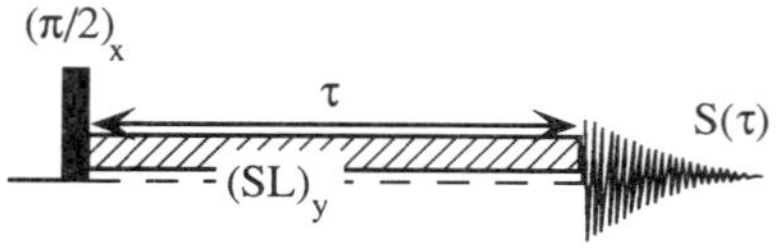

Figure 4.11 Principle of a spin-lock experiment leading to the determination of the relaxation time in the rotating frame ($T_{1\rho}$). $(\mathrm{SL})_y$ stands for the spin-lock period which corresponds to the application of a r.f. field along the y axis of the rotating frame

The transverse magnetization spin-locked along B_1 must end up at its thermal equilibrium value, that is, zero. The corresponding evolution is exponential with a time constant denoted by $T_{1\rho}$ (relaxation time in the rotating frame), very close (if not identical) to T_2 (see below, Section 4.2.3). In practice, the signal is measured (and subsequently Fourier transformed) for a set of τ values in successive experiments, and obeys the equation

$$S(\tau) = S_0 \exp(-\tau/T_{1\rho}) \tag{4.20}$$

where S_0 is the signal amplitude immediately after the $(\pi/2)_x$ pulse. An immediate advantage of the method is the absence of precession during the evolution period and thus, among other things, the absence of J-modulation.

4.1.3 CROSS-RELAXATION AND CHEMICAL EXCHANGE. RELATED TWO-DIMENSIONAL TECHNIQUES: NOESY, HOESY, ROESY

Experimental methods outlined in the preceding section concern systems which obey Bloch equations and which consequently exhibit monoexponential evolutions according to time constant T_1 for longitudinal magnetization and T_2 for transverse magnetization. Although in some instances Bloch equations are either strictly valid or represent a good approximation, there exist situations where biexponentiality shows up obviously, and it would be a shame not to exploit this feature. Such a situation occurs whenever two spins A and X interact by a time-dependent mechanism. Such a process is dubbed *cross-relaxation* and means that any modification of the A magnetization induces a modification of X magnetization which adds up to the specific evolution of X magnetization; the symmetrical process, between X and A of course, holds as well. More formally, this coupling can be expressed via the famous Solomon equations, which are written below for the two longitudinal magnetizations I_z^{A} and I_z^{X} (as will be seen later, any coupling between longitudinal and transverse magnetizations is forbidden):

$$\begin{aligned} \mathrm{d}I_z^{\mathrm{A}}/\mathrm{d}t &= -R_1^{\mathrm{A}}(I_z^{\mathrm{A}} - I_{\mathrm{eq}}^{\mathrm{A}}) - \sigma(I_z^{\mathrm{X}} - I_{\mathrm{eq}}^{\mathrm{X}}) \\ \mathrm{d}I_z^{\mathrm{X}}/\mathrm{d}t &= -R_1^{\mathrm{X}}(I_z^{\mathrm{X}} - I_{\mathrm{eq}}^{\mathrm{X}}) - \sigma(I_z^{\mathrm{A}} - I_{\mathrm{eq}}^{\mathrm{A}}) \end{aligned} \tag{4.21}$$

R_1^{A} represents the specific longitudinal relaxation rate of spin A ($R_1^{\mathrm{A}} = 1/T_1^{\mathrm{A}}$). σ is the cross-relaxation rate which reflects the coupling between the two magnetizations alluded to above; $I_{\mathrm{eq}}^{\mathrm{A}}$ stands for the equilibrium magnetization. It can be seen that Solomon equations resemble Bloch equations to which the cross-relaxation rate σ has been appended. In fact, $I_{\mathrm{eq}}^{\mathrm{A}}$ represents the polarization of nucleus A (which accounts for the difference in energy level populations, and is therefore proportional to γ_A) rather than its magnetization (see Section 4.2.3) which includes the magnetic moment and which depends on γ_A^2. When dealing with homonuclear systems, this distinction is irrelevant; however, in the case of heteronuclear systems, we shall use the term 'magnetization' which shall be understood as 'polarization'. σ may have two origins:

- the dipolar interaction between the two nuclei A and X, modulated by molecular motions. This contribution is very interesting with regard to the information it contains and will be treated more explicitly in Section 4.2.3. For the moment, it is important to acknowledge that it is proportional to $(r_{AX})^{-6}$, where r_{AX} is the internuclear distance between A and X.
- a chemical exchange process such as spins moving from site A to site X and conversely from site X to site A. If we define as τ the residence time in each site (this implies identical concentrations for A and X, thus giving the situation precisely considered for the two spins A and X), then $\sigma = -k = -1/\tau$. Moreover, the exchange rate k must be added to each specific rate R_1^A and R_1^X.

We shall be concerned here with the determination of the three dynamic parameters R_1^A, R_1^X and σ involved in equations (4.21), deferring to a forthcoming section their interpretation at a molecular level. A first approach consists of examining the whole evolution of I_z^A and I_z^X, and, by a non-linear analysis, extracting the three considered parameters. Indeed, because we are dealing with two simultaneous differential equations, this evolution is biexponential in nature (for both I_z^A and I_z^X. After solving the relevant equations, we obtain

$$\begin{aligned} [I_{eq}^A - I_z^A(t)]/(2I_{eq}^A) &= a_1 \exp(\lambda_1 t) + a_2 \exp(\lambda_2 t) \\ [I_{eq}^X - I_z^X(t)]/(2I_{eq}^X) &= x_1 \exp(\lambda_1 t) + x_2 \exp(\lambda_2 t) \end{aligned} \tag{4.22}$$

where λ_1 and λ_2 are the roots of the characteristic equation associated with the two differential equations (4.21):

$$\lambda_{1,2} = (-R_+ \pm X)/2 \tag{4.23}$$

where

$$R_\pm = R_1^A \pm R_1^X \quad \text{and} \quad X = \sqrt{R_-^2 + 4\sigma^2}.$$

The coefficients a_1, a_2, x_1 and x_2 depend not only on relaxation parameters but also on the *initial conditions* $I_z^A(0)$ and $I_z^X(0)$:

$$\begin{aligned} a_{1,2} &= \left\{\frac{[I_{eq}^A - I_z^A(0)]}{2I_{eq}^A}\right\}\left(\frac{1}{2} \mp \frac{R_-}{2X}\right) \mp \left\{\frac{[I_{eq}^X - I_z^X(0)]}{2I_{eq}^X}\right\}\left(\frac{I_{eq}^X}{I_{eq}^A}\right)\left(\frac{\sigma}{X}\right) \\ x_{1,2} &= \left\{\frac{[I_{eq}^X - I_z^X(0)}{2I_{eq}^X}\right\}\left(\frac{1}{2} \pm \frac{R_-}{2X}\right) \mp \left\{\frac{[I_{eq}^A - I_z^A(0)}{2I_{eq}^A}\right\}\left(\frac{I_{eq}^A}{I_{eq}^X}\right)\left(\frac{\sigma}{X}\right) \end{aligned} \tag{4.24}$$

The above relationships are quite general in the sense that they can accommodate any kind of initial conditions: non-selective inversion, selective inversion of one of the magnetizations (Figure 4.12) or any intermediate situation. However, the

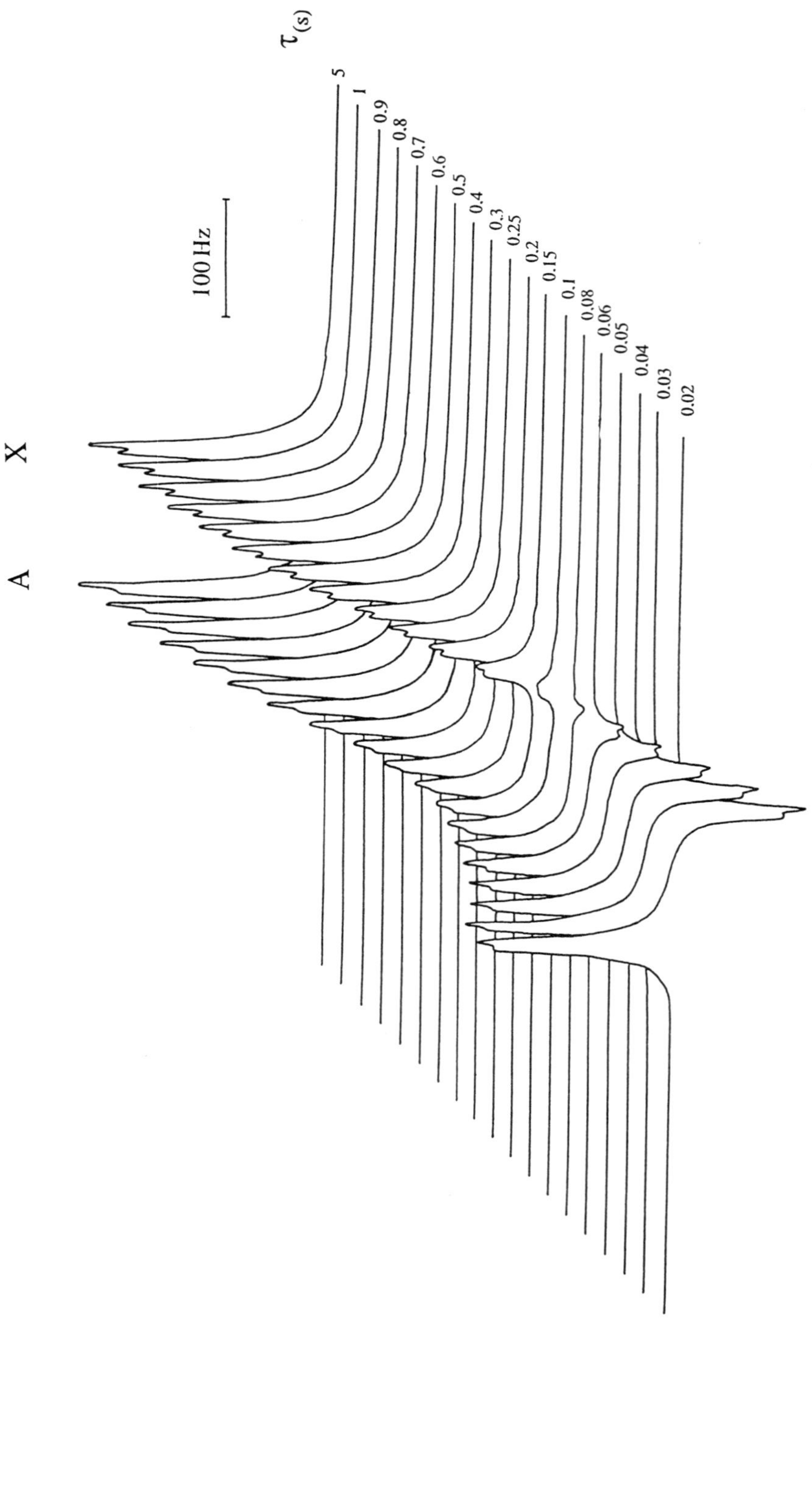

Figure 4.12 Evolution of two nuclear magnetizations obeying equations (4.21) after the selective inversion of one of the magnetizations. One observes a biexponential behavior for both of the magnetizations

complexity of formulae (4.23) and (4.24) has prompted the design of more direct methods based on the *initial behavior* after an appropriate perturbation. Again, we start from equations (4.21) and we assume that t is short enough for ensuring that a first-order expansion is adequate:

$$\begin{aligned} I_z^A(t) &= I_z^A(0) + t(dI_z^A/dt)_{t=0} \\ I_z^X(t) &= I_z^X(0) + t(dI_z^X/dt)_{t=0} \end{aligned} \tag{4.25}$$

It is convenient, for a conceptual and practical point of view, to express equations (4.21) in the form of the *inital slopes* of the reduced quantities already introduced in (4.22):

$$\begin{aligned} \mathcal{S}_A &= \frac{d}{dt}\left(\frac{I_{eq}^A - I_z^A}{2I_{eq}^A}\right)_{t=0} = -R_1^A\frac{I_{eq}^A - I_z^A(0)}{2I_{eq}^A} - \sigma\frac{I_{eq}^X - I_z^X(0)}{2I_{eq}^A} \\ \mathcal{S}_X &= \frac{d}{dt}\left(\frac{I_{eq}^X - I_z^X}{2I_{eq}^X}\right)_{t=0} = -R_1^X\frac{I_{eq}^X - I_z^X(0)}{2I_{eq}^X} - \sigma\frac{I_{eq}^A - I_z^A(0)}{2I_{eq}^X} \end{aligned} \tag{4.26}$$

These initial slopes are readily evaluated, as we can always measure the signal corresponding to I_{eq}^A (or I_{eq}^X) and as the relevant instrumental factor is identical to the one which prevails in the measurement of I_z^A (or I_z^X). Anyhow, initial conditions can be devised for determining separately one of the three relaxation parameters R_1^A, R_1^X or σ. The simplest experiment consists of *selectively* inverting one of the two magnetizations. Consider first the selective inversion of A magnetization for which the following initial conditions hold: $I_z^A(0) = -I_{eq}^A$ and $I_z^X(0) = I_{eq}^X$. This yields for the initial slopes (with $I_{eq}^A/I_{eq}^X = \gamma_A/\gamma_X$)

$$\begin{aligned} &\mathcal{S}_A \text{ (A selectivity inverted)} = -R_1^A \\ &\mathcal{S}_X \text{ (A selectivity inverted)} = -\sigma(\gamma_A/\gamma_X). \end{aligned} \tag{4.27a}$$

The complementary experiment (selective inversion of X magnetization) leads to

$$\begin{aligned} &\mathcal{S}_A \text{ (X selectivity inverted)} = -\sigma(\gamma_X/\gamma_A) \\ &\mathcal{S}_X \text{ (X selectivity inverted)} = -R_1^X. \end{aligned} \tag{4.27b}$$

Another popular way of measuring the cross-relaxation rate σ relies upon the saturation (by continuous irradiation) of one of the two nuclei, say X. Referring again to the first of equations (4.21) with $I_z^X \equiv 0$ (saturation of spin X) and with $dI_z^A/dt = 0$, which implies that a stationary state has been reached prior to the measurement, (this means that the irradiation of spin X has been applied for a time sufficiently long with respect to relaxation times), we obtain a new value for the longitudinal magnetization of spin A which will be denoted as I_{stat}^A.

$$I_{stat}^A = I_{eq}^A[1 + (\gamma_X/\gamma_A)(\sigma/R_1^A)] \tag{4.28}$$

Hence, provided that I_{eq}^A is known and that R_1^A has been determined by means of an independent experiment, I_{stat}^A provides the cross-relaxation rate σ. This stems from the *nuclear Overhauser effect* or NOE (Overhauser was the first scientist to recognize that, by a related method, electron spin polarization could be transferred to nuclear spins). This effect is usually quantified by the so-called NOE factor η

$$\eta = (I_{stat}^A - I_{eq}^A)/I_{eq}^A = (\gamma_X/\gamma_A)(\sigma/R_1^A) \tag{4.29}$$

In practice, I^A_{stat} and I^A_{eq} must be determined under identical instrumental conditions. The decoupler channel of the spectrometer is generally used for saturating the spin X. Homonuclear and heteronuclear systems have to be distinguished. In the former case, a weak (selective) secondary r.f. field centered on the X resonance is used for measuring I^A_{stat}, whereas I^A_{eq} is obtained by the same experiment with the irradiation frequency shifted to a zone devoid of resonances; the results of these two experiments are stored in different memory blocks (Figure 4.13(a)). In the heteronuclear case, broadband X decoupling (which amounts to saturating X resonances) is used to measure I^A_{stat}. Since instrumental conditions as close as possible should be used for the measurement of I^A_{eq}, the decoupling frequency is moved to a significantly different value (as in the homonuclear case) and the modulation scheme (which insured effective irradiation on a proper frequency band) is simply switched off so that no residual NOE can occur. The normal decoupling scheme is reset for signal acquisition (so as to obtain a decoupled spectrum); it can be mentioned that the nuclear Overhauser effect does not affect transverse magnetization. Again I^A_{stat} and I^A_{eq} are stored in different memory blocks for comparison purposes. The spectra of Figure 4.14 illustrate this procedure.

It is perfectly clear that, although the above experiments appear especially gratifying, they should be repeated for each pair of (A, X) nuclei (with the requirement of tedious experimental adjustments, since selective pulses must be

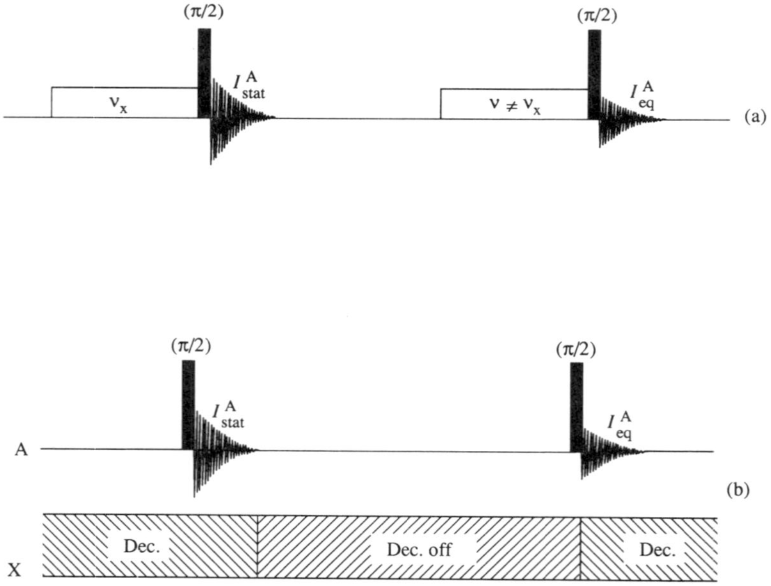

Figure 4.13 (a) Determination of the NOE factor in the homonuclear case. The frequency of the selective irradiation is moved at a value different from ν_X for the measurement of I^A_{eq}. (b) Determination of the NOE factor in the heteronuclear case. The broadband decoupler is applied a long time prior to the measurement of I^A_{stat}. For the measurement of I^A_{eq} and in order to maintain identical conditions, the decoupling frequency is moved far away from its normal value and the modulation scheme is switched off (Dec. off)

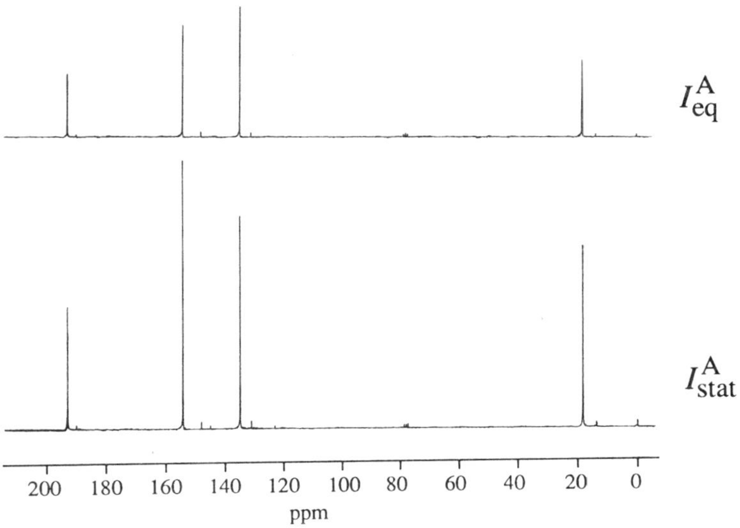

Figure 4.14 Determination of the heteronuclear Overhauser effect for the *trans*-crotonaldehyde molecule (experiment of Figure 4.13b). Carbon-13 is observed while the proton is irradiated only during the signal acquisition (top) or continuously (bottom). The comparison of line intensities in the two spectra yields the NOE factor η for each carbon

employed). Heteronuclear Overhauser effect measurements constitute an exception since they concern X nuclei in their globality. As a matter of fact, experimental procedures can be simplified and automated through the two-dimensional techniques which will be described in the forthcoming sections.

The NOESY sequence

This two-dimensional method (J. Jeener *et al.*, *J. Chem. Phys.*, **71**, 4546 (1979)) provides in a single experiment all cross-relaxation rates. The acronym (Nuclear Overhauser Effect SpectroscopY) originates from the one-dimensional experiments described above, and is mostly limited to the study of homonuclear cross-relaxation. As explained in detail in Chapter 3, the analysis of a signal $S(t_1, t_2)$ by means of a double Fourier transformation leads to a two-dimensional map in the frequency domains (ν_1, ν_2), whose cross peaks (off-diagonal peaks for which $\nu_1 = \nu'_1$ and $\nu_2 = \nu'_2$) generally contain correlation information regarding two nuclei resonating at frequencies ν'_1 and ν'_2. Concerning the NOESY sequence depicted in Figure 4.15, this correlation arises from cross-relaxation and indicates either a dipolar interaction (thus a spatial proximity of the two considered nuclei) or chemical exchange between the two relevant sites. Let us emphasize again that the advantage of this two-dimensional technique stems from the global view of all dipolar or exchange correlations existing within the molecular system under investigation.

The preparation period allows for the return to equilibrium of nuclear magnetization. The evolution period provides a labeling according to the resonance

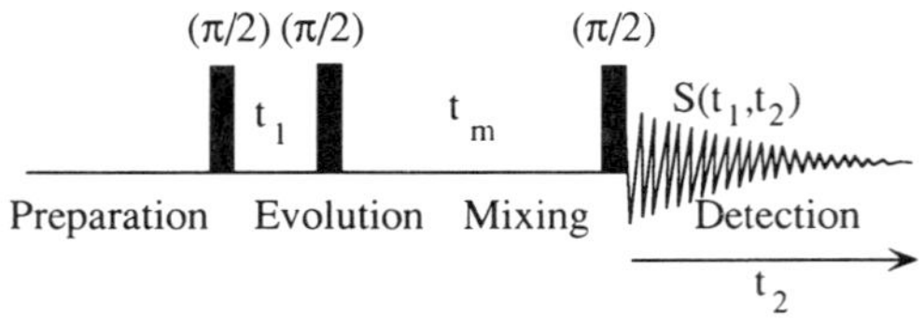

Figure 4.15 Basic scheme of the NOESY sequence which provides essentially the homonuclear cross-relaxation rate from a double Fourier transform of the signal $S(t_1, t_2)$

frequency of each nucleus by repeating the experiment for incremented values of t_1 according to $k_1*\mathrm{Dw}$, where the integer k_1 takes the values between 0 and $(n-1)$ and where Dw stands for the sampling rate (dwell time) appropriate to the spectral window to be investigated (see Chapter 3). Let us first consider the situation at the end of the first part of the sequence ($\pi/2 - t_1 - \pi/2$) by assuming that the two $\pi/2$ pulses possess the same phase (this means that both of them act along the same axis of the rotating frame, say x). The z component of the magnetization, corresponding to a resonance frequency ν' (in the rotating frame), can be readily evaluated from the schematic representation of Figure 4.16.

If we limit ourselves to longitudinal components of magnetization (we shall see later that an appropriate phase cycling eliminates the transverse components), we can write the initial conditions before the mixing period t_m for the two spins A and X:

$$\begin{aligned} I_z^{\mathrm{A}}(0) &= -I_{\mathrm{eq}}^{\mathrm{A}} \cos(2\pi\nu_{\mathrm{A}} t_1) \\ I_z^{\mathrm{X}}(0) &= -I_{\mathrm{eq}}^{\mathrm{X}} \cos(2\pi\nu_{\mathrm{X}} t_1). \end{aligned} \tag{4.30}$$

ν_{A} and ν_{X} stand for the resonance frequencies (ν') of the two spins A and X. (Relaxation or signal attenuation during t_1 is neglected). During the so-called *mixing time* t_m the system evolves according to Solomon equations, and we shall further assume that t_m is sufficiently short so that we can rely on equations (4.25) and (4.26), which describe the 'initial behavior'. The last $\pi/2$ pulse of the sequence of Figure 4.15 is simply a read pulse which converts longitudinal magnetization into observable transverse magnetization, yielding the free induction decay physically detected during t_2. Again disregarding attenuation factors during t_2 (governed by

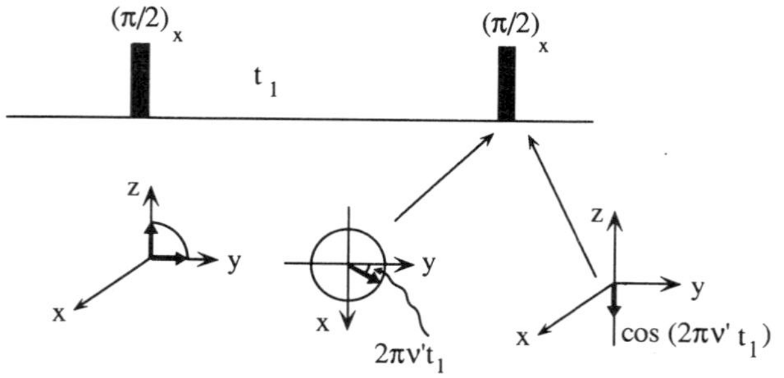

Figure 4.16 Evolution of the magnetization corresponding to a precession of frequency ν' during the first stage of sequence 4.15

the effective transverse relaxation T_2^*), we can write the actual signal, which is a function of both the t_1 and t_2 variables, as

$$\begin{aligned}S(t_1, t_2) = \{&-I_{eq}^{A}\cos(2\pi\nu_A t_1) + t_m[-R_1^{A}(-I_{eq}^{A}\cos(2\pi\nu_A t_1) - I_{eq}^{A})\\&- \sigma(-I_{eq}^{X}\cos(2\pi\nu_X t_1) - I_{eq}^{X})]\}\cos(2\pi\nu_A t_2)\\&+ \{-I_{eq}^{X}\cos(2\pi\nu_X t_1) + t_m[-R_1^{X}(-I_{eq}^{X}\cos(2\pi\nu_X t_1) - I_{eq}^{X})\\&- \sigma(-I_{eq}^{A}\cos(2\pi\nu_A t_1) - I_{eq}^{A})]\}\cos(2\pi\nu_X t_2).\end{aligned} \tag{4.31}$$

Expressions between braces represent the longitudinal magnetizations of spins A and X immediately before the read pulse; they subsequently precess at frequencies ν_A and ν_X respectively, hence the factors $\cos(2\pi\nu_A t_2)$ and $\cos(2\pi\nu_X t_2)$. Since we are dealing with an homonuclear system, we can put $I_{eq}^{A} = I_{eq}^{X} = I_0$ and recast accordingly the first term of (4.31) recognizing that we can deduce the second term by simply interchanging the labels A and X:

$$\begin{aligned}S_A(t_1, t_2) = I_0\{&(-1 + t_m R_1^{A})\cos(2\pi\nu_A t_1)\cos(2\pi\nu_A t_2)\\&+ t_m\sigma\cos(2\pi\nu_X t_1)\cos(2\pi\nu_A t_2)\\&+ t_m(R_1^{A} + \sigma)\cos(2\pi\nu_A t_2)\}.\end{aligned} \tag{4.32}$$

A double Fourier transform with respect to t_1 and t_2 yields the peaks whose intensities and frequencies are reported in Table 4.1.

As shown in Figure 4.17, this experiment yields diagonal peaks whose intensity depends on the specific relaxation rate of each nucleus $(-1 + t_m R_1^{A})$ and $(-1 + t_m R_1^{X})$ and two cross peaks, symmetrical with respect to the diagonal, whose

Table 4.1 Characteristics and origins of peaks obtained after a double Fourier transform of $S_A(t_1, t_2)$, given by a equation (4.32), and its homolog for spin X, $S_X(t_1, t_2)$

Frequencies		Type	Intensity	Origin
$\nu_1 = \nu_A$;	$\nu_2 = \nu_A$	diagonal	$(-1 + t_m R_1^{A})I_0$	$S_A(t_1, t_2)$
$\nu_1 = \nu_X$;	$\nu_2 = \nu_A$	cross	$t_m\sigma I_0$	$S_A(t_1, t_2)$
$\nu_1 = 0$;	$\nu_2 = \nu_A$	axial	$t_m(R_1^{A} + \sigma)I_0$	$S_A(t_1, t_2)$
$\nu_1 = \nu_X$;	$\nu_2 = \nu_X$	diagonal	$(-1 + t_m R_1^{X})I_0$	$S_X(t_1, t_2)$
$\nu_1 = \nu_A$;	$\nu_2 = \nu_X$	cross	$t_m\sigma I_0$	$S_X(t_1, t_2)$
$\nu_1 = 0$;	$\nu_2 = \nu_X$	axial	$t_m(R_1^{X} + \sigma)I_0$	$S_X(t_1, t_2)$

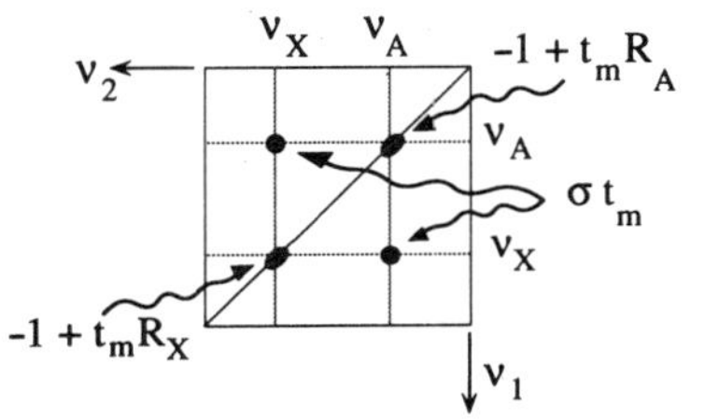

Figure 4.17 Meaning of the diagonal and cross peaks in a NOESY two-dimensional map

intensity leads directly to σ. Axial peaks are cancelled by appropriate phase cycling (see below). Since $t_m R_1^A$ and $t_m R_1^X$ are generally smaller than 1, diagonal and cross peaks are of opposite sign if σ is positive (corresponding to short correlation times; see the second part of this chapter) and of the same sign if long correlation times prevail (slowly tumbling molecules). In any event, the existence of cross peaks indicates a non-negligible dipolar interaction between the involved spins, thus an actual spatial proximity (Figure 4.18).

It must however be borne in mind that, for a system involving a large number of spins, the existence of cross peaks may be misleading even in the case of relatively short mixing times. This is because the approximation of initial behavior is no longer valid, especially if cross-relaxation rates become important (in absolute value), as occurs for large biomolecules which, due to their size, reorient on a slow time scale. In order to understand this phenomenon, let us consider three spins A, M and X and let us suppose that σ_{AM} and σ_{MX} are important while σ_{AX} is

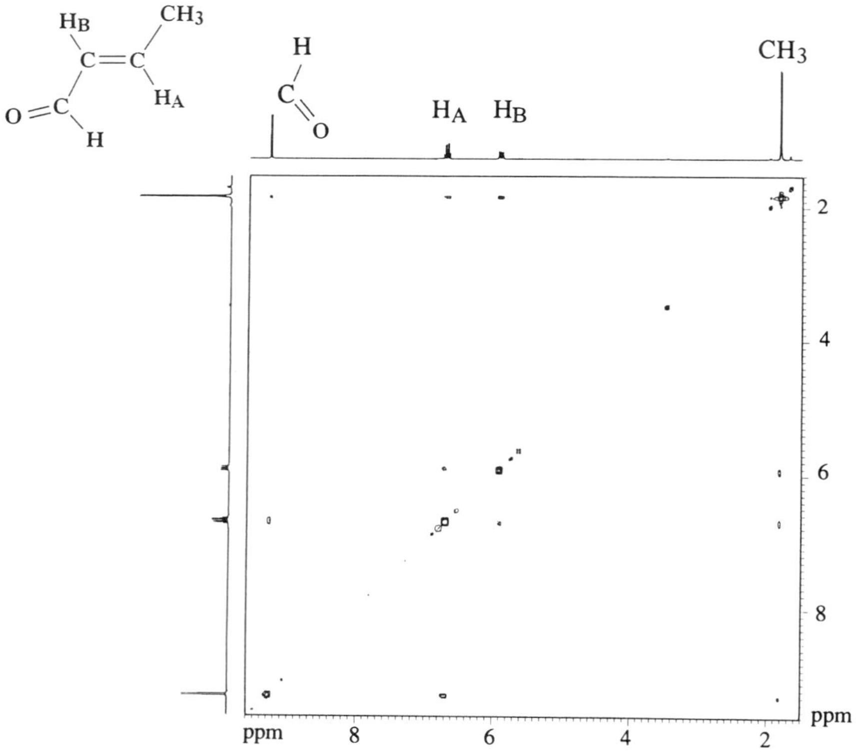

Figure 4.18 The NOESY sequence applied to the *trans*-crotonaldehyde molecule, which can be compared with the COSY experiment shown in Chapter 1 (Figure 1.34). In this latter case, cross peaks arise from *J* couplings, thus from chemical bonds joining the two nuclei concerned. In the present case, cross peaks simply reflect spatial proximity. Notice the absence of correlation between the proton H_B and the aldehyde proton (the correlation between the latter and the methyl protons is due to a relayed process and has no geometrical meaning)

negligible; thus no cross-peak between A and X should be visible. It turns out that, for a mixing time long enough, the perturbation undergone by M from the AM interaction will affect X via σ_{MX}, leading finally to a cross peak between A and X. The latter, however, does not reflect any spatial proximity; this phenomenon is known under the terminology of 'spin diffusion' and can be avoided by choosing mixing times as short as some tens of milliseconds.

We shall now, in respect of the NOESY sequence, further exemplify the concept of phase cycling introduced in Section 3.6. It may be recalled that it consists of adding free precession signals of experiments which differ only by the phase of the r.f. pulses and possibly by the receiver phase. Let us first suppose that the phases of the three r.f. pulses of sequence 4.15 are identical for a first experiment

$$(\pi/2)_x - t_1 - (\pi/2)_x - t_m - (\pi/2)_x - (\mathrm{Acq})_+$$

and become for a second experiment

$$(\pi/2)_{-x} - t_1 - (\pi/2)_x - t_m - (\pi/2)_x - (\mathrm{Acq})_-$$

where $(\mathrm{Acq})_+$ and $(\mathrm{Acq})_-$ correspond to the detection period (t_2) and symbolize the sign attributed to the acquired signals, which is simply the receiver phase. It is obvious that the sign of $I_z^{\mathrm{A}}(0)$ and $I_z^{\mathrm{X}}(0)$ [see equation (4.30)] and the sign of the detected signal are modified. It can then be seen, from equations (4.31) and (4.32), that only the sign of axial peaks has been altered. The addition of the fids resulting from these two experiments thus constitutes a very simple way for the elimination of axial peaks.

The obtention of a quadrature signal in t_1 can be achieved through another phase cycle which can be superimposed onto the one given above. This can be accomplished provided that *physical* quadrature detection exists in t_2. This will be represented by $[\cos(2\pi\nu_2' t_2) + i\sin(2\pi\nu_2' t_2)]$ and not by a simple cosine function as has been done until now for the sake of simplicity regarding magnetization transfer processes. For that purpose, let us envision the r.f. phases of the two experiments below:

$$(\pi/2)_x - t_1 - (\pi/2)_x - t_m - (\pi/2)_x - t_2$$
$$(\pi/2)_x - t_1 - (\pi/2)_{-y} - t_m - (\pi/2)_y - t_2$$

the receiver phase being unchanged.

Now, with quadrature detection in the t_2 dimension as indicated above, let us assign (arbitrarily) the real part of the detection to $\cos(2\pi\nu_2' t_2)$ and the imaginary part to $\sin(2\pi\nu_2' t_2)$. This amounts to assigning the y axis to the real part and the x axis to the imaginary part. The first of the experiments considered above yields for all terms of (4.31) (except those corresponding to axial peaks)

$$\cos(2\pi\nu_1' t_1)[\cos(2\pi\nu_2' t_2) + i\sin(2\pi\nu_2' t_2)]$$

where ν_1' and ν_2' stand for ν_A or ν_X, whereas the second experiment yields

$$\sin(2\pi\nu_1' t_1)[(\sin(2\pi\nu_2' t_2) - i\cos(2\pi\nu_2' t_2)].$$

In the above expression the phases of $\pi/2$ pulses and the signs of projections along the x and y axes of Figure 4.15 together with the convention retained for real and imaginray parts in t_2 have been taken into account. Adding these two contributions, as is usually done in a phase cycling procedure, yields

$$[\cos(2\pi\nu'_1 t_1) - i\sin(2\pi\nu'_1 t_1)][\cos(2\pi\nu'_2 t_2) + i\sin(2\pi\nu'_2 t_2)] = \exp(-2i\pi\nu'_1 t_1)\exp(2i\pi\nu'_2 t_2).$$

A 'quadrature detection' is actually achieved in t_1. Moreover, the opposite signs in the exponential arguments ensure a final result of type n (see Section 3.5.1). The two phase cyclings given above can then be combined so as to provide (i) the cancellation of axial peaks, (ii) quadrature detection in both dimensions. A double complex Fourier transform can then be applied and the result displayed in the form of an amplitude spectrum, with, however, the inconveniences mentioned in Section 3.5.1. Table 4.2 summarizes the corresponding four-step phase cycle.

Pure absorption spectra, retaining t_1 quadrature detection, can be obtained by the TPPI procedure (Section 3.5.3), which consists in adding $\pi/2$ to the phase φ_1 at each new value of t_1 value (remember that in this case the t_1 increment should be half the size it is in the conventional experiment above described). From the phase cycling scheme of Table 4.2, only the suppression of axial peaks needs to be retained. On the other hand, it can be shown (G. Bodenhausen *et al.*, *J. Magn. Reson.*, **58**, 370 (1984)) that multiquanta coherences, which may occur during t_m if there exists a J coupling between the two nuclei involved in cross-relaxation processes, are eliminated by cycling through x, y, $-x$, $-y$ the phase of the read pulse (φ_3) concomitantly with the receiver phase. This procedure, known under the name of CYCLOPS (of widespread use; see Section 3.6), presents the further advantage of compensating for the instrumental imperfections of the physical quadrature detection in t_2. The relevant phase cycling scheme is given in Table 4.3.

Finally, it can be mentioned that possible zero-quantum coherences (Section 2.2.3) cannot disappear through the cycling of φ_3 and φ_4. They may manifest themselves by unwanted cross peaks, corresponding to coherence transfers by J coupling. Since they evolve during t_m according to sine or cosine functions, they can be attenuated, eventually suppressed, by a small random variation of the mixing time t_m.

Table 4.2 Minimal phase cycling of the NOESY sequence (Figure 4.15) with quadrature detection in both dimensions (leading to an amplitude spectrum) and with cancellation of axial peaks. φ_1, φ_2 and φ_3 are the phases of the three ($\pi/2$) pulses whereas φ_4 indicates the sign to be applied to the physically detected signal (receiver phase)

φ_1	φ_2	φ_3	φ_4
x	x	x	+
$-x$	x	x	−
x	$-y$	y	+
$-x$	$-y$	y	−

Table 4.3 CYCLOPS superimposed on the elimination of axial peaks in the NOESY sequence. This phase cycling implies quadrature detection in t_1 by the TPPI procedure along with the obtention of pure obsorption spectra

φ_1	φ_2	φ_3	φ_4
x	x	x	x
$-x$	x	x	$-x$
x	x	y	y
$-x$	x	y	$-y$
x	x	$-x$	$-x$
$-x$	x	$-x$	x
x	x	$-y$	$-y$
$-x$	x	$-y$	y

The HOESY sequence

We shall be concerned here with the extension of the NOESY sequence to a heteronuclear AX system. According to a commonly encountered situation, A and X can be thought as a proton and a carbon-13 atom, respectively. The acronym originates from Heteronuclear Overhauser effect Spectroscopy (P. L. Rinaldi, *J. Am. Chem. Soc.*, **105**, 5167 (1983); C. Yu and G. C. Levy, *J. Am. Chem. Soc.*, **105**, 6994 (1983)). Referring to the above example, it should be emphasized that the observations of ^{13}C resonances will be favored for two reasons: (i) the carbon-13 spectrum under proton decoupling is much simpler than the proton spectrum (which would in any event be dominated by resonances of protons bonded to a carbon-12 about a hundred times larger than resonances corresponding to protons bonded to a carbon-13 (see Section 1.3.5)); and (ii) the proton–carbon dipolar interaction, which is to be determined, represents the dominant relaxation mechanism for carbon-13 whereas proton–proton dipolar interactions constitute the overwhelming mechanism of proton relaxation. Thus, regarding the determination of the heteronuclear cross-relaxation rate, we can expect more reliability from carbon-13.

The pulse sequence is sketched in Figure 4.19. It differs from the NOESY sequence (i) by he read pulse, which is applied solely to X nuclei, the fid

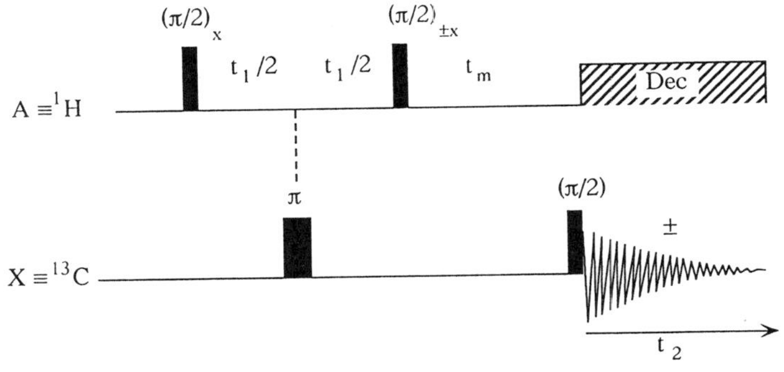

Figure 4.19 Basic scheme of the HOESY sequence

acquisition being performed under decoupling of A nuclei; and (ii) by a 180° pulse applied to X in the middle of the evolution period. This pulse refocuses the J_{AX} coupling effects during t_1 (see Section 2.2.6) and amounts to decoupling carbon-13 from the proton. Without this precaution and owing to the magnitude of these J couplings (greater than 100 Hz), the A magnetization at the beginning of t_m would be considerably reduced.

In spite of experimental differences, responses of the HOESY sequence are of similar nature to those of the NOESY sequence. An almost identical phase cycling can be invoked: a 180° alternation of the phase of the second ($\pi/2$) pulse applied to A concomitantly with the sign of the acquisition leads to the cancellation of axial peaks. As in the NOESY sequence, the CYCLOPS procedure can be superimposed in order to compensate for quadrature artifacts in t_2. Similary, the TPPI incrementation can be employed at the level of the first ($\pi/2$) pulse (applied to spin A) for ultimately obtaining a pure absorption spectrum.

The real interest of the method lies in the visualization, by means of a single experiment, of all the protons (in the ν_1 dimension) which interact by dipolar coupling with a given carbon (ν_2 dimension), indicating the spatial proximity of the corresponding atoms within a molecule, or between two distinct molecules (for instance in the context of solute–solvent interactions). These latter effects will be especially apparent in the case of a carbon not directly bonded to a proton, since they will not be hidden by the otherwise overwhelming one-bond dipolar interaction (due to the short carbon–proton distance). The HOESY two-dimensional map of a surfactant in aqueous solution illustrates these features (Figure 4.20).

The ROESY sequence

We return here to a homonuclear spin system in which we aim to determine the *transverse* cross-relaxation rates, by contrast with the NOESY experiment, which yields *longitudinal* cross-relaxation rates. It can be recalled that the latter may be positive or negative, depending on the molecular mobility, and thus go to zero (this situation can occur at certain measuring frequencies for slowly tumbling molecules such as proteins or nucleic acids). In contrast, transverse cross-relaxation rates are always positive (see below). In that sense, a ROESY experiment which yields transverse cross-relaxation rates (A. A. Bothner-By *et al.*, *J. Am. Chem. Soc.*, **106**, 811 (1984); A. Bax and D.G. Davis, *J. Magn. Reson.*, **63**, 207 (1985)) can be thought as complementary of a NOESY experiment.

For theoretical reasons explained in Section 4.2.3, transverse cross-relaxation rates do not actually affect a conventional transverse relaxation experiment using spin echoes. Rather, they become visible in a *spin-lock* experiment. To maximizing their effects, the two magnetizations of interest should be taken along the locking r.f. field in opposite directions. Hence, a possible design for a one-dimensional experiment is as follows: a selective r.f. pulse takes the X magnetization toward $-z$; it is followed by a non-selective pulse which brings both magnetizations along the spin-locking field so permitting cross-talk during the mixing time t_m (Figure 4.21). During the mixing time t_m, magnetizations evolve according to the relaxation rates in the rotating frame $R_{1\rho}^{A}$ and $R_{1\rho}^{X}$, and also according to the transverse

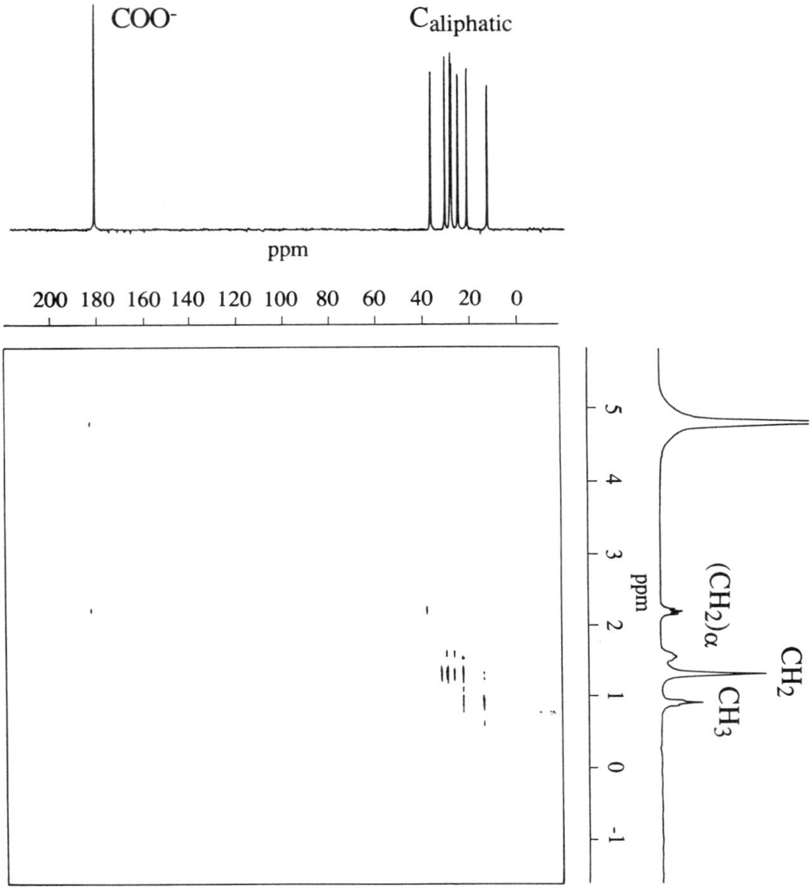

Figure 4.20 HOESY diagram of micellized sodium octanoate for a mixing time t_m of 2 s. Horizontal scale (ν_1): proton chemical shifts. Vertical scale (ν_2): carbon chemical shifts. The correlation peaks indicate proton–carbon spatial proximity. An interesting point concerns the correlation between the carbon and the water protons which allows one to visualize the contact between the solvent and the surfactant polar head

cross-relaxation rate denoted by σ_ρ (or $\sigma_\perp$). This is homologous to the relaxation parameters which govern the longitudinal evolution magnetizations (R_1^A, R_1^X and σ). We can therefore expect having similar behavior, with, however, relaxation rates of different values in conditions of slow molecular mobility (Section 4.2.3).

The two-dimensional counterpart (Figure 4.22) can be deduced from the sequence of Figure 4.21 by substituting for $(\mathrm{Sel(X)}(\pi/2)_x)$ a non-selective $\pi/2$ pulse followed by an evolution period. Subsequently to the evolution period, the x components of A and X magnetizations are respectively $I_{eq}^A \cos(2\pi\nu_A t_1)$ and $I_{eq}^X \cos(2\pi\nu_x t_1)$; they stand as initial conditions for the mixing period, as the spin-lock field, for the first step of the phase cycle, is applied along the x axis of the rotating frame. All other components are defocused or dispersed by the

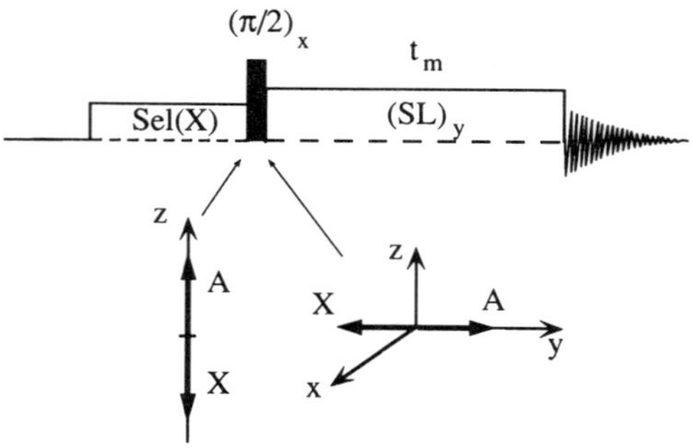

Figure 4.21 One-dimensional version of a ROESY experiment. Sel(X) stands for a selective pulse which takes the X magnetization toward $-z$. During the application of a spin locking (SL) r.f. field along y and for a time t_m, exchange takes place between the two magnetizations by transverse cross relaxation

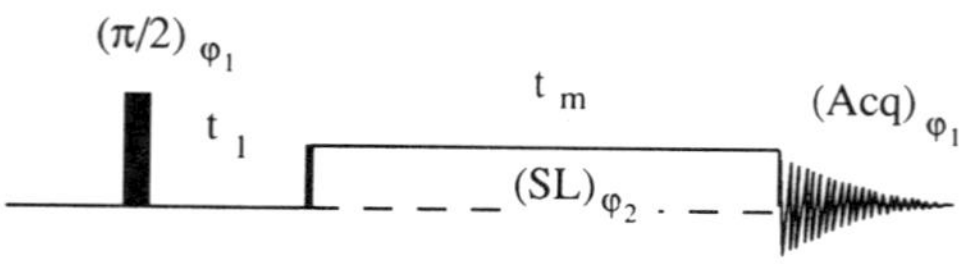

Figure 4.22 Basic scheme of a two-dimensional ROESY experiment. A possible phase cycling is as follows: $\varphi_1 = x, x, -x, -x, y, y, -y, -y$ along with $\varphi_2 = x, -x, x, -x, y, -y, y, -y$. In order to reduce unwanted effects due to the small amplitude of the spin-lock field, the relevant period (t_m) can be bracketed by two standard $\pi/2$ hard pulses (C. Griesinger and R.R. Ernst, *J. Magn. Reson.*, **75**, 261 (1987))

inhomogeneity of the r.f. field (see the discussion about field gradients in Section 2.2.6). Hence, x magnetizations are the only ones to be considered; they evolve under relaxation phenomena and qualitatively lead to the same type of result as a NOESY experiment: a cross peak reveals a non-zero transverse cross-relaxation rate. The phase cycling indicated in the legend of Figure 4.22 is quite simple. The four chemical steps $(x, -x, y, -y)$ applied to the initial $(\pi/2)$ pulse and to the receiver are superimposed onto a two-step phase cycle of the spin lock field (e.g. $+x, -x$). These two latter phases would be equivalent for an ideal instrument and in the absence of off-resonance effects. This phase alternation may partially compensate for instrumental artifacts. It should be noted that the phase cycling given here does not allow for quadrature in t_1. To achieve this goal, the easiest way is to use the TPPI scheme which in addition leads to pure absorption spectra.

4.2 Molecular Interpretation of Nuclear Spin Relaxation

4.2.1 SPECTRAL DENSITIES

Up to now, we have only considered r.f. fields as the only means of acting on nuclear spins with the prerequisite that the relevant frequency is close to the one of the involved transitions. This r.f. field is of course created by an appropriate device outside the sample. By extension, we may wonder whether *local fields*, taking place within the sample would not play a similar role. They would induce transitions whose effect would be to take the nuclear magnetization back to its equilibrium configuration in the same way as the damping forces acting on a periodically moving body. These local fields must be time-dependent so that, when performing a spectral decomposition, we can find a non-zero component at frequencies equal to those of the various transitions which can exist within the energy level diagram of the considered nuclear spin system. For that reason, the efficiency of local fields (denoted by $b(t)$) can be appreciated by quantities named *spectral densities* of the form

$$\mathcal{J}(\omega) = \int_0^{\infty} \overline{b(t)b(0)} \exp(-i\omega t)\, \mathrm{d}t \tag{4.33}$$

It can be recognized that $\mathcal{J}(\omega)$ is the Fourier transform of $\overline{b(t)b(0)}$ and thus provides the amplitude (in the frequency domain) of this latter quantity at the frequency $\nu = \omega/2\pi$. The bar denotes an average over the *ensemble of spin systems belonging to the sample*, in accordance with the concept of nuclear magnetization which also originates from all the spin systems in the sample. On the other hand, a transition can be induced by a fluctuating local field at the condition that a certain degree of *coherence* is present. This can be understood on the basis that the external r.f. field evolving as a sine or a cosine function is itself totally coherent. Indeed, the quantity $\overline{b(t)b(0)}$, named the correlation function, is indicative of this coherence. Its Fourier transform determines the ability of the interaction responsible for $b(t)$ to induce a transition.

The fluctuations of local magnetic fields originate from molecular motions which, because they are by nature somewhat complicated, should be modeled so that we can arrive at a simple interpretation of spectral densities and, consequently, of relaxation parameters. A widely used model for molecular reorientation is the *rotational diffusion*, which leads to an exponential correlation function of the form:

$$\overline{b(t)b(0)} = \overline{b^2} \exp(-t/\tau_c) \tag{4.34}$$

where $\overline{b^2}$ is the average value of the square of the local field and τ_c is called the *correlation time*. This quantity can be considered as the time taken by a molecule to rotate by an angle of one radian. For a non-viscous liquid, τ_c is in the range 10^{-10}–10^{-12} s. In that case, the real part of the spectral density (4.33) can easily be calculated:

$$\mathcal{J}(\omega) = \overline{b^2} \frac{\tau_c}{1 + \omega^2 \tau_c^2} \tag{4.35}$$

Whenever the interaction (mechanism) responsible for relaxation phenomena is well defined (for instance the chemical shift anisotropy mechanism, the dipolar interaction, etc.; see below), we are allowed to substitute to the local field a tensorial quantity which is meaningful at the molecular level. We shall denote by T_{uv} (with u, v = x, y or z) the elements of the relevant *cartesian tensor* (see Section 1.4.1) which in any instance can be converted into *irreducible tensors*, as explained shortly. Irreducible tensors are especially convenient when switching from the laboratory frame to a molecular frame where they can be expressed in terms of structural or dynamical parameters; such an operation is performed according to a methodology based on the Wigner matrices. Another advantage of irreducible tensors stems from the evaluation of correlation functions. The irreducible tensors with which we shall be concerned are of rank 2 and will be denoted as $F^{(\mathrm{m})}$, where m (which is called the projection) can take the values −2, −1, 0, 1, 2. It can be shown that in an isotropic medium (hence solids or crystal liquids are excluded) correlation functions involving any arbitrary projection can be expressed as the correlation function of irreducible tensors of zero projection:

$$\begin{aligned}\int_0^\infty \overline{F^{(\mathrm{m}')}(t)F^{(-\mathrm{m})}(0)}\exp(-i\omega t)\,\mathrm{d}t &= (-1)^{\mathrm{m}}\delta_{\mathrm{mm}'}\int_0^\infty \overline{F^{(0)}(t)F^{(0)}(0)}\exp(-i\omega t)\,\mathrm{d}t \\ &= (-1)^{\mathrm{m}}\delta_{\mathrm{mm}'}\mathcal{J}(\omega) \qquad (4.36)\end{aligned}$$

where $\mathcal{J}(\omega)$ is a spectral density bearing the same properties as that defined by (4.35).

In order to express a cartesian tensor in terms of irreducible tensors, we can resort to spherical harmonics of rank 2 which themselves belong to the class of irreducible tensors of rank 2. It may be recalled that these functions constitute the angular part of the familiar d atomic orbitals, and are expressed as a function of spherical coordinates (Figure 4.23):

$$\begin{aligned}\mathrm{Y}_2^{(\pm 2)} &= \sqrt{15/8\pi}(1/2)\sin^2\theta\exp(\pm 2i\varphi) \\ \mathrm{Y}_2^{(\pm 1)} &= \mp\sqrt{15/8\pi}\sin\theta\cos\theta\exp(\pm i\varphi) \qquad (4.37) \\ \mathrm{Y}_2^{(0)} &= \sqrt{15/8\pi}(\sqrt{3/2})(\cos^2\theta - 1/3)\end{aligned}$$

We can notice that $\mathrm{Y}_2^{(0)}$ involves the quantity $\cos^2\theta$, which is proportional to Z^2, whereas T_{ZZ} behaves (or transforms) also as Z^2. Hence we are allowed to write

$$\sqrt{15/8\pi}\sqrt{3/2}\cos^2\theta = \mathrm{k}T_{\mathrm{ZZ}}$$

where the factor k accounts for the specificity of the considered cartesian tensor. Likewise, with the same factor k

$$\sqrt{15/8\pi}\sqrt{3/2}\sin^2\theta\cos^2\varphi = \mathrm{k}T_{\mathrm{XX}}$$

$$\sqrt{15/8\pi}\sqrt{3/2}\sin^2\theta\sin^2\varphi = \mathrm{k}T_{\mathrm{YY}}$$

Hence

$$\sqrt{15/8\pi}\sqrt{3/2} = 3\mathrm{k}T_0$$

where T_0 is a third of the tensor trace: $T_0 = (T_{\mathrm{XX}} + T_{\mathrm{YY}} + T_{\mathrm{ZZ}})/3$. It can be recalled

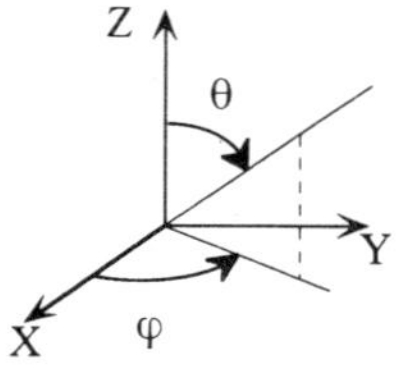

Figure 4.23 Polar angles of the spherical coordinates

that this invariant governs NMR spectra in isotropic media (see Section 1.4.3). Now, because $F^{(0)}$, the irreducible associated with the considered cartesian tensor, is homologous to $Y_2^{(0)}$, we can establish the following correspondence

$$F^{(0)} = T_{ZZ} - T_0 \tag{4.38}$$

The specific coefficient of the considered tensor (in other words, its magnitude) is of course included in $F^{(0)}$. Employing similar geometrical considerations and taking into account the coefficients of the spherical harmonics of rank 2, we are able to establish the complementary relationships:

$$\begin{aligned}
F^{(-1)} - F^{(1)} &= \sqrt{8/3}\,T_{XZ} \\
F^{(-1)} + F^{(1)} &= -i\sqrt{8/3}\,T_{YZ} \\
F^{(-2)} - F^{(2)} &= -i\sqrt{8/3}\,T_{XY} \\
F^{(-2)} + F^{(2)} &= \sqrt{2/3}(T_{XX} - T_{YY})
\end{aligned} \tag{4.39}$$

From the above expressions, we are now in the position to express any element of a cartesian tensor as a function of the associated irreducible tensors of rank 2:

$$\begin{aligned}
T_{XX} &= T_0 - (1/2)F^{(0)} + \sqrt{3/8}(F^{(2)} + F^{(-2)}) \\
T_{YY} &= T_0 - (1/2)F^{(0)} - \sqrt{3/8}(F^{(2)} + F^{(-2)}) \\
T_{ZZ} &= T_0 + F^{(0)} \\
T_{XY} &= i\sqrt{3/8}(F^{(-2)} - F^{(2)}) \\
T_{XZ} &= \sqrt{3/8}(F^{(-1)} - F^{(1)}) \\
T_{YZ} &= i\sqrt{3/8}(F^{(1)} + F^{(-1)}).
\end{aligned} \tag{4.40}$$

These expressions, combined with (4.36), provide in a straightforward manner the Fourier transforms of the correlation function of all cartesian tensor elements (it may be recalled that the Fourier transform of a constant is zero).

$$\begin{aligned}
\int_0^\infty \overline{T_{XX}(t)T_{XX}(0)}\exp(-i\omega t)\,dt &= \int_0^\infty \overline{T_{YY}(t)T_{YY}(0)}\exp(-i\omega t)\,dt \\
&= \int_0^\infty \overline{T_{ZZ}(t)T_{ZZ}(0)}\exp(-i\omega t)\,dt = \mathcal{J}(\omega).
\end{aligned} \tag{4.41a}$$

$$\begin{aligned}
\int_0^\infty \overline{T_{XY}(t)T_{XY}(0)}\exp(-i\omega t)\,dt &= \int_0^\infty \overline{T_{XZ}(t)T_{XZ}(0)}\exp(-i\omega t)\,dt \\
&= \int_0^\infty \overline{T_{YZ}(t)T_{YZ}(0)}\exp(-i\omega t)\,dt = (3/4)\mathcal{J}(\omega).
\end{aligned} \tag{4.41b}$$

The next step is to express these tensorial quantities in a molecular frame, as usable information (either dynamic or structural) can be extracted only from quantities expressed in that frame. We shall limit the following discussion to an isotropic medium and (i) to an *axially symmetric tensor* in the case of an arbitrary reorientational motion, (ii) to an *arbitrary tensor* (without any particular symmetry) if the *motion is isotropic* (any direction in the molecule undergoes the same type of rotational motion).

Let x, y, z be the molecular frame in which the considered tensor is diagonal (such a frame is usually called the principal axis system or PAS), the axes being labeled so that $|T_{zz}| \geqslant |T_{yy}| \geqslant |T_{xx}|$. In the following, the tensor will be characterized by $T_z \equiv T_{zz}$, $T_0 = (T_{xx} + T_{yy} + T_{zz})/3$ and the asymmetry parameter $\eta = (T_{xx} - T_{yy})/T_{zz}$ (η is of course zero in the case of an axially symmetric tensor). In order to proceed further, we must switch to irreducible tensors, which will be denoted $f^{(m)}$ in the molecular frame and which can be constructed along the same lines as the irreducible tensor $F^{(m)}$ in the laboratory frame (see (4.38) and (4.39)). Because the cartesian tensor is here diagonal, we are left with

$$f^{(0)} = T_z - T_0$$
$$f^{(1)} = f^{(-1)} = 0$$
$$f^{(2)} = f^{(-2)} = (1/\sqrt{6})\eta T_z$$

Since, at the outset, we have to calculate

$$\mathcal{J}(\omega) = \int_0^{\infty} \overline{F^{(0)}(t)F^{(0)}(0)} \exp(-i\omega t)\,\mathrm{d}t$$

$F^{(0)}$ needs to be expressed according to $f^{(0)}$, $f^{(2)}$ and $f^{(-2)}$. As mentioned above, the interest of irreducible tensors lie, among other things, in the ease with which they can be expressed in one or another frame. $F^{(0)}$ is deduced from $f^{(0)}$, $f^{(2)}$ and $f^{(-2)}$ through appropriate elements of Wigner matrices which here reduce to spherical harmonics:

$$F^{(0)} = \sqrt{4\pi/5}[Y_2^{(0)}(\beta)f^{(0)} + Y_2^{(2)}(\alpha, \beta)f^{(2)} + Y_2^{(-2)}(\alpha, \beta)f^{(-2)}].$$

where α and β are the angles which define the molecular frame with respect to the laboratory frame, β being the angle between the two z and Z axes (Z, as usual, coinciding with B_0). We should also be concerned by the angle α; however, it turns out that: (i) when dealing with an axially symmetric tensor $f^{(2)}$ and $f^{(-2)}$ are obviously zero, and (ii) when dealing with an isotropic motion the Fourier transforms involving $Y_2^{(2)}$ or $Y_2^{(-2)}$ are equal to the Fourier transform involving $Y_2^{(0)}$. Hence, in the two limiting cases considered in this chapter, the angle α does not matter.

The above discussion leads to the extension of (4.35) when the relaxation mechanism can be specified in terms of a cartesian tensor defined in the *molecular frame* of its principal axis (where it is diagonal) by T_z (the T_{zz} element), the third of its trace $T_0 = (T_{xx} + T_{yy} + T_{zz})/3$ (an invariant which eventually governs line positions in the NMR spectrum) and the asymmetry parameter $\eta = (T_{xx} - T_{yy})/T_{zz}$ (with $|T_{zz}| \geqslant |T_{yy}| \geqslant |T_{xx}|$; of course this condition does not need to be satisfied in the case of an axially symmetrical tensor for which $\eta = 0$). We arrive at

$$\mathcal{J}(\omega) = (1/10)[(T_z - T_0)^2 + \eta^2 T_z^2/3]\tilde{J}(\omega) \quad (4.42a)$$

for an isotropic reorientation, and at

$$\mathcal{J}(\omega) = (1/10)(T_z - T_0)^2\tilde{J}(\omega) \quad (4.42b)$$

for an axially symmetric tensor whatever the type of reorientation. $\tilde{J}(\omega)$ is called *reduced spectral density* and has the form

$$\tilde{J}(\omega) = 4\pi\int_{-\infty}^{+\infty} \overline{Y_2^{(0)}[\beta(t)]Y_2^{(0)}[\beta(0)]} \exp(-i\omega t)\,\mathrm{d}t \quad (4.43)$$

where β is the angle between the z and the Z axis (Z coinciding with the B_0 direction) and $Y_2^{(0)}$ is the spherical harmonics of the form

$$Y_2^{(0)}(\beta) = \sqrt{5/16\pi}(3\cos^2\beta - 1)$$

Whenever the correlation function involved in (4.43) obeys a simple exponential law defined by a characteristic time τ_c [see (4.34)], the real part of the reduced spectral density exhibits the simple form

$$\tilde{J}(\omega) = \frac{2\tau_c}{1 + \omega^2\tau_c^2} \tag{4.44}$$

Variables are seen to be separated in formulae (4.42), the dynamic features being accounted for in the reduced spectral density, whereas the significance of a given relaxation mechanism arises from the magnitude of the relevant interaction (represented by T_z, T_0 or η). Since, as will be shown shortly, relaxation rates (inverse of relaxation times, $R_1 = 1/T_1$, $R_2 = 1/T_2$) can be expressed as linear combinations of spectral densities, contributions from different mechanisms (m) have simply to be added in the final result:

$$R_1 = \sum_{\mathrm{m}}(R_1)_{\mathrm{m}}; \quad R_2 = \sum_{\mathrm{m}}(R_2)_{\mathrm{m}}$$

Because this will be useful, an important result [formulae (4.41)] should be mentioned to the reader who could have skipped the above theoretical developments: when expressed in the laboratory frame, the Fourier transform of the correlation function of diagonal elements of the cartesian tensor is equal to $\mathcal{J}(\omega)$, while it reduces to (3/4) $\mathcal{J}(\omega)$ for off-diagonal elements.

4.2.2 PHENOMENOLOGICAL APPROACH OF RELAXATION PHENOMENA. RANDOM FIELDS, CHEMICAL SHIFT ANISOTROPY (CSA) AND SCALAR RELAXATION OF THE SECOND KIND

As in the previous chapters, we will avoid a quantum mechanical treatment if the system can be described by the three magnetization components M_x, M_y, and M_z. This is valid for a single spin 1/2 nucleus or if we disregard interactions involving explicitly two spins (e.g. the dipolar interaction).

Random fields

This is a mechanism for which the nature of the interactions has not to be specified. It stems from the assumption that each elementary magnetic momentum $\boldsymbol{\mu}$ is subjected to a magnetic field $\mathbf{b}(t)$ randomly varying in time whose components b_x, b_y and b_z are not correlated, although they evolve in a similar fashion:

$$\begin{gathered}\overline{b_x(t)b_y(t)} = \overline{b_x(t)b_z(t)} = \overline{b_y(t)b_z(t)} = 0\\ \overline{b_x^2(t)} = \overline{b_y^2(t)} = \overline{b_z^2(t)} = \overline{b^2}.\end{gathered} \tag{4.45}$$

The following interactions can be included in this mechanism:

(i) intermolecular dipolar interactions when they concern two different spin systems;

(ii) spin–spin interactions (direct or indirect) between the considered nuclear spin and the spin of an unpaired electron (as occurs where paramagnetic species are present in the sample). This important topic will not be treated here and the reader is referred to specialized textbooks. From a practical point of view, it can be mentioned that this mechanism may be extremely efficient and prone to mask the other contributions. As a matter of fact, any paramagnetic impurity should be removed from the sample whenever meaningful relaxation parameters are desired. This is especially the case for the dissolved oxygen. Oxygen must be taken away by degassing procedures (freeze–pump–thaw techniques) or by argon or nitrogen substitution;

(iii) the so-called spin rotation mechanism, which arises from the coupling between two angular momenta: that of the considered nuclear spin, and the one describing the rotation of the molecule. This mechanism is generally unimportant except for small molecules or for rapidly rotating molecular moieties (e.g. a methyl grouping). It can easily be identified because, in contrast to all other mechanisms, it manifests itself by an increase of relaxation rates with temperature.

In order to study the evolution of nuclear magnetization whenever the nuclear spins are subjected to random fields, we shall proceed in the frame (x', y', z) rotating at the angular velocity $\omega_0 = \gamma B_0$ (see Section 2.1.3) which has the virtue of removing any precession effect. Consequently, we are left with the sole random fields in the equation of motion pertaining to an elementary magnetic momentum

$$\frac{d}{dt}\boldsymbol{\mu} = \gamma\boldsymbol{\mu} \wedge \mathbf{b}$$

which can be explicited for each component

$$\begin{aligned} \frac{d}{dt}\mu_{x'} &= \gamma(\mu_{y'}b_z - \mu_z b_{y'}) \\ \frac{d}{dt}\mu_{y'} &= \gamma(\mu_z b_{x'} - \mu_{x'}b_z) \\ \frac{d}{dt}\mu_z &= \gamma(\mu_{x'}b_{y'} - \mu_{y'}b_{x'}) \end{aligned} \tag{4.46}$$

with

$$\begin{aligned} b_{x'} &= b_x \cos(\omega_0 t) - b_y \sin(\omega_0 t) \\ b_{y'} &= b_x \sin(\omega_0 t) + b_y \cos(\omega_0 t). \end{aligned} \tag{4.47}$$

First, we shall focus on the temporal evolution of μ_z, recognizing that it yields the z component of nuclear magnetization through an ensemble average, $M_z(t) = \overline{\mu_z(t)}$,

$$\mu_z(t) = \mu_z(0) + \int_0^t \frac{d\mu_z}{dt'}dt'$$

or, inserting the expansion of $d\mu_z/dt'$ given in (4.46),

$$\mu_z(t) = \mu_z(0) + \gamma\int_0^t [\mu_{x'}(t')b_{y'}(t') - \mu_{y'}(t')b_{x'}(t')]\,dt' \tag{4.48}$$

$\mu_{x'}(t')$ and $\mu_{y'}(t')$ can be expressed accordingly

$$\mu_{x'}(t') = \mu_{x'}(0) + \int_0^{t'} \frac{d\mu_{x'}}{dt''} dt''$$

$$\mu_{y'}(t') = \mu_{y'}(0) + \int_0^{t'} \frac{d\mu_{y'}}{dt''} dt''$$

We shall now make the assumptions that: (i) t is relatively short and (ii) $\boldsymbol{\mu}$ evolves more slowly than $\mathbf{b}$. With these assumptions and owing to the fact that $t'' \leqslant t' \leqslant t$, we can approximate $\mu_{x'}(t'')$, $\mu_{y'}(t'')$ and $\mu_z(t'')$ by $\mu_{x'}(0)$, $\mu_{y'}(0)$ and $\mu_z(0)$ respectively, hence:

$$\mu_{x'}(t') = \mu_{x'}(0) + \gamma \int_0^{t'} [\mu_{y'}(0) b_z(t'') - \mu_z(0) b_{y'}(t'')] \, dt''$$

$$\mu_{y'}(t') = \mu_{y'}(0) + \gamma \int_0^{t'} [\mu_z(0) b_{x'}(t'') - \mu_{x'}(0) b_z(t'')] \, dt''$$

These latter expressions can be inserted into (4.48) and further subjected to an ensemble average for obtaining $M_z(t)$

$$M_z(t) = M_z(0) + \gamma^2 \int_0^t \int_0^{t'} [\overline{\mu_{y'}(0) b_z(t'') b_{y'}(t')} - \overline{\mu_z(0) b_{y'}(t'') b_{y'}(t')}] \, dt' \, dt''$$

$$- \gamma^2 \int_0^t \int_0^{t'} [\overline{\mu_z(0) b_{x'}(t'') b_{x'}(t')} - \overline{\mu_{x'}(0) b_z(t'') b_{x'}(t')}] \, dt' \, dt''$$

It is quite obvious that the elementary magnetic momentum at time zero does not depend on random fields which will exist at times t' and t''. Moreover, two distinct components of the random field are not correlated (see (4.45)). Therefore, the first term involved in the above equation can be decomposed according to

$$\overline{\mu_{y'}(0) b_z(t'') b_{y'}(t')} = \overline{\mu_{y'}(0)} \; \overline{b_z(t'') b_{y'}(t')} = M_{y'}(0) \overline{b_z(t'') b_{y'}(t')} = 0$$

Similar decompositions can be dealt with for other terms. Two of them yield a non-zero result which gives

$$M_z(t) = M_z(0) - \gamma^2 M_z(0) \int_0^t f(t') \, dt'$$

with

$$f(t') = \int_0^{t'} [\overline{b_{y'}(t'') b_{y'}(t')} + \overline{b_{x'}(t'') b_{x'}(t')}] \, dt''. \tag{4.49}$$

This should allow us to retrieve the Bloch equation pertaining to longitudinal magnetization. To this end, let us differentiate $M_z(t)$

$$\frac{dM_z}{dt} = -\gamma^2 M_z(0) \frac{d}{dt} \int_0^t f(t') \, dt'$$

and let us denote F the primitive of $f (f = dF/dt)$, so as to acknowledge the fact that

$$\frac{d}{dt} \int_0^t f(t') \, dt' = \frac{d}{dt} [F(t) - F(0)] = \frac{dF(t)}{dt} = f(t)$$

Thus dM_z/dt can be written as

$$\frac{dM_z}{dt} = -\gamma^2 M_z(0) \int_0^t [\overline{b_{y'}(t'') b_{y'}(t)} + \overline{b_{x'}(t'') b_{x'}(t)}] \, dt''$$

Replacing the variable t'' by $\tau = t' - t''$ and assuming that a correlation function depends solely upon the difference of the two involved instants [that is, $\overline{b_{y'}(t)b_{y'}(t'')} = \overline{b_{y'}(t - t'')b_{y'}(0)}$], we arrive at

$$\frac{\mathrm{d}M_z}{\mathrm{d}t} = -\gamma^2 M_z(0)\int_0^t [\overline{b_{y'}(\tau)b_{y'}(0)} + \overline{b_{x'}(\tau)b_{x'}(0)}]\,\mathrm{d}\tau \tag{4.50}$$

At this stage, it is convenient to return to the laboratory frame through equations (4.47). Noticing further that $b_{x'}(0) = b_x(0)$ and $b_{y'}(0) = b_y(0)$ and using again the fact that b_x and b_y are correlated, we are able to express $\mathrm{d}M_z/\mathrm{d}t$ as:

$$\frac{\mathrm{d}M_z}{\mathrm{d}t} = -\gamma^2 M_z(0)\int_0^t [\overline{b_x(\tau)b_x(0)} + \overline{b_y(\tau)b_y(0)}]\cos(\omega_0\tau)\,\mathrm{d}\tau.$$

We must also recognize that the actual evolution of M_z stems from random fields which afford longitudinal magnetization to recover the equilibrium value. Hence, we must rather consider the deviation of M_z with respect to its equilibrium value, that is $M_z - M_0$. On the other hand, t has been chosen so that nuclear magnetization changes slowly, whereas random fields may evolve on a faster time scale. Hence, in the above equation, $M_z(0)$ can reasonably be substituted by $[M_z(t) - M_0]$:

$$\frac{\mathrm{d}M_z}{\mathrm{d}t} = -\gamma^2(M_z - M_0)\int_0^t [\overline{b_x(\tau)b_x(0)} + \overline{b_y(\tau)b_y(0)}]\cos(\omega_0\tau)\,\mathrm{d}\tau.$$

Moreover, because correlation functions decrease rapidly, the upper limit in the above integral can be replaced by infinity. Finally, for the sake of recasting the above expression with the usual notation of Fourier transformation, we shall write it in the form

$$\frac{\mathrm{d}M_z}{\mathrm{d}t} = -(M_z - M_0)[\gamma^2\,\mathrm{Re}\int_0^\infty [\overline{b_x(t)b_x(0)} + \overline{b_y(t)b_y(0)}]\,\mathrm{e}^{-i\omega_0 t}\,\mathrm{d}t] \tag{4.51a}$$

where Re stands for 'real part of'. This expression featuring the *spectral densities* defined by (4.33) is just the familiar Bloch equation

$$\frac{\mathrm{d}M_z}{\mathrm{d}t} = -\left(\frac{1}{T_1}\right)_{\mathrm{rf}}(M_z - M_0)$$

with (4.51b)

$$\left(\frac{1}{T_1}\right)_{\mathrm{rf}} = (R_1)_{\mathrm{rf}} = \gamma^2\,\mathrm{Re}\,[2\mathcal{J}_{\mathrm{rf}}(\omega_0)].$$

with the subscript r.f. meaning 'random fields'.

A similar approach can be applied to the transverse components of nuclear magnetization $M_{x'}$ and $M_{y'}$ (in the rotating frame). The relevant calculations involve such terms as $\int_0^t\int_0^{t'} \overline{b_{y'}(t'')b_{y'}(t')}\,\mathrm{d}t''\,\mathrm{d}t'$ which lead to $\mathcal{J}_{\mathrm{rf}}(\omega_0)$ and $\int_0^t\int_0^{t'} \overline{b_z(t'')b_z(td')}\,\mathrm{d}t''\,\mathrm{d}t'$, which evidently does not require to switch to the laboratory frame, and which leads to a spectral density at zero frequency. Carrying out all the calculations yields again the Bloch equations

$$\frac{\mathrm{d}M_{x',y'}}{\mathrm{d}t} = -\left(\frac{1}{T_2}\right)_{\mathrm{rf}} M_{x'y'}$$

with (4.52)

$$\left(\frac{1}{T_2}\right)_{\mathrm{rf}} = (R_2)_{\mathrm{rf}} = \gamma^2\,\mathrm{Re}\,[\mathcal{J}_{\mathrm{rf}}(\omega_0) + \mathcal{J}_{\mathrm{rf}}(0)].$$

Because $\mathcal{J}(0) \geqslant \mathcal{J}(\omega_0)$ [see for instance (4.44)], $(R_2)_{rf} \geqslant (R_1)_{rf}$. This leads to the important property that generally *the transverse relaxation time is shorter than the longitudinal relaxation time*.

Chemical shift anisotropy (CSA)

This relaxation mechanism originates from the tensorial nature of the shielding effect, and can be treated through the concept of random fields by noticing that the effective induction at a given nucleus can be expressed in the laboratory frame as (see Section 1.4.1):

$$b_X = \sigma_{XZ} B_0$$
$$b_Y = \sigma_{YZ} B_0$$
$$b_Z = (1 - \sigma_{ZZ}) B_0$$

Switching to a molecular frame has the consequence that one must account for the molecular reorientation, which entails a time modulation of the tensor elements σ_{XZ}, σ_{YZ} and σ_{ZZ}. Therefore, this mechanism should be active in respect of nuclear spin relaxation. Using the random field formalism, we can write from (4.51) that $(R_1)_{csa}$ is the sum of the Fourier transforms of the correlation functions pertaining to γb_x and γb_y or, in other words, of correlation functions of the tensor off-diagonal elements (namely $\gamma\sigma_{XZ} B_0$ and $\gamma\sigma_{YZ} B_0$). From (4.41b), which tells us that the relevant spectral density is $(3/4)\mathcal{J}(\omega)$, we are led to

$$(R_1)_{csa} = \left(\frac{1}{T_1}\right)_{csa} = (3/2)\gamma^2 B_0^2 \mathcal{J}_{csa}(\omega_0) \tag{4.53a}$$

and by a similar approach

$$(R_2)_{csa} = \left(\frac{1}{T_2}\right)_{csa} = \gamma^2 B_0^2[(3/4)\mathcal{J}_{csa}(\omega_0) + \mathcal{J}_{csa}(0)] \tag{4.53b}$$

The next step is to express the spectral density $\mathcal{J}_{csa}(\omega)$ as a function of the shielding tensor in its principal *molecular* axis system (where it is diagonal). In that molecular frame, we shall define the following quantity according to the notations commonly employed:

$$\begin{aligned} \Delta\sigma &= \sigma_{zz} - (\sigma_{xx} + \sigma_{yy})/2 \\ \eta_{csa} &= (3/2)(\sigma_{xx} - \sigma_{yy})/\Delta\sigma \end{aligned} \tag{4.54}$$

(with $|\sigma_{zz}| \geqslant |\sigma_{yy}| \geqslant |\sigma_{xx}|$).

We thus obtain for an isotropic reorientation

$$(R_1)_{csa} = \frac{1}{15}(\gamma B_0)^2(\Delta\sigma)^2\left(1 + \frac{\eta_{csa}^2}{3}\right)\tilde{J}(\omega_0) \tag{4.55a}$$

$$(R_2)_{csa} = \frac{1}{15}(\gamma B_0)^2(\Delta\sigma)^2\left(1 + \frac{\eta_{csa}^2}{3}\right)\left[\frac{2\tilde{J}(0)}{3} + \frac{\tilde{J}(\omega_0)}{2}\right] \tag{4.55b}$$

If $\eta = 0$ (there, the condition $(\sigma_{zz}) \geqslant (\sigma_{xx})$ does not need to be verified), the following formulae apply also in the case of an anisotropic reorientation:

$$(R_1)_{csa} = \frac{1}{15}(\gamma B_0)^2(\Delta\sigma)^2\tilde{J}(\omega_0) \tag{4.56a}$$

$$(R_2)_{csa} = \frac{1}{15}(\gamma B_0)^2(\Delta\sigma)^2\left[\frac{2\tilde{J}(0)}{3} + \frac{\tilde{J}(\omega_0)}{2}\right] \tag{4.56b}$$

It can be seeen that, in all cases, relaxation rates are directly proportional to $(\Delta\sigma)^2$. Because $\Delta\sigma$ reflects the anisotropy of the shielding tensor and because chemical shifts originate from the shielding effect, the terminology 'chemical shift anisotropy' is generally used for denoting this relaxation mechanism. It can be further noted that these relaxation rates depend in two ways upon the measurement frequency: via $\tilde{J}(\omega_0)$, but also via the factor $(\gamma B_0)^2 = \omega_0^2$. Moreover, for an exponential correlation function characterized by a single correlation time τ_c, $\tilde{J}(\omega)$ is expressed as $2\tau_c/(1 + \omega^2\tau_c^2)$ [see (4.44)] and reduces to $2\tau_c$ within *extreme narrowing* conditions ($\omega^2\tau_c^2 \ll 1$). Even in that case, the two relaxation rates R_1 and R_2 are different; $(R_2)_{csa}/(R_1)_{csa} = 7/6$.

With the aim of providing illustrative figures, we shall consider two examples for which this relaxation mechanism may be important because of a large chemical shift anisotropy value: the carbon-13 of a carbonyl group and the phosphorus-31 of a phosphate group. For these two examples, we shall add a realistic dipolar contribution arising from a remote proton and assume that the molecular motion is isotropic [that is, $\tilde{J}(\omega) = 2\tau_c/(1 + \omega^2\tau_c^2)$]. Numerical values of longitudinal relaxation times are gathered in Table 4.4 for different values of B_0 (these values being identified by the corresponding proton resonance frequency).

Table 4.4 Longitudinal relaxation times (in s) of the carbon-13 of a carbonyl group and of the phosphorus-31 of a phosphate group for different B_0 values indicated by the corresponding proton resonance frequency (i.e. $B_0 = 4.7$ T for $\nu(^1H) = 200$ MHz)

	$\nu(^1H)$		
	200 MHz	400 MHz	600 MHz
C=O; $\Delta\sigma = 250$ ppm			
$d_{CH} = 1.85$ Å			
$\tau_c = 5 \times 10^{-9}$ s	0.61	0.61	0.60
$\tau_c = 5 \times 10^{-11}$ s	11.6	4.75	2.4
P=O; $\Delta\sigma = 150$ ppm			
$d_{PH} = 1.85$ Å			
$\tau_c = 5 \times 10^{-9}$ s	0.87	1.33	1.5
$\tau_c = 5 \times 10^{-11}$ s	6.5	3.7	2.2

Scalar relaxation of the second kind

This mechanism operates whenever a spin A possessing a long intrinsic relaxation time is *J*-coupled with a spin X whose relaxation is much faster (typically A is a spin 1/2 nucleus whereas X is a quadrupolar nucleus). The relevant interaction between the two spins is familiar and can be expressed as

$$J_{AX}(I_x^A I_x^X + I_y^A I_y^X + I_z^A I_z^X)$$

It must be realized that the time-dependent quantities are here I_x^X, I_y^X and I_z^X which evolve on a short time scale (T_2^X for the transverse components and T_1^X for the longitudinal components). Formally, the above interaction can be treated as random fields provided that we identify $J_{AX}I_x^X$, $J_{AX}I_y^X$ and $J_{AX}I_z^X$ with γb_x, γb_y and γb_z respectively. Consequently, J_{AX} must be expressed in rad s^{-1} (since I_x^X, I_y^X and I_z^X are dimensionless), and the following correlation functions have to be evaluated:

$$\overline{I_x^X(t)I_x^X(0)},\ \overline{I_y^X(t)I_y^X(0)},\ \overline{I_y^X(t)I_z^X(0)}$$

We shall first proceed with the correlation function involving I_z^X, whose time evolution arises from the X longitudinal relaxation time and can thus be expressed as $I_z^X(t) = I_z^X(0)\exp(-t/T_1^X)$. Because we are dealing with operators, we can rely on the following relationships

$$\overline{I_x^X(0)^2} = \overline{I_y^X(0)^2} = \overline{I_z^X(0)^2} = \frac{1}{3}(I^X)^2 = \frac{I_X(I_X + 1)}{3}$$

where I_X is the spin number of nucleus X. We can therefore substitute $\gamma\int_0^\infty \overline{b_z(t)b_z(0)}\,dt$, which is involved in the calculation of $(R_2)_{rf}$, by

$$J_{AX}^2\int_0^\infty \overline{I_z^X(t)I_z^X(0)}\,dt = (J_{AX}^2/3)I_X(I_X + 1)T_1^X.$$

The evolution of the two other correlation functions, those involving I_x^X and I_y^X requires that the precession be taken into account. The calculation is better carried out with the help of raising and lowering operators $I_+^X = I_x^X + iI_y^X$ and $I_-^X = I_x^X - iI_y^X$.

It can be recognized that by expanding $\overline{I_x^X(t)I_x^X(0) + I_y^X(t)I_y^X(0)}$ we are left with $\overline{I_+^X(t)I_-^X(0)}$ because, for symmetry reasons, $\overline{I_+(t)I_-(0)} = \overline{I_-(t)I_+(0)}$. The evolution of I_+^X depends also on precession, hence

$$I_+^X(t) = I_+^X(0)\exp(-t/T_2^X)\exp(i\omega_X t)$$

where $\omega_X/2\pi$ is the precession frequency of nucleus X. The spectral density relevant to transverse components can thus be expressed as

$$\frac{J_{AX}^2}{2}\int_0^\infty \overline{I_+^X(t)I_-^X(0)}\,e^{-i\omega_A t}\,dt = \frac{J_{AX}^2}{2}\int_0^\infty \overline{I_+^X I_-^X}\,e^{-t/T_2^X}\,e^{-i(\omega_A-\omega_X)t}\,dt$$
$$= \frac{J_{AX}^2}{3}I_X(I_X + 1)\frac{T_2^X}{1 + (\omega_A - \omega_X)^2(T_2^X)^2}.$$

By reference to (4.51), ω_0 has been changed into ω_A, because A is the observed nucleus. Finally, scalar relaxation of the second kind (sc) contributes so the transverse and longitudinal relaxation rates as indicated below. The expressions of $(R_1^A)_{SC}$ and $(R_2^A)_{SC}$ follow from the insertion of the above results into (4.51b) and (4.52)

$$
\begin{aligned}
(R_1^{\mathrm{A}})_{\mathrm{SC}} &= \frac{2J_{\mathrm{AX}}^2}{3} I_{\mathrm{X}}(I_{\mathrm{X}}+1) \frac{T_2^{\mathrm{X}}}{1+(\omega_{\mathrm{X}}-\omega_{\mathrm{A}})^2 (T_2^{\mathrm{X}})^2} \\
(R_2^{\mathrm{A}})_{\mathrm{SC}} &= \frac{J_{\mathrm{AX}}^2}{3} I_{\mathrm{X}}(I_{\mathrm{X}}+1) \left[\frac{T_2^{\mathrm{X}}}{1+(\omega_{\mathrm{X}}-\omega_{\mathrm{A}})^2 (T_2^{\mathrm{X}})^2} + T_1^{\mathrm{X}} \right]
\end{aligned}
\tag{4.57}
$$

(where J_{AX} has to be expressed in rad s^{-1}).

Because of the term $(\omega_{\mathrm{X}} - \omega_{\mathrm{A}})^2(T_2^{\mathrm{X}})^2$ and because T_2^{X} cannot lie in the submicrosecond range, $(R_1^{\mathrm{A}})_{\mathrm{sc}}$ is generally negligible except when A and X resonance frequencies are close to each other. Hence, only the transverse relaxation rate of the observed nucleus (A) is usually concerned in scalar relaxation of the second kind. The J_{AX} coupling cannot be directly determined owing to the coalescence of the relevant multiplet by a phenomenon analogous to chemical exchange. However, if T_1^{X} is independently determined, the measurement of $(R_2^{\mathrm{A}}) \approx T_1^{\mathrm{X}} I_{\mathrm{X}}(I_{\mathrm{X}}+1) J_{\mathrm{AX}}^3/3$ (if scalar relaxation is overwhelming) does yield the J_{AX} coupling. Conversely, the knowledge of J_{AX} enables one to determine indirectly the (usually very short) longitudinal relaxation time of the quadrupolar nucleus, whose direct measurement would be very difficult. As examples of such determinations, we can mention deuterium (after labeling) or nitrogen-14 (both nuclei are of spin 1) with many applications in the field of local mobility in biomolecules.

4.2.3 QUANTUM MECHANICAL APPROACH OF RELAXATION PHENOMENA: DIPOLAR AND QUADRUPOLAR INTERACTIONS

As soon as the spin system cannot be treated by the vectorial model (such a situation has been encountered in Chapter 2 for some states of the system of two J-coupled nuclei), we must turn to quantum mechanics. As already mentioned in Chapter 2, the appropriate tool is the density operator, whose evolution equation is relatively easy to handle and which enables one to evaluate the time evolution of any quantity.

For example, we can recognize that the quadrupolar interaction, for a nucleus of spin greater than 1/2, implies *more than two eigenstates*. This is akin to a system involving several spins 1/2 (for instance, a two-spin-1/2 system possesses four eigenstates). Hence, it is not surprising that a quadrupolar nucleus cannot be treated according to a classical vectorial model, and should also be considered with the help of the density operator.

We will not give here a detailed account of the quantum mechanical treatment; rather, we shall indicate the major steps of the approach and suggest the way in which it could be carried out. First, let us recall (Chapter 2) the evolution of the density operator $\hat{\sigma}$:

$$
\frac{\mathrm{d}\hat{\sigma}}{\mathrm{d}t} = 2i\pi[\hat{\sigma}, \hat{\mathscr{H}}(t)]
$$

where $\hat{\mathscr{H}}(t)$, expressed in Hz, includes:

- the static hamiltonian $\hat{\mathscr{H}}_0$ which governs the intensity and position of lines in the NMR spectrum and which, in a more general way, governs precession phenomena;
- the hamiltonian $\hat{\mathscr{H}}_1(t)$, responsible for relaxation phenomena, and which accounts for all time-dependent interactions. In any instance, it can be cast in the form

$$\mathcal{H}_1(t) = \sum_{\mathrm{r}} \sum_{m=-2}^{2} (\hat{A}_{\mathrm{r}}^{m})^{\dagger} F_{\mathrm{r}}^{m}(t). \tag{4.58}$$

where r refers to a given relaxation mechanism (interaction) and where the symbol † means 'transpose complex conjugate'. Both $\hat{A}_{\mathrm{r}}^{m}$ and F_{r}^{m} are irreducible tensors. However, $\hat{A}_{\mathrm{r}}^{m}$ is only made of spin operators whereas F_{r}^{m} depends only on molecular properties. As irreducible tensors, they possess the following properties:

$$\begin{aligned}(\hat{A}_{\mathrm{r}}^{m})^{\dagger} &= (-1)^{m}\hat{A}_{\mathrm{r}}^{-m} \\ F_{\mathrm{r}}^{-m} &= (-1)^{m}(F_{\mathrm{r}}^{m})^{*}\end{aligned} \tag{4.59}$$

It turns out that, for the dipolar interaction and for quadrupolar interaction involving an electric field gradient tensor for axial symmetry, the molecular functions F_{r}^{m} are proportional to spherical harmonics of rank 2 [see (4.37)]:

$$F_{\mathrm{r}}^{m} = \mathrm{K}_{\mathrm{D,Q}}\mathrm{Y}_{2}^{m}(\theta, \varphi) \tag{4.60}$$

$\mathrm{K_D} = -(\mu_0/4\pi)\sqrt{6/5\pi}\hbar\gamma_{\mathrm{A}}\gamma_{\mathrm{X}}/r_{\mathrm{AX}}^{3}$ is the coefficient appropriate for the dipolar interaction, r_{AX} being the internuclear distance, μ_0 the vacuum permeability and the other symbols having their usual meaning. $\mathrm{K_Q} = \sqrt{6/5\pi}\,\mathrm{e}QV_{zz}/[4\hbar I(2I-1)]$ is the coefficient appropriate for the quadrupolar interaction; eQ stands for the nuclear quadrupole moment (invariant for a given nucleus); V_{zz} is the element of the electric field gradient tensor pertaining to its *molecular* symmetry axis z; I is the spin number. These relationships arise from the expressions of the relevant interactions given in Section 1.4.2 to which are added the non-secular terms. θ and φ are the polar angles which define the orientation of a molecular direction with respect to the laboratory frame: this molecular direction is r_{AX} for the dipolar interaction and z (the tensor symmetry axis) for the quadrupolar interaction. The operators $\hat{A}_{\mathrm{r}}^{m}$, given in Table 4.5 present some similarities for these two mechanisms.

Using methods of the same kind as those developed in the previous section, but within the frame of a quantum mechanical formalism, it is possible to derive the evolution equation for any quantity that will be denoted by $\langle M \rangle$. Here $\langle M \rangle$ can stand for the expectation value of a magnetization component or for the expectation value of an operator product associated with a particular state of the spin system (e.g. an antiphase doublet; see Chapter 2). Relying on the properties of the density operator (Section 2.2.1) one could arrive at

$$\frac{\mathrm{d}}{\mathrm{d}t}\langle M \rangle = -iTr\{\hat{\sigma}[\mathcal{H}_0, \hat{M}]\} - \sum_{\mathrm{r,r'}} \sum_{m=-2}^{2} Tr\{(\hat{\sigma} - \hat{\sigma}_{\mathrm{eq}})[\hat{A}_{\mathrm{r'}}^{m}, [(\hat{A}_{\mathrm{r}}^{m})^{\dagger}, \hat{M}]]\}\mathcal{J}^{\mathrm{rr'}}(m\omega_0) \tag{4.61}$$

The first term in the right hand side of (4.61) refers to precession and is of little

Table 4.5 Spin operators $\hat{A}_{\mathrm{r}}^{m}$ for dipolar and quadrupolar mechanisms

	Dipolar interaction	Quadrupolar interaction
$A_{\mathrm{r}}^{\pm 2}$	$(\hat{I}_{\pm}^{\mathrm{A}}\hat{I}_{\pm}^{\mathrm{X}})/2$	$\hat{I}_{\pm}^{2}/2$
$A_{\mathrm{r}}^{\pm 1}$	$\mp(\hat{I}_{\pm}^{\mathrm{A}}\hat{I}_{z}^{\mathrm{X}} + \hat{I}_{z}^{\mathrm{A}}\hat{I}_{\pm}^{\mathrm{X}})/2$	$\mp(\hat{I}_{\pm}\hat{I}_{z} + \hat{I}_{z}\hat{I}_{\pm})/2$
A_{r}^{0}	$[4\hat{I}_{z}^{\mathrm{A}}\hat{I}_{z}^{\mathrm{X}} - (\hat{I}_{+}^{\mathrm{A}}\hat{I}_{-}^{\mathrm{X}} + \hat{I}_{-}^{\mathrm{A}}\hat{I}_{+}^{\mathrm{X}})]/\sqrt{24}$	$[4\hat{I}_{z}^{2} - (\hat{I}_{+}\hat{I}_{-} + \hat{I}_{-}\hat{I}_{+})]/\sqrt{24}$

concern here. Conversely, the second term is associated with relaxation phenomena. It can be noticed that this term can be decomposed into the product of a quantity involving spin operators ($Tr\{\ldots\}$, which we shall discuss shortly) and of quantity related solely to dynamic properties, namely $\mathcal{J}^{\mathrm{rr'}}(\omega)$. It can be shown that this generalized spectral density does not depend upon m provided that the medium is isotropic (the only situation considered in the present context) and can thus be expressed as

$$\mathcal{J}^{\mathrm{rr'}}(\omega) = \int_0^{\infty} \overline{F_{\mathrm{r}}^{(0)}(t)F_{\mathrm{r'}}^{(0)}(0)} \exp(-\, i\omega t)\,\mathrm{d}t. \tag{4.62}$$

We have again to deal with the Fourier transform of a correlation function $(\overline{F_{\mathrm{r}}^{(0)}(t)F_{\mathrm{r'}}^{(0)}(0)})$, as in equation (4.36), with optionally the consideration of two distinct relaxation mechanisms ($\mathrm{r} \neq \mathrm{r'}$). This situation leads to the concept of *cross-correlation* spectral densities. Examples of such cross-correlations are provided (i) by two different dipolar interactions (r standing for the dipolar interaction between spins A and X and r′ for the dipolar interaction between spins A and Y), (ii) by a dipolar interaction (r standing, for instance, for the dipolar interaction between spins A and X) and a CSA relaxation mechanism (r′ standing, for instance, for the CSA mechanism at nucleus A). These cross-correlation spectral densities will be envisaged later. For now, we shall concentrate upon autocorrelation spectral densities ($\mathrm{r} = \mathrm{r'}$) which in many practical situations play essential role. A last point concerns the frequency appearing in the spectral density $\mathcal{J}^{\mathrm{rr'}}(\omega)$. It can be shown that it is the frequency given by $\omega_{ij}/2\pi = (E_j - E_i)$, E_j and E_i being the energies, expressed in Hz, of two eigenstates such that $\langle i|\hat{A}_{\mathrm{r}}^{m}|j\rangle$ is non-zero [hence $\mathcal{J}^{\mathrm{rr'}}(m\omega_0)$ in (4.61)].

Since most of the time the spectral density (4.62) involves spherical harmonics Y_2^0 [see (4.60)], it can be expressed in terms of reduced spectral densities [see (4.43)]. The only difficulty in establishing the evolution equation of any quantity $\langle M \rangle$ lies in the calculation of the commutators of (4.61). An interesting method for achieving this goal consists in defining a set of orthogonal and normalized operators (or product operators) as has been done in Chapter 2 for studying the evolution of a spin system in a multipulse experiment. These operators must be adapted to the actual relaxation problem and lead to the so-called magnetization modes

$$v_i(t) = \langle \hat{V}_i - \hat{V}_{i,\mathrm{eq}} \rangle \tag{4.63}$$

where $\hat{V}_i$ is one of the operators of the relevant set. For instance, longitudinal relaxation in a two-spin system can be handled with $\hat{V}_1 = \hat{I}_z^{\mathrm{A}}$, $\hat{V}_2 = \hat{I}_z^{\mathrm{X}}$ and $\hat{V}_3 = 2\hat{I}_z^{\mathrm{A}}\hat{I}_z^{\mathrm{X}}$. It can then be shown from equation (4.61) that the evolution of magnetization modes obeys the simultaneous first order kinetic equations (ignoring possible precessional motions).

$$\frac{\mathrm{d}}{\mathrm{d}t} v_i(t) = \sum_j \Gamma_{ij} v_j(t) \tag{4.64}$$

with

$$\Gamma_{ij} = \sum_{\mathrm{r,r'}} \sum_{m=-2}^{2} T_{\mathrm{r}}\{[\hat{A}_{\mathrm{r}}^{-m}, \hat{V}_i][\hat{A}_{\mathrm{r'}}^{-m}, \hat{V}_j]\} \mathcal{J}^{\mathrm{rr'}}(m\omega_0) \tag{4.65}$$

The only task is therefore the calculation of the commutators of (4.65). The elements Γ_{ij} constitute the relaxation matrix, which can be shown to be symmetrical. The number of magnetization modes is *a priori* identical to the number of operators (or product operators) necessary to describe any arbitrary state of the system. For a system of n spin 1/2 nuclei, this number is equal to 2^{2n} (see Chapter 2). Fortunately, the relaxation matrix can be factorized, thus reducing the size of the problem. First it can be easily demonstrated that the coupling term Γ_{ij} cancels if the two modes v_i and v_j correspond to quantities of different nature (i.e. populations and coherences). This explains why longitudinal and transverse relaxations can be treated separately. Concerning coherences, as can be deduced from the Redfield theory, Γ_{ij} again disappears if v_i and v_j correspond to distinct precessional frequencies. For instance, there is no relaxation coupling term between the transverse magnetizations of two nuclei A and X.

Dipolar relaxation in a two-spin system. Solomon equations

The equations governing the longitudinal relaxation of two spin 1/2 nuclei have already been alluded to in Section 4.1.3 in order to understand the NOESY experiment. Referring to the previous section, they can be viewed as arising from the two magnetization modes $\langle I_z^A - I_{eq}^A\rangle$ and $\langle I_z^X - \hat{I}_{eq}^X\rangle$, provided that the CSA mechanism can be ignored (otherwise, as we shall see in the next section, these two modes are further coupled with $\langle 2I_z^A I_z^X\rangle$). These evolution equations can be established according to the methods described in the previous section; they lead to the well-known Solomon equations already given in equations (4.21) and recalled here for the reader's convenience (of course, $\mathrm{d}/\mathrm{d}t\langle I_z^A\rangle \equiv \mathrm{d}/\mathrm{d}t\langle I_z^A - I_{eq}^A\rangle$. It can be stressed again that, although the term 'magnetization' is used, one is rather dealing with polarizations, proportional to γ and not to γ^2.)

$$\begin{aligned}\frac{\mathrm{d}}{\mathrm{d}t}\langle I_z^A\rangle &= -R_1^A(\langle I_z^A\rangle - I_{eq}^A) - \sigma(\langle I_z^X\rangle - I_{eq}^X)\\ \frac{\mathrm{d}}{\mathrm{d}t}\langle I_z^X\rangle &= -R_1^X(\langle I_z^X\rangle - I_{eq}^X) - \sigma(\langle I_z^A\rangle - I_{eq}^A).\end{aligned} \tag{4.66}$$

along with the relaxation parameters expressed in terms of reduced spectral densities. The specific relaxation rate R_1^A can be decomposed according to the dipolar contribution (d) and to contributions arising from other relaxation mechanisms: $R_1^A = (R_1^A)_d + (R_1^A)_{others}$ (with a similar decomposition for R_1^X), whereas the cross-relaxation rate σ originates solely from the dipolar interaction:

$$(R_1^A)_d = (1/20)(\mu_0/4\pi)^2(\gamma_A\gamma_X\hbar/r_{AX}^3)^2[6\tilde{J}(\omega_A + \omega_X) + 3\tilde{J}(\omega_A) + \tilde{J}(\omega_A - \omega_X)] \tag{4.67}$$

$$\sigma = (1/20)(\mu_0/4\pi)^2(\gamma_A\gamma_X\hbar/r_{AX}^3)^2[6\tilde{J}(\omega_A + \omega_X) - \tilde{J}(\omega_A - \omega_X)] \tag{4.68}$$

Whenever the two spins A and X become indistinguishable (e.g. equivalent nuclei), they are dubbed 'like spins' and we are forced to consider the sum of their longitudinal magnetizations $\langle I_z^A\rangle + \langle I_z^X\rangle$, which thus recovers according to a single relaxation rate whose dipolar contribution is the sum of $R_1^A(\equiv R_1^X)$ and σ. This yields with $\omega_0 = \omega_A \equiv \omega_X(\gamma_A \equiv \gamma_X)$

$$(R_1^{like})_d = (1/20)(\mu_0/4\pi)^2(\gamma^2\hbar/r^3)^2[12\tilde{J}(2\omega_0) + 3\tilde{J}(\omega_0)]. \tag{4.69}$$

It can be mentioned that, if two nuclei of identical isotopic nature are subjected to non-selective pulses, $(R_1^{like})_d$ should also prevail for related reasons. In extreme narrowing conditions ($\omega^2\tau_c^2 \ll 1$), spectral densities become frequency-independent, yielding $(R_1^{like})_d/(R_1^A)_d = 3/2$, a property which is known as the '3/2 effect'.

Finally, if the spin 1/2 nucleus A interacts by dipolar coupling with a nucleus X of spin I_X greater than 1/2, the formula (4.67) must be rewritten as follow

$$(R_1^A)_d = (1/15)I_X(I_X + 1)(\mu_0/4\pi)^2(\gamma_A\gamma_X\hbar/r_{AX}^3)^2[6\tilde{J}(\omega_A + \omega_X) + 3\tilde{J}(\omega_A) + \tilde{J}(\omega_A - \omega_X)]. \tag{4.70}$$

We now go back to the widely considered nucleus Overhauser effect factor generally denoted by η and defined as

$$\eta = \frac{I_{stat}^A - I_{eq}^A}{I_{eq}^A} = (\gamma_X/\gamma_A)(\sigma/(R_1^A)$$

The practical way of measuring η has been given in Section 4.1.3. It may be recalled that I_{stat}^A (for stationary A magnetization) is the new 'equilibrium' magnetization of spin A under continuous irradiation of spin X:

$$I_{stat}^A = I_{eq}^A\left(1 + \frac{\sigma}{R_1^A}\frac{\gamma_X}{\gamma_A}\right).$$

As far as R_1^A has been independently measured, the NOE factor η leads to the determination of the cross-relaxation rate σ. From equation (4.67) and (4.68), it can be acknowledged that the factor η is simply equal to $(\gamma_X/2\gamma_A)$ provided that (i) extreme narrowing conditions prevails (each reduced spectral density is equal to $2\tau_c$) and (ii) contributions to R_1^A other than dipolar can be considered as negligible (in that case $R_1^A = 2\sigma$). Therefore, in extreme narrowing conditions, the value of the NOE factor tells us if other relaxation mechanisms are involved. In that case, a value smaller than $(\gamma_X/2\gamma_A)$ is found (e.g. when the CSA mechanism contributes significantly or when paramagnetic species are present). In a general way, the nuclear Overhauser effect affords a sensitivity improvement whenever A nuclei are observed under continuous broadband decoupling. Conversely, if it can be assumed that dipolar interactions are predominant for R_1^A, and if extreme narrowing conditions no longer hold, the ratio R_1^A/σ is smaller than 2 and the separate determinations of R_1^A and σ (via the η factor) provide the reduced spectral densities as a function of the frequency. The evolution of spectral densities with frequency is nevertheless appreciated through several measurements performed at different values of the static field B_0.

Another interesting feature is that continuous irradiation of spin X suppresses the biexponential behavior in the evolution of spin A longitudinal magnetization since, with $\langle I_z^X\rangle \equiv 0$, the Solomon equation relating to A becomes

$$\frac{d}{dt}\langle I_z^A\rangle = -R_1^A(\langle I_z^A\rangle - I_{stat}^A).$$

This property has led to a wealth of dynamic information originating from the intrinsic relaxation rates of carbon-13 and nitrogen-15: if extreme narrowing conditions prevail (small molecules) and if the dipolar mechanism can be considered as predominant, R_1^A yields immediately the correlation time associated with the reorientation of the vector CH or NH, provided that the considered carbon or nitrogen is directly bonded to proton(s).

A salient feature of the longitudinal cross-relaxation concerns the homonuclear case for which σ becomes

$$\sigma_{\text{homonuclear}} = \left(\frac{1}{20}\right)\left(\frac{\mu_0}{4\pi}\right)^2\left(\frac{\gamma^2\hbar}{r_{AX}^3}\right)^2[6\tilde{J}(2\omega_0) - \tilde{J}(0)] \tag{4.71}$$

In extreme narrowing conditions (short τ_c; small molecules), the value of σ is relatively small ($\sigma = 1/20(\mu_0/4\pi)^2(\gamma^2\hbar/r_{AX}^3)^2(10\tau_c)$). For slow tumbling molecules (large τ_c), the value of τ_c is such that $\tilde{J}(\omega_0)$ is shortened by the denominator $(1 + \omega_0^2\tau_c^2)$ whereas $\tilde{J}(0)$ (equal to $2\tau_c$) may become preponderant, depending on the value of ω_0. In that case, σ is negative and its modulus is rather large, leading to intense cross peaks in NOESY spectra. This feature is illustrated by the NOESY two-dimensional of a protein (Figure 4.24).

This explains the impact of the method for structure determinations in large biomolecules whereas, paradoxically, its interest remains dubious for small molecules. Intermediate situations, for which $6\tilde{J}(\omega_0)$ is close to $\tilde{J}(0)$, are especially

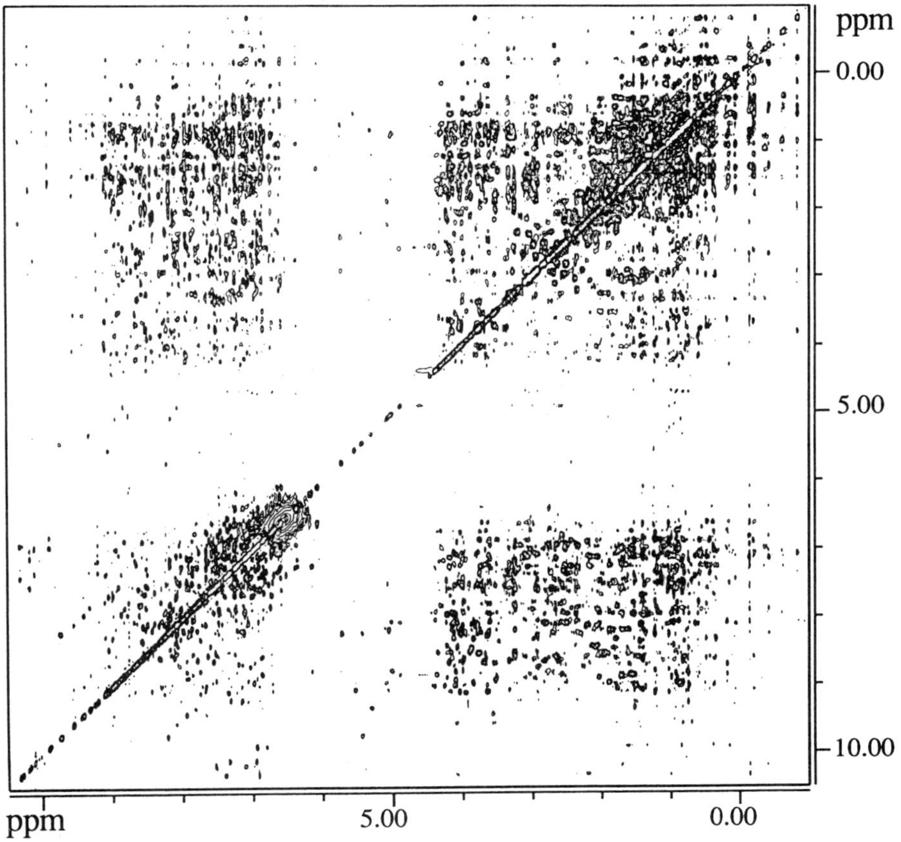

Figure 4.24 The NOESY spectrum of lysozyme at a concentration of 2 mM in H_2O (with suppression of the water resonance, see Section 5.1). Note the wealth of information that can be extracted from such a diagram

frustrating since σ gets close to zero and cross peaks disappear in NOESY spectra. For many molecules, this occurs for a proton measurement frequency around 200–400 MHz. This difficulty can be circumvented by cross-relaxation measurements in the rotating frame (see below).

We turn now to transverse relaxation. The dipolar contribution to specific transverse relaxation rates can again be calculated by the methods given above:

$$(R_2^A)_d = (1/20)(\mu_0/4\pi)^2(\gamma_A\gamma_X\hbar/r_{AX}^3)^2$$
$$[3\tilde{J}(\omega_A + \omega_X) + 3\tilde{J}(\omega_A)/2 + 3\tilde{J}(\omega_X) + \tilde{J}(\omega_A - \omega_X)/2 + 2\tilde{J}(0)] \quad (4.72)$$

Because, as was stated earlier in a general way, two quantities are relaxation-coupled only if their precession frequencies are identical (or nealy identical), transverse cross-relaxation rates are most of the time negligibly small. They must however be accounted for in the case of two equivalent spins (same chemical shift) and enter the effective transverse relaxation rate:

$$(R_2^{\text{like spins}})_d = (1/20)(\mu_0/4\pi)^2(\gamma^2\hbar/r_{AX}^3)^2[3\tilde{J}(2\omega_0) + 15\tilde{J}(\omega_0)/2 + 9\tilde{J}(0)/2] \quad (4.73)$$

In fact, owing to the experimental difficulties of measuring transverse relaxation time (and the necessity to eradicate the contributions of B_0 inhomogeneities), the above expressions are merely useful whenever B_0 inhomogeneities are negligible with respect to the natural T_2 broadening. This occurs of course for short T_2 (or large R_2), as for slowly tumbling molecules (e.g. large biomolecules). In that case, $\tilde{J}(0)$ can be considered as predominant (for the same reasons as above) and the linewidth provides the correlation time directly (as long as no other relaxation mechanism, such as CSA, is involved).

Transverse cross-relaxation terms are regained in the spin-lock experiment (the ROESY experiment; see Figures 4.21 and 4.22) since the corresponding frequencies become identical (equal to zero in the rotating frame). Denoting by x the axis of the rotating frame along which the spin-lock field is applied, we can write the two simultaneous evolution equations as

$$\frac{d}{dt}\langle I_x^A\rangle = -R_{1\rho}^A\langle I_x^A\rangle - \sigma_\rho\langle I_x^X\rangle$$
$$\frac{d}{dt}\langle I_x^X\rangle = -R_{1\rho}^X\langle I_x^X\rangle - \sigma_\rho\langle I_x^A\rangle. \quad (4.74)$$

Calculations worked out with equation (4.61) to which is appended the effect of the spin lock field give rise to the expressions of $R_{1\rho}(=1/T_{1\rho})$ and σ_ρ given below. In these expressions, $\omega_1 = \gamma B_1$ is related to the spin lock field amplitude B_1. In most situations, $\omega_1/2\pi$ is of the order of a few KHz and can be identified with the zero frequency in the relevant spectral densities (in the case of very slow motions only, the exact value of ω_1 should be taken into account). On the other hand, the experiment is by nature homonuclear so that $\gamma_A = \gamma_X = \gamma$ and $\omega_A\omega_X = \omega_0$. Finally, contributions other than dipolar must be added to $R_{1\rho}$, whereas σ_ρ, as for the longitudinal cross-relaxation rate, depends solely on the dipolar interaction.

$$R_{1\rho}^A = R_{1\rho}^X = (1/20)(\mu_0/4\pi)^2(\gamma^2\hbar/r_{AX}^3)^2[3\tilde{J}(2\omega_0) + 9\tilde{J}(\omega_0)/2 + 5\tilde{J}(2\omega_1)/2] \quad (4.75)$$

$$\sigma_\rho = (1/20)(\mu_0/4\pi)^2(\gamma^2\hbar/r_{AX}^3)^2[3\tilde{J}(\omega_0) + 2\tilde{J}(2\omega_1)] \quad (4.76)$$

We can note the resemblance between R_2 and $R_{1\rho}$ [expressions (4.72) and (4.75)] for $\omega_A = \omega_X$ on the one hand and $\omega_1 = 0$ on the other hand. This is not surprising, as these two rates depict the decay of transverse magnetization. Interestingly, σ_ρ (sometimes called transverse cross-relaxation rate) is seen to be *always positive*, by contrast with the longitudinal cross-relaxation rate. Hence there is no risk of cancellation and this explains why rotating frame relaxation is preferred in spite of spurious transfers (thus spurious cross peaks) of the Hartmann–Hahn type (see the TOCSY experiment in the next chapter) when A and X are J coupled. Anyhow, the transverse cross-relaxation rate provides additionnal dynamic (or structural) information owing to its particular dependency upon the measurement frequency, which is different from that of the longitudinal cross-relaxation rate and from that of R_1. These three parameters are plotted in Figure 4.25 as a function of the proton measurement frequency for the simplest reduced spectral density $\tilde{J}(\omega) = 2\tau_c/(1 + \omega^2\tau_c^2)$.

To close this sub-section, we give in Tables 4.6 and 4.7 the relaxation time values of some solvents. They serve merely as examples aimed at providing orders of magnitude. They are not claimed to be reference values as they correspond to non-degassed solutions, obtained at ambient temperature with a low-field spectrometer, so that the CSA mechanism is expected to play a minor role (except for the carbonyl of acetone); extreme narrowing conditions are likewise expected to hold.

Finally, in order to make usable the formula of dipolar relaxation, we give below the numerical values of the relevant constants (in the MKSA system):

$$(1/20)(\mu_0/4\pi)^2(\gamma_A\gamma_X\hbar/r_{AX}^3)^2 = 28.4654 \times 10^9 (A \equiv X \equiv {}^1H)$$
$$= 1.7996 \times 10^9 (A \equiv {}^1H;\ X \equiv {}^{13}C)$$

For both numerical values, r_{AX} has been assumed to be equal to 1 Å. If another interatomic distance r is involved, one must divide the above number by r^6 (r being expressed in Å). If another pair of nuclei is involved, the gyromagnetic

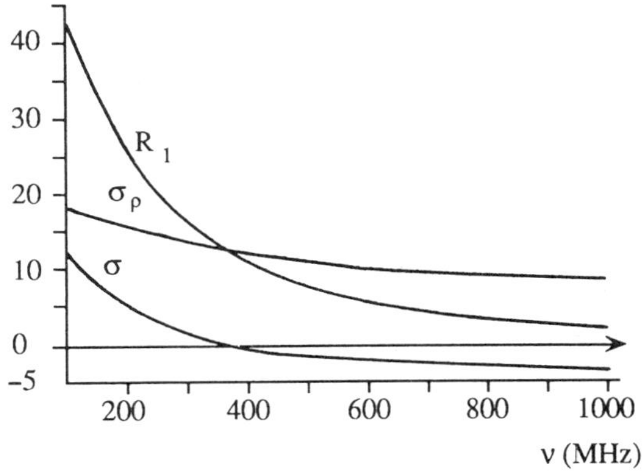

Figure 4.25 Evolution of the longitudinal relaxation rate $(R_1^{like})_d$, and of the longitudinal and transverse cross-relaxation rates σ and σ_ρ for a correlation time of 10^{-9} s and a H–H distance of 1.4 Å

Table 4.6 Proton longitudinal relaxation times of some common solvents (oxygen-17 and deuterium relaxation times of water are also indicated)

Solvent	Formula	T_1 (s)	
water	H_2O	3.6	
		0.51	(^2H)
		0.007	(^{17}O)
methanol	$C\underline{H}_3OH$	4.2	
acetone	$(CH_3)_2CO$	4.3	
benzene	C_6H_6	2.2	
dimethyl sulfoxide	C_2H_6SO	2.2	
chloroform	$CHCl_3$	5.4	
ethanol	$C\underline{H}_3CH_2OH$	3.1	
	$CH_3C\underline{H}_2OH$	2.6	
cyclohexane	C_6H_{12}	0.8	
dioxane	$C_4H_8O_2$	2.0	
methylene chloride	CH_2Cl_2	6.0	
diethyl ether	$(C\underline{H}_3CH_2)_2O$	6.2	
	$(CH_3C\underline{H}_2)_2O$	6.1	

Table 4.7 Carbon-13 longitudinal relaxation times of some solvents

Solvent	Formula	T_1 (s)
methanol	CH_3OH	15.1
acetone	$(\underline{C}H_3)_2CO$	33.1
	$(CH_3)_2\underline{C}O$	23.8
benzene	C_6H_6	26.8
dimethyl sulfoxide	C_2H_6SO	5.8
chloroform	$CHCl_3$	22.6
carbon disulfide	CS_2	35.8
ethanol	$\underline{C}H_3CH_2OH$	8.0
	$CH_3\underline{C}H_2OH$	6.2
cyclohexane	C_6H_{12}	19.7
dioxane	$C_4H_8O_2$	11.2
methylene chloride	CH_2Cl_2	26.0
diethyl ether	$(\underline{C}H_3CH_2)_2O$	15.9
	$(CH_3\underline{C}H_2)_2O$	21.5

ratio values must be changed according to the resonance frequencies (see Table 1.1). For instance, the dipolar relaxation between two fluorine nuclei implies that we multiply 28.4654×10^9 by $(94.077/100)^2$.

Dipolar and CSA relaxation mechanisms in a two-spin system. Generalized Solomon equation for longitudinal relaxation

As explained before, the set of magnetization modes necessary to a full dedscription of longitudinal magnetization in a system of two spin 1/2 nuclei includes

$\langle I_z^A - I_{eq}^A \rangle$, $\langle I_z^X - I_{eq}^X \rangle$ and $\langle 2I_z^A I_z^X \rangle$. In the previous section, only the two former quantities have been considered for two reasons: (i) $\langle 2I_z^A I_z^X \rangle$ is zero at thermal equilibrium and remains at this value unless it is created by some special multipulse preparation; and (ii) $\langle 2I_z^A I_z^X \rangle$ does not cross-talk (is not coupled by relaxation) with $\langle I_z^A \rangle$ and $\langle I_z^X \rangle$ in the absence of a sufficient CSA contribution. In fact, working out the general theory of relaxation summarized above enables one to show that such a coupling exists via a *cross-correlation spectral density* depending on both the dipolar and the CSA mechanism. Unlike cross-relaxation terms, which involve a single relaxation mechanism (e.g. the dipolar interaction), cross-relaxation terms can be considered as a generalization of the latter and reflect the correlation of the molecular motions which modulate two different types of interactions [see (4.62)]. In more complicated systems, such terms may arise from two different dipolar interactions. Here, the only cross-correlation which can occur is between the dipolar AX interaction and the CSA mechanism (for a complete theoretical description, see M. Goldman, *J. Magn. Reson.*, **60**, 437 (1984)). Its existence is of course implied by a non-negligible CSA contribution at either the A or the X nucleus; with the advent of high-field spectrometers such a situation becomes more and more likely and deserves some consideration. A cross-correlation of CSA with dipolar spectral density, sometimes called the interference term, can be expressed according to the familiar reduced spectral density. [From now on, we shall assume an axially symmetric shielding tensor and possibly an anisotropic reorientation; the case of a non-axially symmetric shielding tensor accompanied by an isotropic reorientation can also be considered by substituting the expression given in (4.56) by the one given in (4.55)].

$$\mathcal{J}^{\mathrm{CSA(A),d(AX)}}(\omega) = -(1/5)\sqrt{2/3}(\gamma_A B_0)(\Delta J_A)(\mu_0/4\pi)(\gamma_A \gamma_X \hbar / r_{AX}^3)\tilde{J}(\omega) \quad (4.77)$$

and may be shown effectively to couple longitudinal magnetization $\langle I_z^A - I_{eq}^A \rangle$ and $\langle I_z^X - I_{eq}^X \rangle$ with the quantity $\langle 2I_z^A I_z^X \rangle$ (sometimes called longitudinal two-spin order). Altogether, the evolution of the three magnetization modes $v_1 = \langle I_z^A - I_{eq}^A \rangle$, $v_2 = \langle I_z^X - I_{eq}^X \rangle$ and $v_3 = \langle 2I_z^A I_z^X \rangle$ is accounted for by simultaneous differential equations

$$\frac{d}{dt} v_i(t) = \sum_{i,j} \Gamma_{ij} v_j(t)$$

with $\Gamma_{ji} = \Gamma_{ij}$ [see equations (4.64) and (4.65)]. The elements Γ_{ij} of the relaxation matrix are given in Table 4.8; they include dipolar and CSA contributions [equations (4.56), (4.67), (4.68), (4.77)] and also, for the sake of completeness, random field contributions. These equations are just an extension of the Solomon equations with the coupling terms between longitudinal magnetizations and the longitudinal two-spin order $\langle 2I_z^A I_z^X \rangle$. It can be acknowledged that the latter can develop under relaxation phenomena and that the evolution of each of the three modes becomes triexponential. In such a situation (or even in more complicated spin systems), the relaxation parameters, possibly involving cross-correlation spectral densities, are extracted from the analysis of multiexponential curves. It may be advantageous to devise particular experiments aimed at enhancing the effect of a given element of the relaxation matrix. (The description of such experiments is, however, beyond the scope of this book).

Table 4.8 Elements of the relaxation matrix corresponding to generalized Solomon equations. Shielding tensors are assumed to be axially symmetric; reduced spectral densities refer to the reorientation of the relevant tensor(s) (dipolar and/or shielding)

$$\Gamma_{11} = -(1/20)(\mu_0/4\pi)^2(\gamma_A\gamma_X\hbar/r_{AX}^3)^2[6\tilde{J}^d(\omega_A+\omega_X)+3\tilde{J}^d(\omega_A)+\tilde{J}^d(\omega_A-\omega_X)] \\ -(1/15)(\gamma_A B_0)^2(\Delta\sigma_A)^2\tilde{J}^{csa(A)}(\omega_A) - 2\gamma_A^2\mathcal{J}^{rf(A)}(\omega_A)$$

$$\Gamma_{12} = -(1/20)(\mu_0/4\pi)^2(\gamma_A\gamma_X\hbar/r_{AX}^3)^2[6\tilde{J}^d(\omega_A+\omega_X)-\tilde{J}^d(\omega_A-\omega_X)]$$

$$\Gamma_{13} = (1/5)(\gamma_A B_0)(\Delta\sigma_A)(\mu_0/4\pi)(\gamma_A\gamma_X\hbar/r_{AX}^3)\tilde{J}^{d,csa(A)}(\omega_A)$$

$$\Gamma_{22} = -(1/20)(\mu_0/4\pi)^2(\gamma_A\gamma_X\hbar/r_{AX}^3)^2[6\tilde{J}^d(\omega_A+\omega_X)+3\tilde{J}^d(\omega_A)+\tilde{J}^d(\omega_A-\omega_X)] \\ -(1/15)(\gamma_X B_0)^2(\Delta\sigma_X)^2\tilde{J}^{csa(X)}(\omega_X) - 2\gamma_X^2\mathcal{J}^{rf(X)}(\omega_X)$$

$$\Gamma_{23} = (1/5)(\gamma_X B_0)(\Delta\sigma_X)(\mu_0/4\pi)(\gamma_A\gamma_X\hbar/r_{AX}^3)\tilde{J}^{d,csa(X)}(\omega_X)$$

$$\Gamma_{33} = (1/20)(\mu_0/4\pi)^2(\gamma_A\gamma_X\hbar/r_{AX}^3)^2[3\tilde{J}^d(\omega_A)+3\tilde{J}^d(\omega_X)] \\ -(1/15)(\gamma_A B_0)^2(\Delta\sigma_A)^2\tilde{J}^{csa(A)}(\omega_A) - (1/15)(\gamma_X B_0)^2(\Delta\sigma_X)^2\tilde{J}^{csa(X)}(\omega_X) \\ -2\gamma_A^2\mathcal{J}^{rf(A)}(\omega_A) - 2\gamma_X^2\mathcal{J}^{rf(X)}(\omega_X)$$

Transverse relaxation of some simple coherences

Whenever molecular tumbling is slow, transverse relaxation rates may become so large that static magnetic field inhomogeneities can be ignored. Linewidths, or more generally coherence decay, are meaningful and can be predicted from the general equations of relaxation [see equations (4.64) and (4.65)]. With regard to multipulse and multidimensional experiments commonly performed for assigning spectra or establishing correlations, the following coherences are of special concern: (i) the usual transverse magnetization, (ii) antiphase coherences and (iii) multiple quantum coherence. Without going into the details of such calculations, we give below the expressions of the relevant relaxation rates, which may be useful for optimizing the type of experiment and/or the durations of the evolution periods involved in these experiments. For simplicity, we shall again rely on a system of two spin 1/2 nuclei with the three common relaxation mechanisms: dipolar, CSA (the shielding tensor being assumed as of axial symmetry) and random fields. The standard transverse relaxation rate (corresponding to in-phase coherence) already given by relationship (4.72) for the dipolar mechanism must be appended by CSA and random field contributions. This yields

$$(R_2^A)_{in} = (1/20)(\mu_0/4\pi)^2(\gamma_A\gamma_X\hbar/r_{AX}^3)^2[3\tilde{J}^d(\omega_A+\omega_X)+3\tilde{J}^d(\omega_A)/2 \\ +3\tilde{J}^d(\omega_X)+\tilde{J}^d(\omega_A-\omega_X)/2+2\tilde{J}^d(0)] \\ +(1/15)(\gamma_A B_0)^2(\Delta\sigma_A)^2[\tilde{J}^{csa(A)}(\omega_A)/2+2\tilde{J}^{csa(A)}(0)/3] \\ +\gamma_A^2[\mathcal{J}^{rf(A)}_{(\omega_A)}+\mathcal{J}^{rf(A)}_{(0)}] \qquad (4.78)$$

Similarly, the decay rate of an antiphase coherence can be calculated from the evolution equation of the relevant product operator (e.g. $\langle 2I_x^A I_z^X\rangle$) (there is no coupling with other quantities)

$$
\begin{aligned}
(R_2^{\mathrm{A}})_{\mathrm{anti}} = {} & (1/20)(\mu_0/4\pi)^2(\gamma_{\mathrm{A}}\gamma_{\mathrm{X}}\hbar/r_{\mathrm{AX}}^3)^2[3\tilde{J}^{\mathrm{d}}(\omega_{\mathrm{A}}+\omega_{\mathrm{X}}) + 3\tilde{J}^{\mathrm{d}}(\omega_{\mathrm{A}})/2 \\
& + \tilde{J}^{\mathrm{d}}(\omega_{\mathrm{A}}-\omega_{\mathrm{X}})/2 + 2\tilde{J}^{\mathrm{d}}(0)] \\
& + (1/15)(\gamma_{\mathrm{A}}\mathrm{B}_0)^2(\Delta\sigma_{\mathrm{A}})^2[\tilde{J}^{\mathrm{csa(A)}}(\omega_{\mathrm{A}})/2 + 2\tilde{J}^{\mathrm{csa(A)}}(0)/3] \\
& + (1/15)(\gamma_{\mathrm{X}}B_0)^2(\Delta\sigma_{\mathrm{X}})^2\tilde{J}^{\mathrm{csa(X)}}(\omega_{\mathrm{X}}) \\
& + \gamma_{\mathrm{A}}^2[\mathcal{J}_{(\omega_{\mathrm{A}})}^{\mathrm{rf(A)}} + \mathcal{J}_{(0)}^{\mathrm{rf(A)}}] + 2\gamma_{\mathrm{X}}^2\mathcal{J}_{(\omega_{\mathrm{X}})}^{\mathrm{rf(X)}}
\end{aligned}
\tag{4.79}
$$

In the course of an actual evolution period, there is obviously continuous exchange between the in-phase and the antiphase configurations, so that an average of (4.78) and (4.79) should be considered for obtaining an effective transverse relaxation rate. It can be recognized that the effective transverse relaxation rate of the A nucleus actually depends on relaxation mechanisms affecting the X nucleus.

Coherences of the type $\langle 2I_x^{\mathrm{A}}I_x^{\mathrm{X}}\rangle$, $\langle 2I_x^{\mathrm{A}}I_y^{\mathrm{X}}\rangle$, $\langle 2I_y^{\mathrm{A}}I_x^{\mathrm{X}}\rangle$ and $\langle 2I_y^{\mathrm{A}}I_y^{\mathrm{X}}\rangle$ are involved in the HMQC experiment during the evolution period (see Section 5.3.2). After appropriate calculations carried out according to (4.64) and (4.65), and by performing an average over these four quantities, we arrive at (the subscript MQ indicates that one is dealing with multiquanta coherences)

$$
\begin{aligned}
(R_2)_{\mathrm{MQ}} = {} & (1/20)(\mu_0/4\pi)^2(\gamma_{\mathrm{A}}\gamma_{\mathrm{X}}\hbar/r_{\mathrm{AX}}^3)^2[3\tilde{J}^{\mathrm{d}}(\omega_{\mathrm{A}}+\omega_{\mathrm{X}}) + 3\tilde{J}^{\mathrm{d}}(\omega_{\mathrm{A}})/2 \\
& + 3\tilde{J}^{\mathrm{d}}(\omega_{\mathrm{X}})/2 + \tilde{J}^{\mathrm{d}}(\omega_{\mathrm{A}}-\omega_{\mathrm{X}})/2] \\
& + (1/15)(\gamma_{\mathrm{A}}B_0)^2(\Delta\sigma_{\mathrm{A}})^2[\tilde{J}^{\mathrm{csa(A)}}(\omega_{\mathrm{A}})/2 + 2\tilde{J}^{\mathrm{csa(A)}}(0)/3] \\
& + (1/15)(\gamma_{\mathrm{X}}B_0)^2(\Delta\sigma_{\mathrm{X}})^2[\tilde{J}^{\mathrm{csa(X)}}(\omega_{\mathrm{X}})/2 + 2\tilde{J}^{\mathrm{csa(X)}}(0)/3] \\
& + \gamma_{\mathrm{A}}^2[\mathcal{J}_{(\omega_{\mathrm{A}})}^{\mathrm{rf(A)}} + \mathcal{J}_{(0)}^{\mathrm{rf(A)}}] + \gamma_{\mathrm{X}}^2[\mathcal{J}_{(\omega_{\mathrm{X}})}^{\mathrm{rf(X)}} + \mathcal{J}_{(0)}^{\mathrm{rf(X)}}]
\end{aligned}
\tag{4.80}
$$

From a practical point of view, if we assign A to a carbon-13 and X to a proton bound to this carbon, and if we assume that $\mathcal{J}^{\mathrm{rf(X)}}$ represents the dipolar contribution of X with other protons in the molecule, we can notice that (4.80) involves this spectral density at zero frequency. Hence, this term may well predominate in the case of slow tumbling and considerably enhance $(R_2)_{\mathrm{MQ}}$. This is not the case in (4.79), where $\mathcal{J}^{\mathrm{rf(X)}}$ appears at the frequency ω_{X} and is much smaller than $\mathcal{J}^{\mathrm{rf(X)}}(0)$. This explains why HSQC heteronuclear correlation experiments (Section 5.2.2), which involve the evolution of A antiphase coherences, are less prone to sensitivity loss by relaxation than HMQC experiments.

Quadrupolar relaxation

Owing to the form of the quadrupolar hamiltonian [equation (4.58) and Table 4.5], we must rely upon equation (4.61) to obtain the evolution equations of the longitudinal and transverse magnetizations of nuclei whose spin I is greater than 1/2 (quadrupolar nuclei). Analytical solutions, leading to well defined relaxation rates, exist in particular situations which will be the only ones considered here. In the same way that the interatomic distance was involved in the scaling factor of dipolar relaxation, the quadrupole coupling constant χ will define the efficiency of the quadrupolar relaxation mechanism. It can be expressed (in Hz) as

$$\chi = eQV_{zz}/h \tag{4.81}$$

where e is the electron charge; Q is also a constant, namely the quadrupolar momentum of the considered nuclear isotope; V_{zz} is either the largest element of the electric field gradient tensor (supposed diagonal, which implies that it is expressed in the molecular principal axis system) or the element pertaining to the symmetry axis if the tensor is axially symmetric. It can be recalled that the *electric field gradient tensor* originates essentially from the electronic distribution; it is therefore a molecular property depending strongly on the symmetry of this latter distribution. Hence, for a distribution of spherical symmetry, the gradient vanishes and the quadrupolar relaxation mechanism is ineffective. As an example, this occurs for an ion solvated by water molecules (e.g. the sodium ion; $I(\text{Na}) = 3/2$). This no longer holds insofar as this ion is part of a molecular (or supramolecular) structure.

The first special case we are going to consider is the one of *extreme narrowing* which, as often stated before, is defined by $(\omega_0\tau_c)^2 \ll 1$, where τ_c is an effective correlation time and where $\omega_0/2\pi$ is the measurement frequency. In that case, the evolutions of longitudinal and transverse magnetizations are exponential and governed by the same relaxation rate

$$(R_1)_q = (R_2)_q = (3\pi^2/10)\{(2I + 3)/[I^2(2I - 1)]\}\chi^2(1 + \eta_q^2/3)\tau_c \tag{4.82}$$

where $\eta_q = (V_{xx} - V_{yy})/V_{zz}$ represents the asymmetry parameter of the electric field gradient tensor still expressed in the molecular principal axis system with $|V_{zz}| \geqslant |V_{yy}| \geqslant |V_{xx}|$. In many instances, the quadrupole constant is so large that the quadrupolar relaxation mechanism is overwhelming with, as a consequence, an important line broadening (arising from a large R_2 value). This line broadening precludes the observation of any multiplet due to J couplings and may further entail severe overlap if several resonances are present in the spectrum.

Whenever extreme narrowing conditions are not verified, the evolution of longitudinal and transverse magnetizations is no longer exponential, with, however, the noticeable exception of nuclei whose spin number I is equal to 1 (deuterium and nitrogen-14). This arises from the fact that the commutator calculations in equation (4.61) do not lead exclusively to I_z for $M = I_z$ or to I_x for $M = I_x$. In fact, it turns out that most of the time, quasiexponential behavior is observed. In such an instance, the apparent relaxation rates can be interpreted according to the approximate expressions given below (B. Halle and H. Wennerström, *J. Magn. Reson.*, **44**, 89 (1989)):

$$\begin{aligned} ((R_1)_q)_{app} &= [(2I + 3)/5][8\mathcal{J}^q(2\omega_0) + 2\mathcal{J}^q(\omega_0)] \\ ((R_2)_q)_{app} &= [(2I + 3)/5][2\mathcal{J}^q(2\omega_0) + 5\mathcal{J}^q(\omega_0) + 3\mathcal{J}^q(0)] \end{aligned} \tag{4.83}$$

where $\omega_0/2\pi$ is the measurement frequency; $\mathcal{J}^q(\omega)$ is a quadrupolar spectral density which, as we have seen in the previous sections, can be expressed according to a reduced spectral density if one is dealing (i) with a tensor of axial symmetry ($\eta_q = 0$) and an anisotropic reorientation *or* (ii) with an isotropic reorientation and a tensor without axial symmetry:

$$\mathcal{J}^q(\omega) = (3\pi^2/20)[1/I^2(2I - 1)]\chi^2(1 + \eta_q^2/3)\tilde{J}(\omega). \tag{4.84}$$

It can be mentioned that equations (4.83) are strictly valid for the case of a quadrupolar nucleus of spin $I = 1$.

4.2.4 INTRODUCTION TO THE DYNAMICAL INTERPRETATION OF REDUCED SPECTRAL DENSITIES

We shall limit ourselves to the modeling of a correlation function of the form $(3\cos^2\beta'(t) - 1)(3\cos^2\beta(0) - 1)$ where β and β' are the angles between a molecular axis and the Z direction of the laboratory frame. Except for a coefficient, this average appears in the expression of the reduced spectral densities $\tilde{J}(\omega)$($\beta' = \beta$ for autocorrelation spectral densities; $\beta' \neq \beta$ for cross-correlation spectral densities) which are involved in those relaxation parameters amendable to a direct interpretation. In spite of its apparently simple form, the calculation of the above correlation function is far from being trivial and requires in any event to postulate a dynamic model.

We shall, in the following, envisage only three models akin to molecular reorientation; this means that we shall disregard translational motions. Moreover, we shall assume that these rotational motions are by nature diffusional and thus obey formal equation of the type $\partial\psi/\partial t = \mathrm{D_r}\Delta\psi$. The molecular function ψ is related to an orientational probability and Δ is the Laplacian operator. The *rotational diffusion coefficient* $\mathrm{D_r}$, in the hypothesis of the simplest brownian motion, depends upon the viscosity η of the medium and upon an effective molecular radius a (if the molecule under investigation can be approximatively identified with a sphere)

$$\mathrm{D_r} = \frac{k_B T}{8\pi a^3 \eta} \tag{4.85}$$

where k_B is the Boltzmann constant and T is the absolute temperature. The rotational diffusion coefficient may be compared to the translational diffusion coefficient, which can be measured according to the methods indicated in Section 4.1.2:

$$\mathrm{D_t} = \frac{k_B T}{6\pi a \eta} \tag{4.86}$$

Isotropic reorientation

Regardless of the considered molecular direction, the corresponding rotation is effected according to a unique diffusion coefficient, given by (4.85). By defining the correlation time as $\tau_c = 1/6\mathrm{D_r}$, we obtain a very simple expression for reduced autocorrelation spectral densities (already given by (4.44)):

$$\tilde{J}(\omega) = \frac{2\tau_c}{1 + \omega^2\tau_c^2} \tag{4.87}$$

The relevant cross-correlation spectral density is of the form

$$\tilde{J}^{\mathrm{rr'}}(\omega) = \frac{1}{2}(3\cos^2\theta_{\mathrm{rr'}} - 1)\frac{2\tau_c}{1 + \omega^2\tau_c^2} \tag{4.88}$$

where $\theta_{rr'}$ is the angle between the relaxation vectors pertaining to each mechanism. It can be recalled here that we mean by relaxation vector (i) the internuclear vector in the case of dipolar relaxation (ii) the vector defining the direction of the tensor symmetry axis for other mechanisms such as CSA or quadrupolar relxation, or (iii) the vector defining the direction corresponding to the largest diagonal element of the considered tensor when axial symmetry is lacking.

The evolution of $\tilde{J}(\omega)$ with the measurement frequency $\nu = \omega/2\pi$ is shown in Figure 4.26 for different values of τ_c. The validity of the extreme narrowing conditions [$\omega^2\tau_c^2 \ll 1$, leading to a quasi-independence of $\tilde{J}(\omega)$ with respect to ω] can be appreciated from such plots.

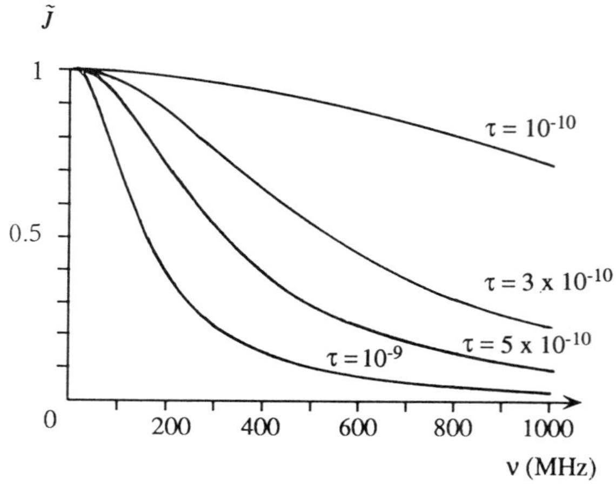

Figure 4.26 $\tilde{J}(\omega)$ (formula (4.87)) plotted as a function of $\nu = \omega/2\pi$ for different values of the correlation time τ_c (in s). The curves have been normalized to 1 for $\nu = 0$

Anisotropic overall reorientation

If the molecule under investigation cannot be identified with a sphere, instead of a single diffusion coefficient, we must resort to a diffusion tensor whose principal axes can be assumed to coincide with those of the inertia tensor. In order to avoid too much complexity, we shall limit ourselves to an axially symmetric diffusion tensor defined by its diagonal elements D_z and $D_x = D_y$. We shall denote by θ the angle between the relaxation vector and the molecular direction corresponding to D_z. The autocorrelation reduced spectral density can be expressed as

$$\tilde{J}(\omega) = \frac{3}{4}\sin^4\theta\frac{2\tau_1}{1+\omega^2\tau_1^2} + \frac{3}{4}\sin^2(2\theta)\frac{2\tau_2}{1+\omega^2\tau_2^2} + \frac{1}{4}(3\cos^2\theta - 1)^2\frac{2\tau_3}{1+\omega^2\tau_3^2} \tag{4.89}$$

with $1/\tau_1 = 2D_x + 4D_z$, $1/\tau_2 = 5D_x + D_z$ and $1/\tau_3 = 6D_x$. The corresponding cross-correlation spectral density depends on the polar angles (θ_r, φ_r) and $(\theta_{r'}, \varphi_{r'})$ which define the orientations (see Figure 4.23) of the two relaxation vectors

$$\begin{aligned}\tilde{J}^{rr'}(\omega) &= \frac{3}{4}\sin^2\theta_r \sin^2\theta_{r'} \cos 2(\varphi_r - \varphi_{r'})\frac{2\tau_1}{1+\omega^2\tau_1^2} \\ &+ \frac{3}{4}\sin(2\theta_r)\sin(2\theta_{r'})\cos(\varphi_r - \varphi_{r'})\frac{2\tau_2}{1+\omega^2\tau_2^2} \\ &+ \frac{1}{4}(3\cos^2\theta_r - 1)(3\cos^2\theta_{r'} - 1)\frac{2\tau_3}{1+\omega^2\tau_3^2} \qquad (4.90)\end{aligned}$$

It can be noted that, if the rotational motion about x and y is slow with respect to the motion about $z(D_x = D_y \ll D_z)$, the correlation time τ_3 is the only one to be affected by the slow motion. The corresponding contribution is modulated according to $(3\cos^2\theta - 1)^2$ and may disappear when θ becomes equal to the magic angle.

'Two-step model' or 'model-free approach'

This is a pragmatic way of accounting for two types of rotational motion: (i) a slow (s) *overall* motion (undergone for instance by a large biomolecule or by an aggregate of surfactant molecules) whose correlation time τ_s is outside extreme narrowing conditions, and (ii) a fast (f) *local* motion (e.g. due to internal rotations) characterized by a correlation τ_f (generally within the extreme narrowing range) and supposed to occur about a local director D (which is itself solely subjected to the slow motion). It turns out that the orientation of the relaxation vector with respect to the director can be restricted; considering for instance the relaxation vector corresponding to $^{13}C-^{1}H$ dipolar interaction (Figure 4.27), we are led to assume that all values of the angle θ, between C–H and D, do not have the same probability. This feature is accounted for by a specific order parameter with the same meaning as in an anisotropic medium (Section 1.4.3):

$$S = 1/2\overline{(3\cos^2\theta - 1)}. \qquad (4.91)$$

Assuming that both motions are diffusional in nature, one arrives at an autocorrelation reduced spectral density of the form (H. Wennerström *et al.*, *J. Am. Chem.*

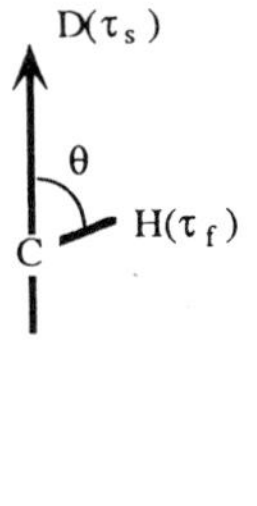

Figure 4.27 Orientation (defined by the angle θ) of a relaxation vector (CH) with respect to a local director D. The relaxation vector is supposed to reorient about D according to a correlation time τ_f whereas the director reorients according to a correlation time $\tau_s(\tau_s \gg \tau_f)$. The orientation of the CH vector with respect to D is characterized by a specific order parameter

Soc., **101**, 6860 (1979); G. Lipari and A. Szabo, *J. Am. Chem. Soc*, **104**, 5146 (1982))

$$\tilde{J}(\omega) = (1 - S^2)2\tau_f + S^2\frac{2\tau_s}{1 + \omega^2\tau_S^2}, \tag{4.92}$$

which, for a cross-correlation spectral density, becomes

$$\tilde{J}^{rr'}(\omega) = \left[\frac{1}{2}(3\cos^2\theta_{rr'} - 1) - S_rS_{r'}\right]2\tau_f + S_rS_{r'}\frac{2\tau_s}{1 + \omega^2\tau_S^2} \tag{4.93}$$

Should the fast motion be outside the extreme narrowing range, $2\tau_f$ must be replaced by $2\tau_f/(1 + \omega^2\tau_f^2)$. The evolution of $\tilde{J}(\omega)$ as a function of the frequency $\omega/2\pi$ is plotted in Figure 4.28 (such a plot is sometimes dubbed the 'dispersion curve'). Breaks occur at $\nu = 1/\tau_s$ and $1/\tau_f$, whence the terminology of 'two-step model'. The interpretation of such spectral densities in terms of τ_s, τ_f and S requires a series of measurements generally performed at different frequencies (multifield measurements).

The interpretation of relaxation data in biomolecules (peptides, proteins, nucleic acids) or in organized systems (micelles, microemulsions) by means of this approach has proved to be quite satisfactory. Experimental results shown in Figures 4.29 and 4.30 demonstrate its validity and also the sensitivity of relaxation spin to the shape of molecular aggregates.

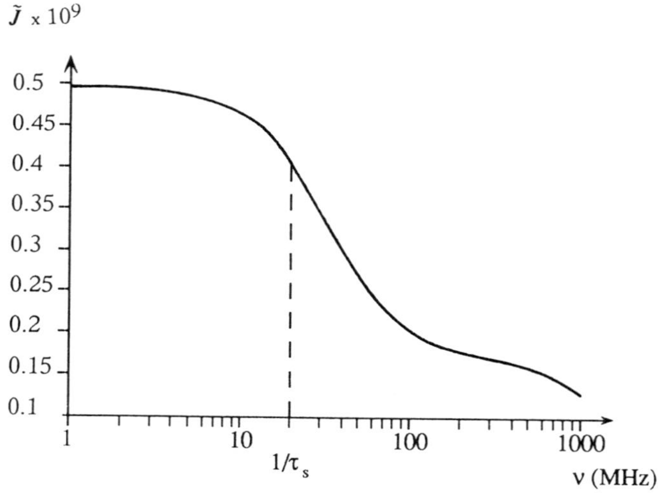

Figure 4.28 Frequency evolution of an autocorrelation spectral density (formula (4.92)) for $\tau_f = 9 \times 10^{-11}$ s, $\tau_s = 5 \times 10^{-9}$ s and $S = 0.18$

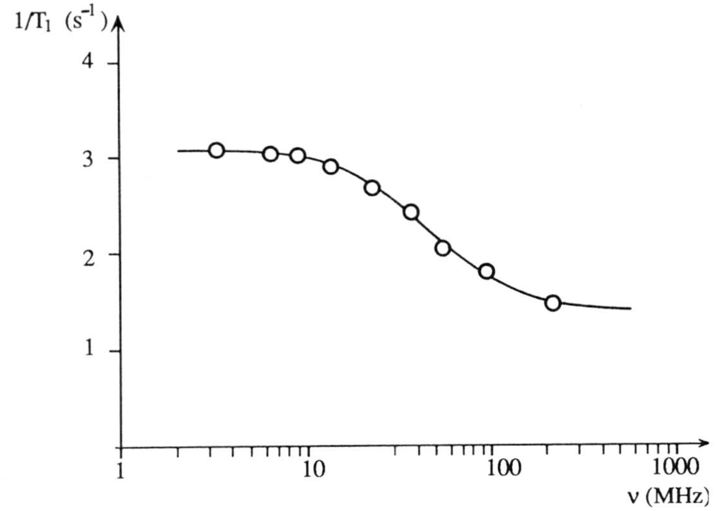

Figure 4.29 Proton relaxation rates of surfactant molecules engaged in spherical micelles, as a function of the measurement frequency. The two steps are associated with the slow motion (overall motion) and the fast local motion, respectively

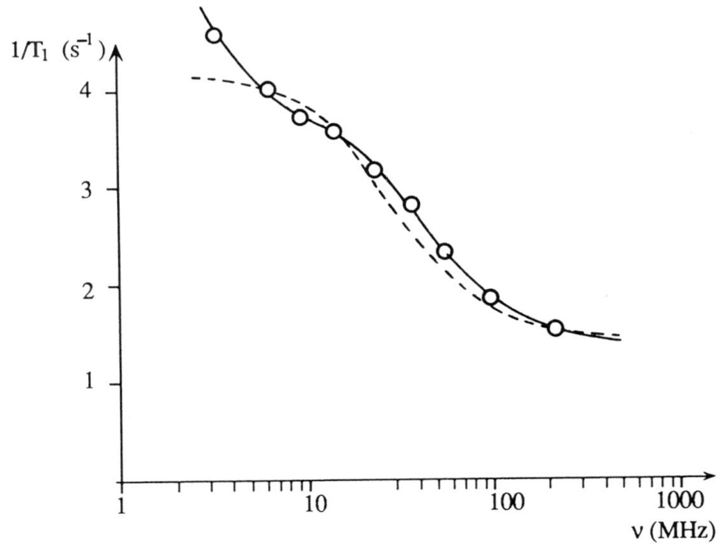

Figure 4.30 Proton relaxation rates of surfactant molecules incorporated in rod-shaped micelles, as a function of the measurement frequency. A third step is necessary for describing the anisotropic motion of the aggregates

Bibliography

A. Abragam: *The pinciples of Nuclear Magnetism*, Clarendon Press, Oxford, 1961, Chapter 8

A. G. Redfield: The theory of relaxation processes, *Adv. Magn. Reson.*, **1**, 1 (1965)
M. E. Rose: *Elementary Theory of Angular Momentum*, Wiley, New York, 1967
W. T. Huntress, Jr.: The study of anisotropic rotation of molecules in liquids by NMR quadrupolar relaxation, *Adv. Magn. Reson.*, **4**, 2 (1970)
J. H. Noggle and R. E. Schirmer: *The Nuclear Overhauser Effect*, Academic Press, New York, 1971
R. Lenk: *Brownian Motion and Spin Relaxation*, Elsevier, Amsterdam, 1977
L. G. Werbelow and D. M. Grant: Intramolecular dipolar relaxation in multispin systems, *Adv. Magn. Reson.*, **9**, 189 (1977)
H. W. Spiess: Rotation of molecules and nuclear spin relaxation, *NMR: Basic Princ. Prog.*, **15**, 55 (1978)
R. L. Vold and R. R. Vold: Nuclear magnetic relaxation in coupled spin systems, *Prog. NMR Spectrosc.*, **12**, 79 (1979)
J. McConnell: *Nuclear Magnetic Relaxation in Liquids*, Cambridge University Press, Cambridge, 1987
D. Neuhaus and M. P. Williamson: *The Nuclear Overhauser Effect in Structural and Conformational Analysis*, VCH, New York, 1989
D. Canet: Construction, evolution and detection of magnetization modes designed for treating longitudinal relaxation of weakly coupled spin 1/2 systems with magnetic equivalence, *Prog. NMR Spectrosc.*, **21**, 237 (1989)
D. Canet and J. B. Robert: Behaviour of the NMR Relaxation Parameters at High Fields, *NMR: Basic Princ. Prog.* **25**, 45 (1990)
J. Kowaleswski: *Annu. Rep. NMR Spectrosc.*, **22**, 306 (1989); **23**, 289 (1991)
D. M. Grant, C. L. Mayne, F. Lin and T. X. Xiang: Spin–lattice relaxation of coupled nuclear spins with applications to molecular motion in liquids, *Chem, Rev.,* **91**, 1591 (1991)
J. J. Delpuech, Ed.: *Dynamics of Solutions and Fluid Mixtures by NMR*, John Wiley & Sons, Chichester, 1995

5 MULTIPULSE AND MULTIDIMENSIONAL NMR

Although still developing, this aspect of nuclear magnetic resonance encompasses some well accepted experiments which have become routine in nature and which will be described in the present chapter. Regarding instrumentation and data processing, these experiments are somewhat elaborate; their justification stems from the necessity to extract specific information from complicated spectra, to establish correlations for assignment and interpretation purposes, or to visualize particular physical properties such as the spin density distribution (imaging). Correlations which can be drawn from dynamic parameters (cross-relaxation or chemical exchange) have been treated in the previous chapter and will not be further considered here. The reader must also be warned that he will not find in this chapter a full description of elaborate three- (or even four-) dimensional methods which are now in common use for structural determinations on biomolecules. However, the relevant sequences are generally built from subunits, each of which plays a well defined role and is derived from one of the basic approaches presented (hopefully) in this chapter.

Most of the objectives which have been mentioned above necessarily require a certain form of spectral selectivity followed by a transfer process. Selectivity may be achieved by an appropriate excitation which affects solely one spin or a group of spins or even a single transition. Another way of reaching this goal rests on the evolution of the system for a incremented time t_1 and on the conversion from time to frequency by means of a Fourier transform ($t_1 \xrightarrow{\text{FT}} \nu_1$). This is of course the basis of two-dimensional experiments; the selectivity extends along the frequency domain ν_1 with each point corresponding to a small frequency range (Figure 5.1).

5.1 Selective Excitation

5.1.1 SOFT PULSES

In order to understand how a 'soft' r.f. pulse works, (that is, a pulse of low B_1 amplitude), we consider the effective radio-frequency field [Figure 2.5 and equation (2.8)] about which nuclear magnetization is due to nutate. It is quite obvious that if the r.f. field amplitude B_1 becomes negligible with respect to $(B_0 - \omega_r/\gamma)$ the effective field coincides with the z axis and cannot act upon the nuclear magnetization. Thus selectivity is achieved, meaning that resonances far away from the excitation frequency will not be affected. In a more formal way, a line at

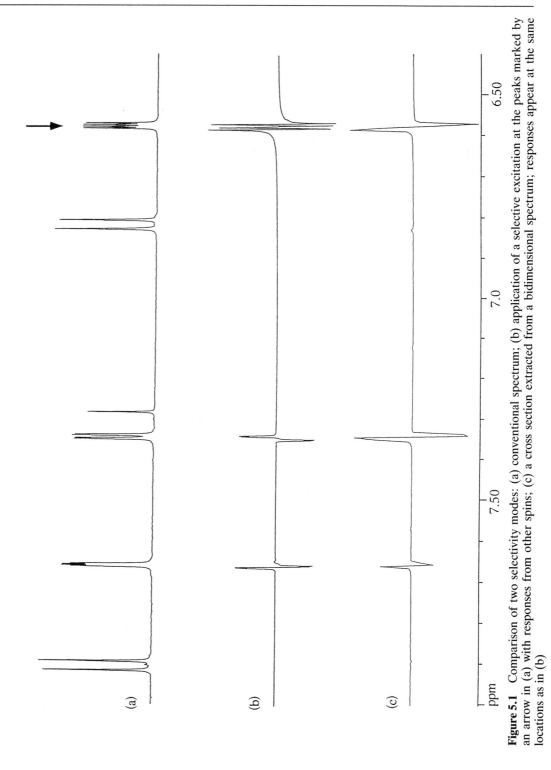

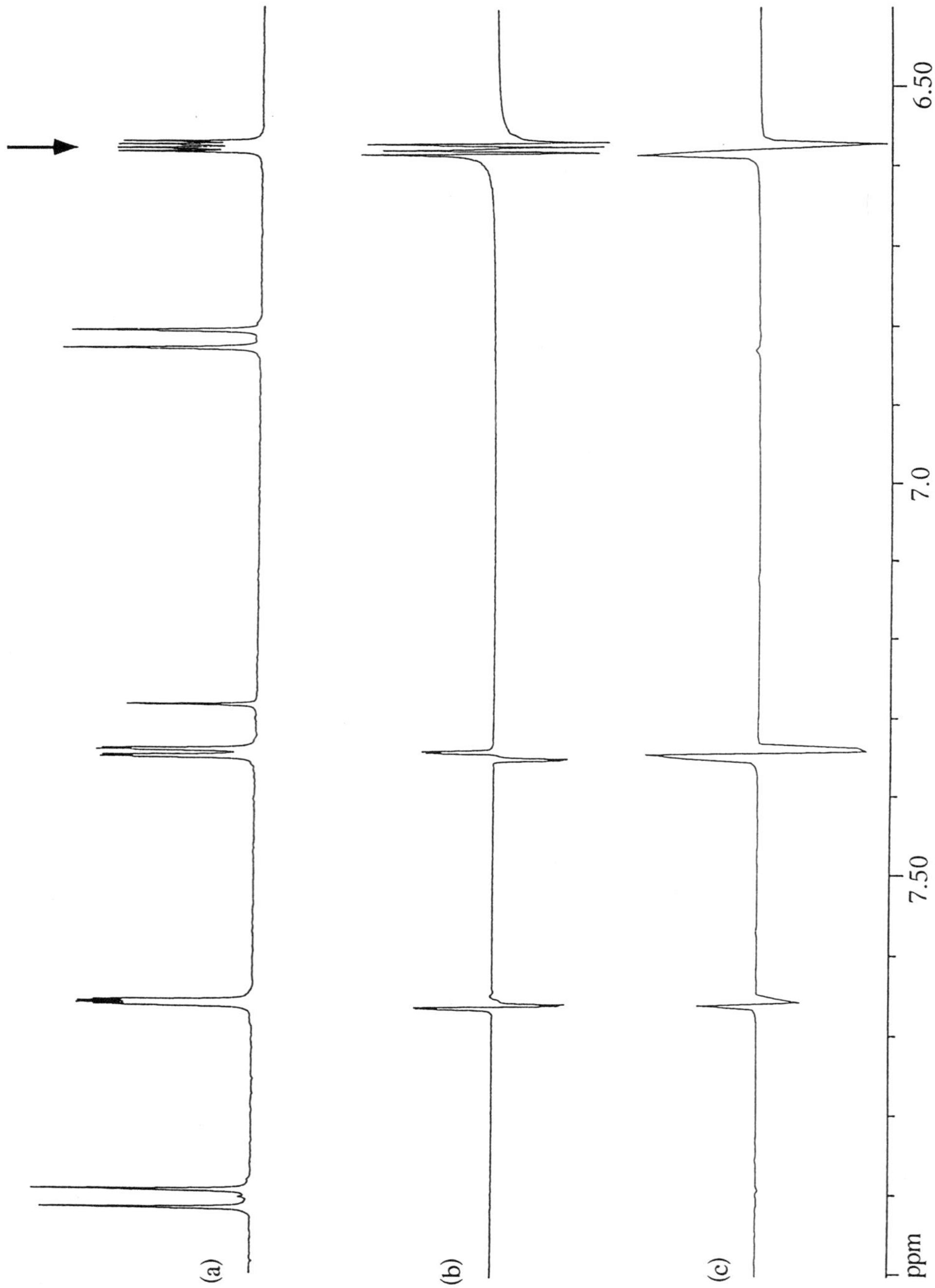

Figure 5.1 Comparison of two selectivity modes: (a) conventional spectrum; (b) application of a selective excitation at the peaks marked by an arrow in (a) with responses from other spins; (c) a cross section extracted from a bidimensional spectrum; responses appear at the same locations as in (b)

frequency $\nu_0 = \gamma B_0/2\pi(1 - \sigma)$(where σ stands for the shielding constant, accounting for the difference in resonance frequencies in the considered spectrum) is not excited as long as the following condition holds

$$|\nu_0 - \nu_r| \gg \frac{\gamma B_1}{2\pi}$$

where $\nu_r = \omega_r/2\pi$ is the carrier frequency.

Conversely, the r.f. field acts fully on resonances whose frequency is close to the carrier frequency. In these conditions, B_1^{eff} is almost identical to B_1 and lies close to the x axis (if the r.f. field is applied along the x axis of the rotating frame) and produces a nutation by an angle $\alpha = \gamma B_1 \tau$, where τ is the time during which the r.f. field is on. This would mean that, according to the r.f. field amplitude B_1, one would be able to select a frequency zone as narrow as desired. However, the physical reality is quite different. On the one hand, relaxation impairs the nutation process if τ becomes long enough, that is if B_1 is purposely made very weak so as to reduce the frequency bandwidth. On the other hand, even if relaxation phenomena could be ignored, a rectangular soft pulse cannot be seen as fully exciting a given frequency zone and leaving resonances outside this frequency zone totally unaltered. The mathematical expression of equation (2.8) helps us to understand this feature: the excitation profile is not rectangular (but merely resembles a sinc function with oscillations or *sidelobes* around the desired excitation frequency zone (see Figure 5.2(a)). This is reminiscent of the Fourier transform of a rectangular function, the soft pulse, which is a sinc function in the frequency domain (see Section 3.3.2). Unfortunately, the concept of Fourier transformation does not apply strictly here because we are not dealing with linear phenomena (the NMR response is approximately linear only within the approximation of small flip angle pulses). However, the Fourier transform of a pulse shape (which is a time function) provides an initial guess of the expected frequency profile and thus may serve as a first guide to devising amplitude (and/or phase) modulation. As a first example, let us consider a small amplitude pulse modulated according to a Gaussian function. Noting that the Fourier transform of a Gaussian function is again a Gaussian function (see Section 3.3.3), we should obtain a frequency profile devoid of sidelobes. This is not really true, as shown in Figure 5.2(b), where two negative lobes are seen on each side of the zone expected to be excited. Moreover, such a pulse shape, in the same way as a rectangular soft pulse, entails a so-called phase gradient: excited resonances exhibit a dephasing which is dependent on their frequency (first order phase alteration). This phase problem arises, of course, from the excitation of the dispersive component (M_x in Figure 5.2, if the soft pulse is applied along the x direction of the rotating frame, M_y being the absorptive component). Gaussian pulses are widely used, and a way to get rid of the absorptive component is to refocus the transverse magnetization by means of a hard π pulse. In general, an alternative method for suppressing the dispersive component is a two-step phase cycle ($\pm$) applied to a $\pi/2$ hard pulse following the soft pulse. The problem of unwanted sidelobes can be alleviated by a half-Gaussian pulse (Figure 5.2(c)), which does not, however, remove the strong dispersive component. As a matter of fact, a simple means of avoiding both sidelobes and the dispersive component, although at the expense of longer

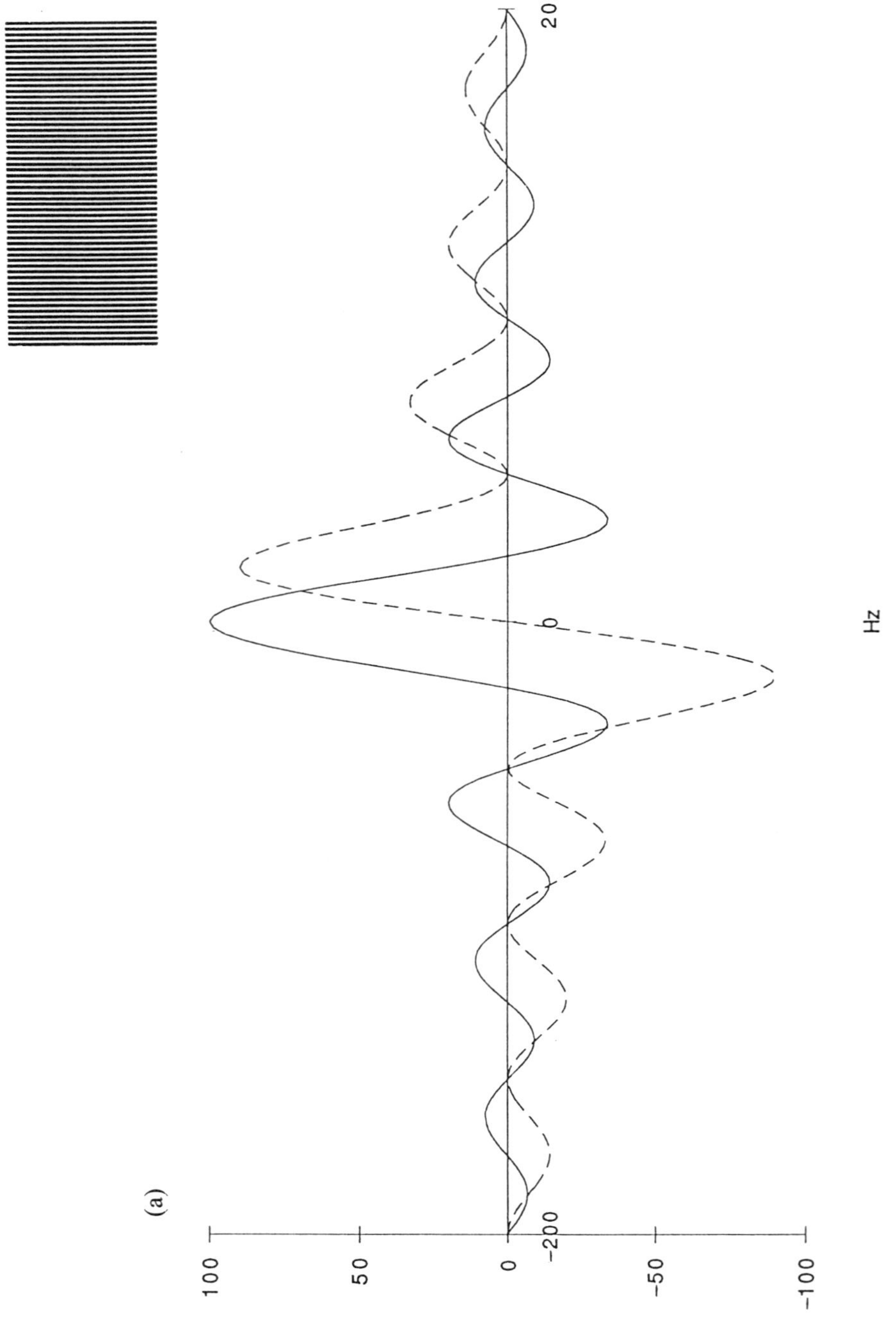
(a)
100
50
0
-50
-100
-200
0
200
Hz

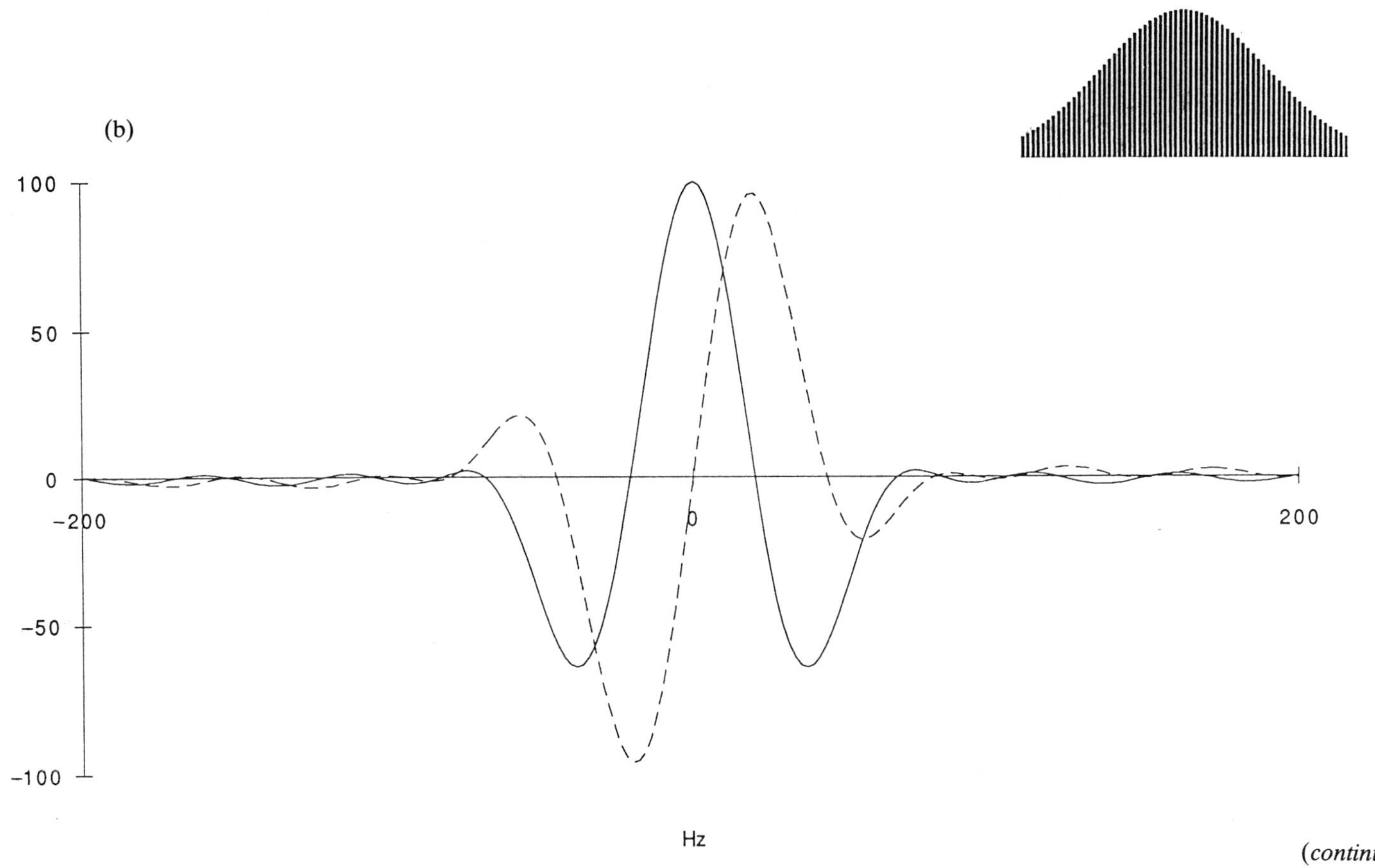

(continued)

Figure 5.2 Excitation profiles (magnetization, relative to its equilibrium value, is indicated in %) for soft pulses of identical duration ($\tau = 20$ ms), corresponding to a 90° flip angle at zero frequency, but of different shapes (schematized in the insert): (a) rectangular; (b) Gaussian (truncated, as the two following pulses, at 95% of its total area); (c) half Gaussian; (d) 270° Gaussian; (e) sinc (truncated at the second zero on each side of the maximum); (f) E-BURP-2 (see text). Each pulse is supposed to be applied along the x direction of the rotating frame in such a way that the desired absorptive component is M_y (full curves), and the unwanted dispersive component is M_x (dashed curves)

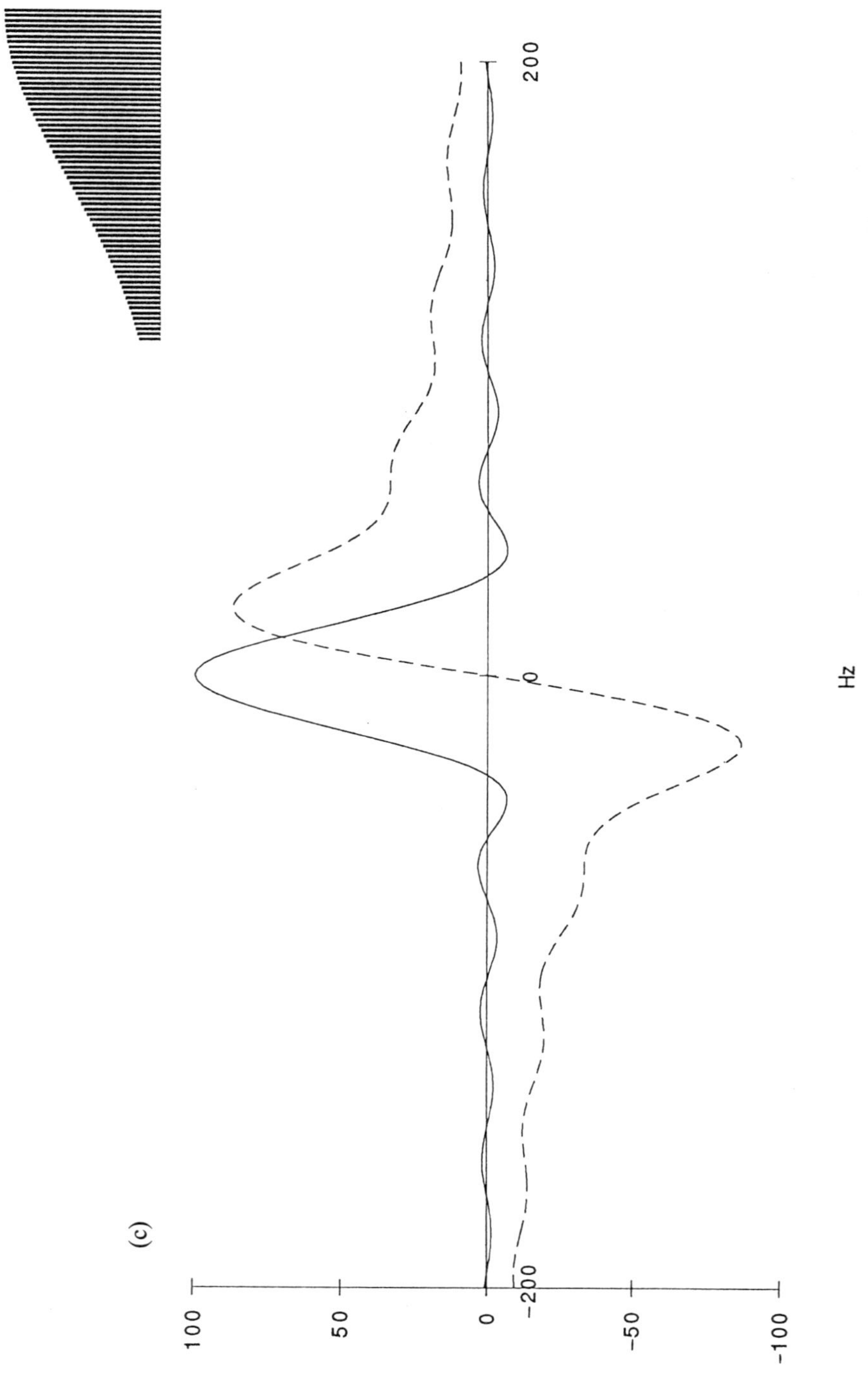
(c)
100
50
0
-50
-100
200
0
-200
Hz

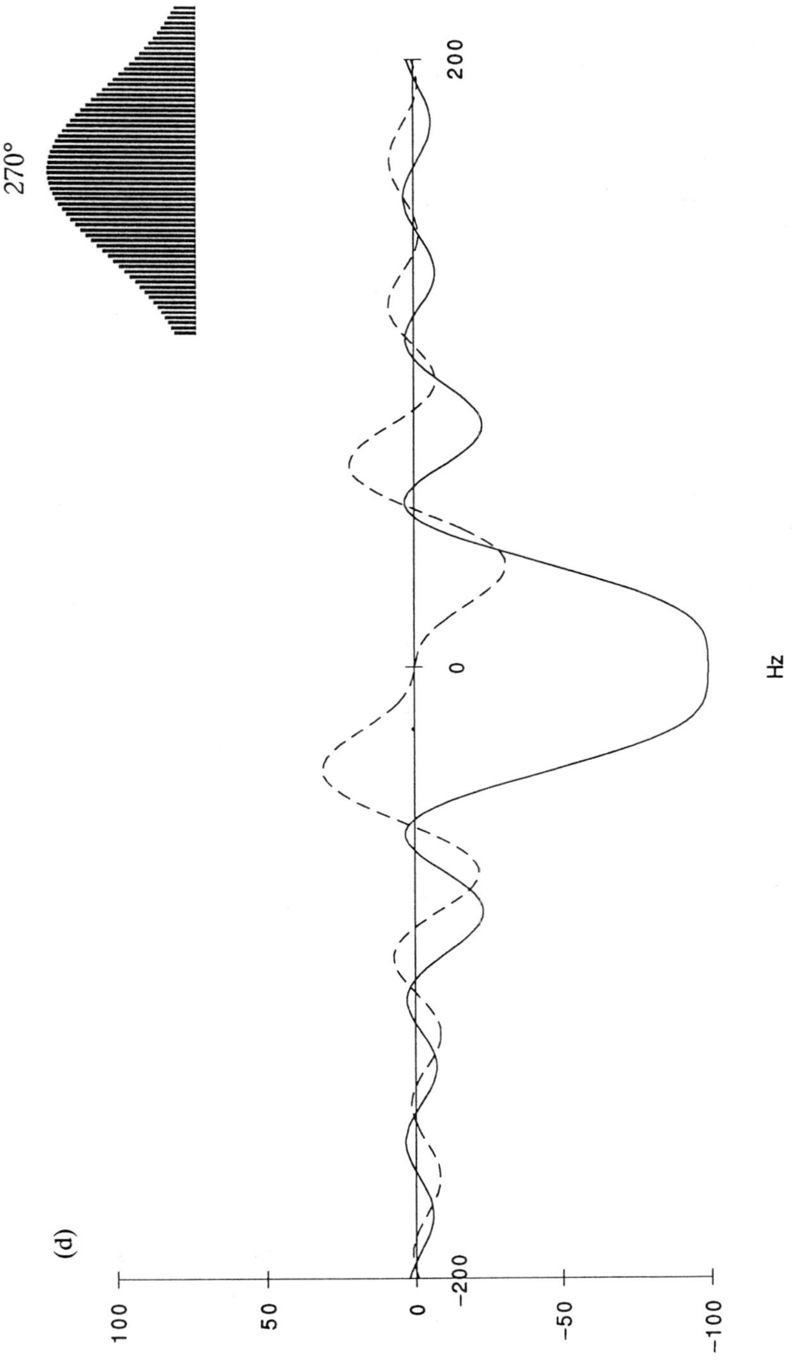

Figure 5.2 *(continued)*

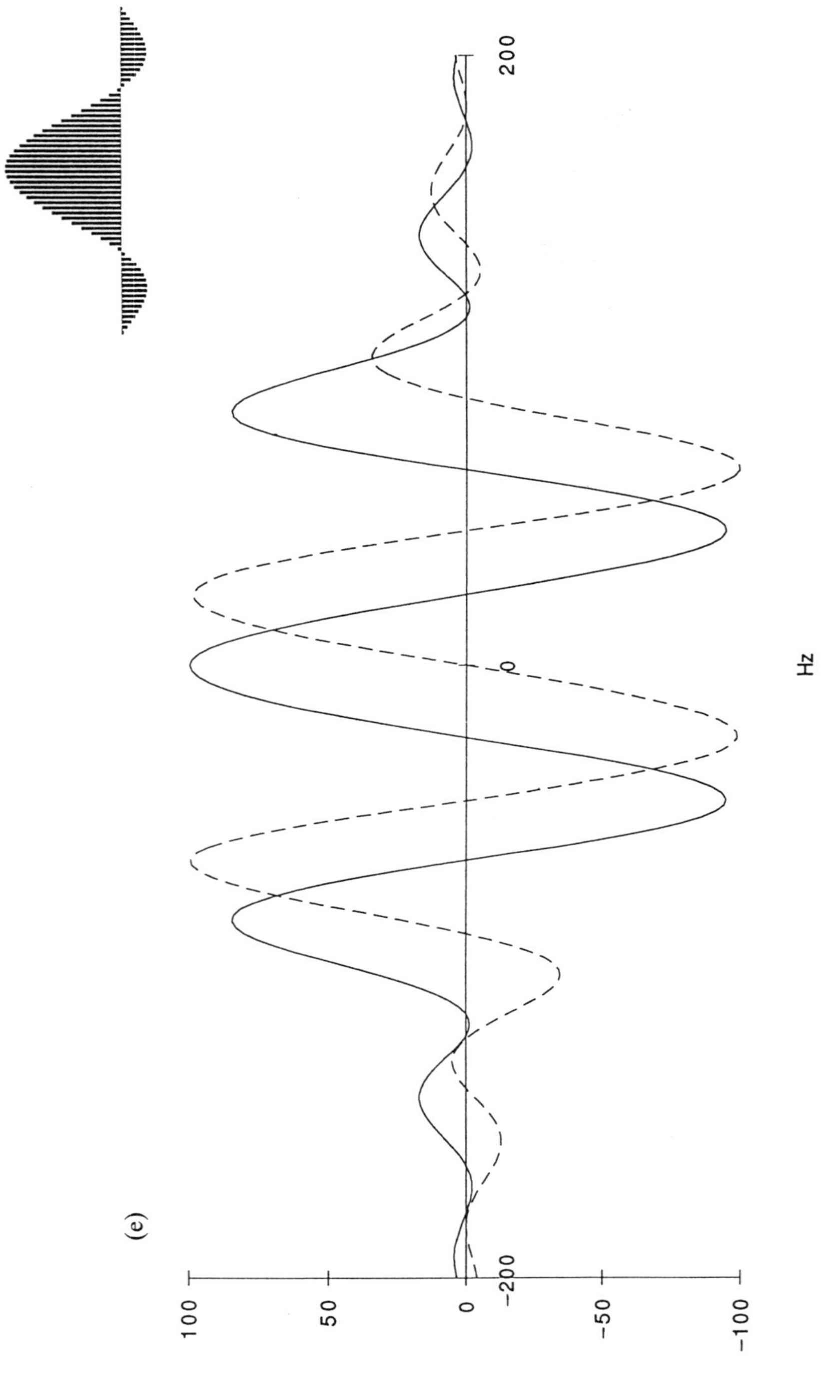
(e)
100
50
0
-50
-100
-200
0
200
Hz

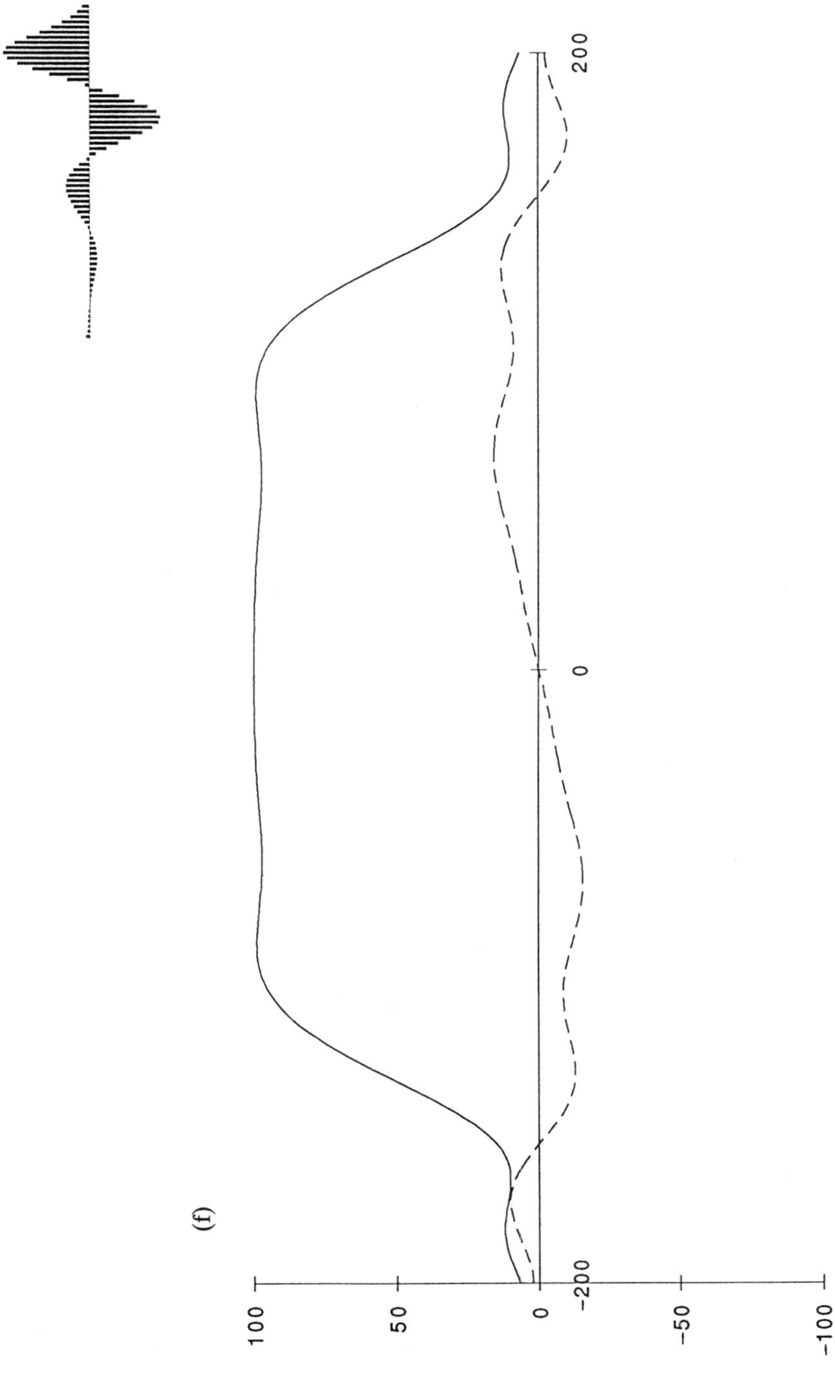

Figure 5.2 *(continued)*

duration, is to apply a 270° Gaussian pulse (Figure 5.2(d)) instead of a 90° pulse (L. Emsley and G. Bodenhausen, *J. Magn. Reson.*, **81**, 211 (1989)). Such a pulse is said to be self-refocused, since the additional 180° flip angle (with respect to a 90° pulse) can be thought of as playing a refocusing role.

Another popular soft pulse in NMR imaging is shaped according to a sinc function and applied in the presence of a B_0 gradient, simply because the Fourier transform of a sinc function (Section 3.3.2) is a rectangle that corresponds, in the imaging context, to the slice which has to be selected (Section 5.4). As for other pulse shapes, the existence of a dispersive component (Figure 5.2(e)) requires a refocusing procedure, which in that case is usually achieved by reversing the gradient polarity. Very pronounced sidelobes of opposite sign can be noticed, ruling out (drastically in that case), the Fourier relationship between the pulse shape and the selectivity profile. The opposite signs of two consecutive lobes make (by chance?) the procedure operational for slice selection purposes provided that the object is sufficiently homogeneous in the relevant direction. Finally, the state-of-the-art is represented by BURP pulses (band-selective uniform response pure-phase; H. Geen and R. Freeman, *J. Magn. Reson.*, **93**, 93 (1991)), whose shape has been optimized by numerical procedures so as to yield a 'top-hat' absorptive component with a negligible dispersive component (Figure 5.2(f)). The function corresponding to the highly specific shape $S(t)$ of the soft pulse is better expressed according to a Fourier series:

$$S(t) = \sum_n [A_n \cos(2\pi nt/\tau) + B_n \sin(2\pi nt/\tau)] \tag{5.1}$$

where τ is the pulse duration. An example of coefficients A_n and B_n is given in Table 5.1.

In general terms, selective excitation is used to reduce the amount of information to a level where straightforward interpretation becomes feasible. This of course includes the selection of a given transition, of a given multiplet, or of a given frequency region, with the immediate benefit of reducing the amount of data and the measuring time in multidimensional NMR experiments.

Another category of applications concerns suppression of the signal of the solvent (generally water) in the spectra of large biomolecules, which are present only at low concentration. Solvent suppression is mandatory in Fourier transform NMR to prevent the signals of interest from being obscured by the analog-to-digital converter (ADC) noise, the ADC resolution being scaled according to the amplitude of the huge solvent signal. One method for getting rid of the solvent signal (which is to some degree a selective excitation) consists in applying a weak r.f. field at the solvent resonance so as to nutate its magnetization for an appropriate period of time (the method is dubbed 'presaturation' and its basis is

Table 5.1 Coefficients which define the shape of one of the BURP family soft pulses (see Equation (5.1); here E-BURP 2, E standing for excitation)

n	0	1	2	3	4	5	6	7	8	9	10
A_n	0.26	0.91	0.45	−1.31	−0.12	0.03	0.01	0.06	0.01	−0.02	−0.01
B_n	0.00	−0.12	−1.79	0.01	0.41	0.08	0.07	0.01	−0.04	−0.01	0.00

that, while nutating, the magnetization decays according to a time constant $T_{1,2}$ which can be easily derived from the Bloch equations $1/T_{1,2} = 0.5(1/T_1 + 1/T_2)$). Thus, to suppress the solvent magnetization, the r.f. field must be applied for a period of sufficient duration with respect to T_1 and T_2. However, this method, which is widely used, suffers from two drawbacks (i) B_0 homogeneity should be at its maximum so that the solvent resonance covers the smallest possible frequency zone, and (ii) owing to the long period of application, resonances of protons exchangeable with water are also saturated (suppressed). To overcome this latter difficulty, it is possible to resort to the so-called J–R (jump and return) method (P. Plateau and M. Guéron, *J. Magn. Reson.*, **104**, 7310 (1982)) which amounts to bringing back the solvent magnetization toward the z axis (where it does not give rise to any detectable signal) while leaving other magnetizations in the xy plane (where they are detected). This is achieved by the simple sequence $(\pi/2)_x - \tau - (\pi/2)_{-x}$, which involves hard pulses and assumes that the solvent is on resonance; its magnetization does not evolve during the time τ and is therefore moved back towards the z axis, whereas magnetizations corresponding to other frequencies remain (in part) in the xy plane. The method has been improved by more sophisticated pulse sequences so as to obtain (i) a broader zone without excitation around the solvent resonance, and (ii) a selectivity profile more favorable for the other resonances (P. J. Hore, *J. Magn. Reson.*, **54**, 539 (1983)). Finally, as already mentioned in Section 2.2.6, the currently preferred method appears to combine a selective r.f. pulse and a gradient pulse so as to defocus the solvent magnetization selectively (M. Von Kienlin *et al.*, *J. Magn. Reson.*, **76**, 169 (1988); M. Piotto *et al.*, *J. Biomol. NMR*, **2**, 661 (1992)).

5.1.2 THE DANTE PULSE TRAIN

This method (G.A. Morris and R. Freeman, *J. Magn. Reson.*, **29**, 433 (1978)) makes use of a train of small flip angle (α) hard pulses, two consecutive pulses being separated by a precession interval τ'. In its basic version, which can be represented schematically as $[(\alpha)_x - \tau]_n$, it is merely equivalent to a rectangular soft pulse, with the major advantage of using the standard transmitter and thus avoiding (i) the need to switch to another transmit channel and (ii) having to cope with phase and amplitude settings between two different channels. To understand how this sequence works, let us consider a DANTE train which produces a total flip angle of 90°, thus with $n\alpha = \pi/2$. It can be noted that, during the interval τ', magnetization at an arbitrary frequency ν precesses by an angle $2\pi\nu\tau'$ (Figure 5.3).

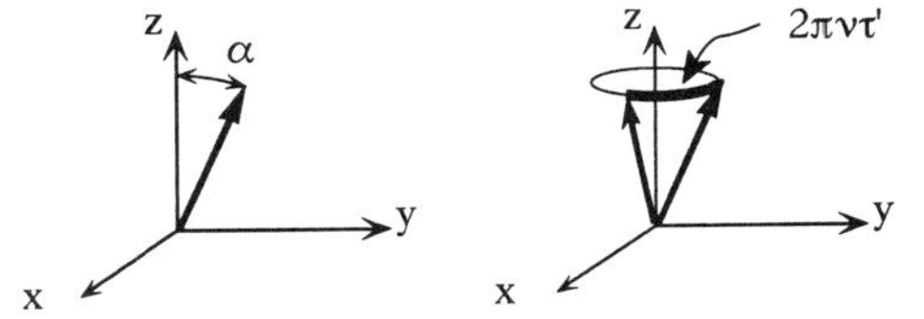

Figure 5.3 Left: nutation by a small flip angle α. Right: during the interval τ', magnetization at frequency ν precesses by an angle $2\pi\nu\tau'$

The next pulse, of flip angle α, acts only on the magnetization component situated in the yz plane and has little effect if this component is relatively small. Conversely, if $\nu = 0$ or $\nu = k/\tau'$ (k being an integer), the corresponding magnetization is always located in the yz plane subsequently to the interval τ', and *the r.f. pulse is fully active. Consequently, r.f. pulses have cumulative effects for magnetizations at frequencies of* $\pm k/\tau'$ *is an (k integer including 0).* Other magnetizations remain in the vicinity of the z axis. Hence selective excitation is effectively achieved at frequencies of $\pm k\tau'$. In fact, numerical simulations show that the excitation profile of a DANTE pulse train of overall duration $n\tau' = \tau$ is very close to that of a single rectangular pulse of duration τ. Interestingly, this analogy suggests that it should be possible to simulate shaped pulses with a DANTE pulse train. Indeed, if we decompose a soft shaped pulse into a series of elementary pulses of identical duration but varying amplitudes, an analogous DANTE sequence can be shown to consist of pulses of constant amplitude but varying duration, so that their areas are identical to those in the soft pulse decomposition (Figure 5.4).

An interesting feature, common to rectangular soft pulses and to a DANTE train of identical small flip angle pulses, is the inversion profile (obtained in the case of a DANTE pulse train for $n\alpha = \pi$). It can be seen (Figure 5.5) that the sidelobes are considerably reduced, suggesting that it should be highly advantageous to rely on the profile of the inverted magnetization.

In fact the following sequence, dubbed DANTE-Z (D. Boudot *et al., J. Magn. Reson.*, **83**, 428 (1989)), takes full advantage of the profile shown in Figure 5.5 and affords much cleaner excitation frequency zones with reduced oscillations and without a phase gradient:

$$[(\alpha)_x - \tau' - (\alpha)_{\pm x} - \tau']_n (\pi/2)(\mathrm{Acq})_\pm \tag{5.2}$$

We shall not enter into detail of the phase cycling aimed at eliminating residual unwanted components, but rather concentrate on its principle; n is chosen according to $2n\alpha = \pi$, so that when the two α pulses are of identical phase we are dealing with a DANTE inverting pulse (Figure 5.5). Conversely, when the two phases are of opposite sign, the pulse train produces essentially a flat profile around the frequency of interest (in practice, the zero frequency). Hence, subtraction of the two signals, after a ($\pi/2$) read pulse, yields for the transverse magnetization an excitation profile almost devoid of sidelobes and without any dispersive component. One feature of this procedure is the ability to produce a clean excitation profile without the need for special instrumentation aimed at the generation of shaped soft pulses.

Another noteworthy application of the DANTE sequence concerns the suppression of huge solvent peaks, a subject alluded to earlier in this section. Suppose that each pulse of the train is a B_1 gradient pulse. Magnetizations apart from the zero frequency remain along the z axis by virtue of the fundamental properties of the DANTE sequence. Conversely, in the zone around the zero frequency, because gradient pulses are used, magnetization defocusing occurs efficiently. The net result is the cancellation of the solvent signal if the carrier frequency has been set to coincide with the solvent resonance. By choosing average flip angles (α) of the

Figure 5.4 Top: a Gaussian soft pulse. Bottom: the analogous situation in the DANTE mode; pulses of constant amplitude but varying duration separated by precession intervals

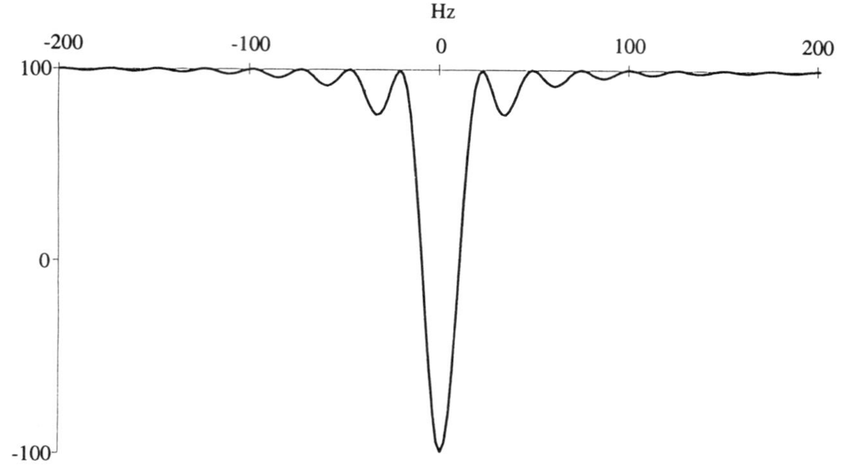

Figure 5.5 The frequency profile (longitudinal magnetization) of an inverting DANTE pulse train (or of an inverting rectangular soft pulse)

order of 2π (the average flip angle refers to the fact that B_1 varies across the sample), the pulse train can be made very short so as to avoid suppression of the effect of protons exchangeable with the solvent (water). Very efficient water suppression is thus possible without noticeable alteration of other resonances in the spectrum, as demonstrated by the profile shown in Figure 5.6.

5.2 Magnetization Transfer, Polarization Transfer and Coherence Transfer

The term magnetization transfer could be reserved for the cross-relaxation phenomena described in the previous chapter although, as explained before, they are actually polarization transfers. These can be homo- or heteronuclear in nature and arise from random processes, so that they are usually designated *incoherent* transfers. We are interested here in *coherent* transfers, which take place because an interaction exists which splits a resonance line into a multiplet, as is the case with the indirect coupling J. Again this property leads to the transfer of a polarization $\boldsymbol{\Delta}$ rather than to a magnetization transfer. $\boldsymbol{\Delta}$ is defined as the population difference between two energy levels whose I_z eigenvalues differ by 1. There is a further ambiguity concerning the terminology of 'coherence transfer' which originates in fact from a particular configuration of the transverse magnetization. We shall focus in the forthcoming sections on the particular aspects of polarization and coherence transfer:

- selective inversion of a particular transition and modifications of the energy level populations (polarization transfer);

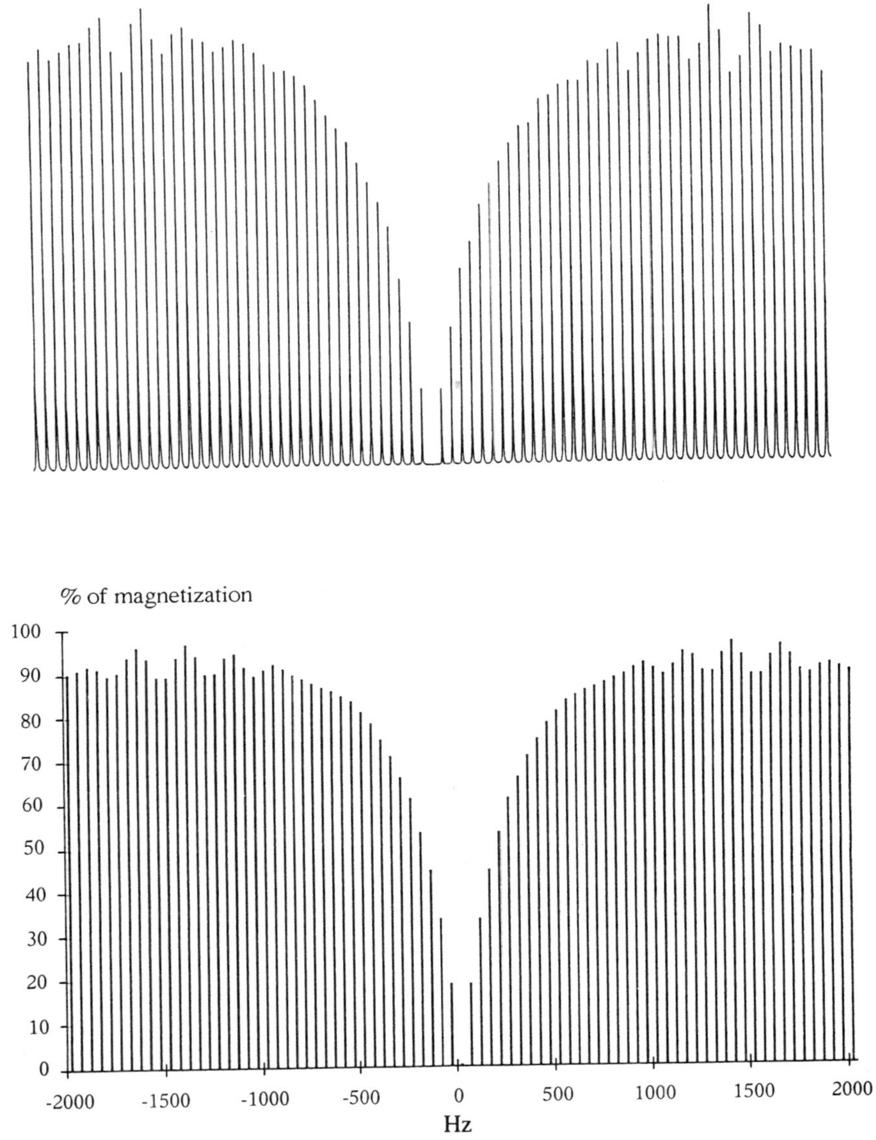

Figure 5.6 A typical selectivity profile obtained by a DANTE train made of B_1 gradient pulses (DEBOG sequence; D. Canet *et al.*, *J. Magn. Reson.*, **A105**, 239 (1993)). Top: experimental; bottom: simulated

- creation of an A antiphase doublet for an AX spin system and transformation into an X antiphase doublet (coherence transfer).

We shall see that these two types of transfer actually proceed from very similar approaches.

5.2.1 SOME EXAMPLES OF POLARIZATION TRANSFER

As a first example, let us consider a heteronuclear two-spin system denoted by AX, A being the nucleus of highest gyromagnetic ratio. We shall consider the consequences of the selective inversion of one of the lines in the A doublet. To this end, we reproduce in Figure 5.7 the energy diagram of Figure 2.11, which shows the populations of energy levels at thermal equilibrium. The modifications of the populations after inversion of one of the A transitions are indicated in the right part of Figure 5.7.

Starting from thermal equilibrium, the intensity of both X transitions is evidently $\boldsymbol{\Delta}_X$. This line intensity does not include the factor arising from the detection procedure but is merely indicative of the polarization denoted here by $\boldsymbol{\Delta}_X$. The global intensity of the X spectrum is thus $2\boldsymbol{\Delta}_X$ or 2 in $\boldsymbol{\Delta}_X$ units. After inversion of the A transition, identified by a dashed arrow in the diagram of Figure 5.7, the intensities of the two X transitions ($\alpha\alpha \rightarrow \alpha\beta$ and $\beta\alpha \rightarrow \beta\beta$) are changed to: $-\boldsymbol{\Delta}_A + \boldsymbol{\Delta}_X$ and $\boldsymbol{\Delta}_A + \boldsymbol{\Delta}_X$, respectively, yielding -3 and $+5$ in $\boldsymbol{\Delta}_X$ units for the $^{13}C{-}^{1}H$ spin system ($\boldsymbol{\Delta}_A \approx 4\boldsymbol{\Delta}_X$). If the X spectrum is acquired under A decoupling conditions, these two transitions coalesce, leading to a net intensity of $2\boldsymbol{\Delta}_X$, therefore without sensivity enhancement. By contrast, if a precession period allows for the refocusing of the two lines in the X doublet (that is, a period of duration $1/2\ J_{AX}$ which brings the two lines into phase), the global intensity becomes $8\boldsymbol{\Delta}_X$. This corresponds to an enhancement by a factor of 4 for a $^{13}C{-}^{1}H$ system, which is simply the ratio γ_A/γ_X, indicating that *polarization transfer from A to X has occurred*. This type of experiment may be dubbed *selective polarization transfer* SPT (K. G. R. Pachler and P. L. Wessels, *J. Magn. Reson.*, **12**, 337 (1973)) and can possibly be analyzed within the frame of product-operator formalism:

Starting from an equilibrium state described by $\gamma_A I_z^A + \gamma_X I_z^X$ (Section 2.2.3), we obtain a state represented by $\gamma_A I_z^A + \gamma_X I_y^X$ after the application of a pulse acting only

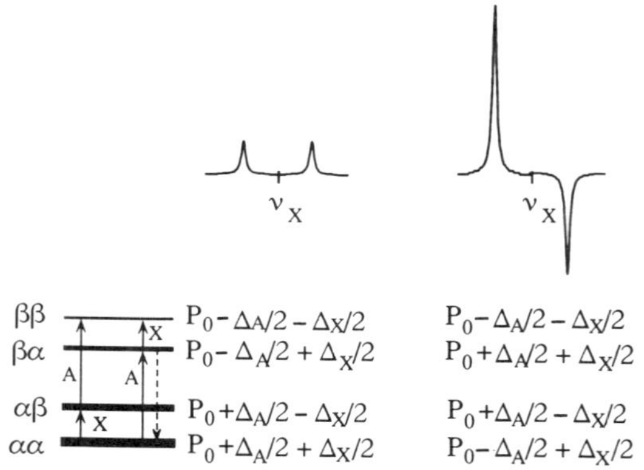

Figure 5.7 Energy diagram of the AX spin system (two spin 1/2 nuclei) with $\gamma_A > \gamma_X$, along with the energy level populations, at thermal equilibrium (left), and after inversion of the line indicated by a dotted arrow (right). The X doublet is represented schematically for each of these situations

on X along the x axis of the rotating frame. This is obviously observed as an X doublet with global intensity proportional to γ_X.

If one of the A transitions has been selectively inverted, the initial state can be represented by $\gamma_A(2I_z^A I_z^X) + \gamma_X I_z^X$. The same pulse as above leads to $\gamma_A(2I_z^A I_y^X) + \gamma_X I_y^X$. The first term corresponds to an antiphase X doublet whose intensity is proportional to γ_A (Chapter 2) whereas the second term corresponds to the conventional in-phase X doublet (as obtained above). The superposition of these two doublets (antiphase and in-phase) results in the situation which was derived from considerations of energy level populations.

The second example of polarization transfer, which we shall present now, involves a quite different process, since it implies the simultaneous application of *spin-locking* radio-frequency fields to both nuclei. This type of transfer is widely used in solid-state NMR with the objective of improving the sensitivity of a nucleus of low gyromagnetic ratio (e.g. carbon-13). It consists of bringing in contact the two magnetizations (of protons and of carbon) by means of spin-lock fields which must fulfill the Hartmann–Hahn condition (S. R. Hartmann and E. L. Hahn, *Phys. Rev.*, **128**, 337 (1973)): $\gamma_C B_1^C = \gamma_H B_1^H$, where B_1^C and B_1^H are the amplitudes of r.f. fields applied to carbon and to proton, respectively. The existence of a dipolar coupling between the two nuclei is the prerequisite for polarization transfer to take place. This experiment is termed 'cross-polarization' and is such that proton polarization is transferred to the carbon-13, thus enhancing the line intensities of the latter nucleus. A complete analysis of the experiment is somewhat complex and beyond the scope of this book. We shall rather focus on what is considered to be the analogous experiment for the liquid state. It is widely employed for homonuclear spin systems, though it can also be applied to heteronuclear systems, and is known under the acronyms HOHAHA (HOmonuclear, HArtmann HAhn; D. G. Davis and A. Bax, *J. Am. Chem. Soc.*, **107**, 2821 (1985)) or TOCSY (TOtal Correlation SpectroscopY; L. Braunschweiler and R. R. Ernst, *J. Magn. Reson.*, **53**, 521 (1983)). As shown below, if the magnetizations of two J-coupled nuclei A and X initially are taken along the spin-lock axis (of the rotating frame) in opposite directions, there then occurs a relentless polarization transfer modulated by $\sin(2\pi Jt)$. Hence, depending on the initial conditions, transfer (between A and X, or X and A) is maximum at $t = 1/2J$. It can be noted that, because one is dealing with a homonuclear spin system, the Hartmann–Hahn condition is automatically fulfilled (provided that the amplitude of the spin-locking field is such that off-resonance effects can be neglected).

A proper understanding of this polarization transfer process requires a quantum mechanical approach. During the spin-lock period, magnetizations are forced along the rotating frame axis associated with the r.f. field so that the effective hamiltonian (expressed in Hz) includes solely the J interaction

$$\mathcal{H} = J_{AX}\mathbf{I}^A\mathbf{I}^X = J_{AX}(I_x^A I_x^X + I_y^A I_y^X + I_z^A I_z^X) \tag{5.3}$$

Note that, because chemical shift effects are absent, the full J coupling hamiltonian must be considered (the first order approximation no longer holds). We shall neglect relaxation phenomena so that the evolution of the spin system is described by the density operator equation (the Liouville–von Neumann equation; see Section 2.2.1)

$$\frac{d\hat{\sigma}}{dt} = 2i\pi[\hat{\sigma}, \mathscr{H}]$$

with the hamiltonian given in (5.3). The evolution of any quantity $\langle G \rangle$ follows from the relationship $\langle G \rangle = Tr(\hat{\sigma}\hat{G})$ (again see Section 2.2.1) which can be inserted in the previous equation to yield

$$\frac{d\langle G \rangle}{dt} = 2i\pi Tr([\mathscr{H}, \hat{G}]\hat{\sigma}) \tag{5.4}$$

The solution of the problem amounts to calculating the commutator in (5.4) for the quantities of interest. Denoting by x the spin-lock axis in the rotating frame and acknowledging the fact that the r.f. field inhomogeneity will defocus all quantities existing prior to the spin-lock interval except $\langle I_x^A \rangle$, $\langle I_x^X \rangle$, $\langle 2I_y^A I_z^X \rangle$ and $\langle 2I_z^A I_y^X \rangle$ (in phase doublets along x and antiphase doublets along y), we can define the following quantities which will prove convenient for solving (5.4)

$$\begin{aligned} S_{\text{in}} &= \langle I_x^A + I_x^X \rangle \\ D_{\text{in}} &= \langle I_x^A - I_x^X \rangle \\ S_{\text{anti}} &= \langle 2I_y^A I_z^X + 2I_z^A I_y^X \rangle \\ D_{\text{anti}} &= \langle 2I_y^A I_z^X - 2I_z^A I_y^X \rangle \end{aligned} \tag{5.5}$$

Evaluating the relevant commutators leads to a four differential equations

$$\begin{aligned} \frac{d}{dt} S_{\text{in}} &= 0 \\ \frac{d}{dt} D_{\text{in}} &= -2\pi J_{AX} D_{\text{anti}} \\ \frac{d}{dt} S_{\text{anti}} &= 0 \\ \frac{d}{dt} D_{\text{anti}} &= -2\pi J_{AX} D_{\text{in}} \end{aligned}$$

and reveals some interesting features: the sum of in-phase and antiphase A and X doublets (denoted respectively by S_{in} and S_{anti}) is invariant and the quantity we are most interested in, D_{in} (the difference between A and X spin-locked magnetizations), obeys a simple second order differential equation

$$\frac{d^2}{dt^2} D_{\text{in}} = -(2\pi J_{AX})^2 D_{\text{in}}$$

whose general solution, assuming that no antiphase coherence is present prior to the spin-lock period, can be expressed as:

$$D_{\text{in}} = D_{\text{in}}(0)\cos(2\pi J_{AX} t) \tag{5.6}$$

This yields for A and X magnetizations along the spin locking direction:

$$\begin{aligned} \langle I_x^A(t) \rangle &= (1/2)\langle I_x^A(0) \rangle[1 + \cos(2\pi J_{AX} t)] + (1/2)\langle I_x^X(0) \rangle[1 - \cos(2\pi J_{AX} t)] \\ \langle I_x^X(t) \rangle &= (1/2)\langle I_x^A(0) \rangle[1 - \cos(2\pi J_{AX} t)] + (1/2)\langle I_x^X(0) \rangle[1 + \cos(2\pi J_{AX} t)] \end{aligned} \tag{5.7}$$

It can be recognized that $t = 1/2J_{AX}$ must have a particular value, as the cosine function becomes equal to -1. In order to understand the polarization transfer process, let us assume that by some selective process $\langle I_x^{\mathrm{A}}(0)\rangle = I_{\mathrm{eq}}$ with $\langle I_x^{\mathrm{X}}(0)\rangle = 0$; we obtain at $t = 1/2J_{\mathrm{AX}}$

$$\langle I_x^{\mathrm{A}}(t)\rangle = 0 \quad \text{and} \quad \langle I_x^{\mathrm{X}}(t)\rangle = I_{\mathrm{eq}}$$

The polarization of spin A has been totally transferred to spin X. Another possible process, alluded to above, consists of bringing the A and X magnetizations along the spin-lock axis in opposite directions so that $\langle I_x^{\mathrm{A}}(0)\rangle = I_{\mathrm{eq}}$ and $\langle I_x^{\mathrm{X}}(0)\rangle = -I_{\mathrm{eq}}$. From (5.7) it can be seen that, at $t = 1/2J_{\mathrm{AX}}$, the polarization transfer process completely reverses the initial situation, that is, $\langle I_x^{\mathrm{A}}(t)\rangle = -I_{\mathrm{eq}}$ and $\langle I_x^{\mathrm{X}}(t)\rangle = I_{\mathrm{eq}}$. An important point (not developed here) concerns doubly selective spin-lock r.f. fields acting simultaneously on A and X (which may be heteronuclei). Because, in that case, the coupling hamiltonian is restricted to $J_{\mathrm{AX}} I_z^{\mathrm{A}} I_z^{\mathrm{X}}$ (which applies when dealing with weakly coupled spins), the characteristic time for maximum polarization transfer becomes $1/J_{\mathrm{AX}}$ instead of $1/2J_{\mathrm{AX}}$ (see R. Konrat *et al.*, *J. Am. Chem. Soc.*, **113**, 9135 (1991)).

We turn now to the two-dimensional counterpart of this experiment, which has found wide application for the delineation of coupling networks in biomolecules. The sequence is depicted in Figure 5.8 together with the example of the TOCSY (or HOHAHA) diagram of a small protein, each row being characteristic of a given amino acid.

As usual, the instrumental time t_1 is used for labeling transverse magnetization components according to their chemical shifts. During the spin-lock period, polarization transfer occurs as explained above. A cross peak in the 2D map reveals the existence of a J coupling between the two considered nuclei. A first advantage of this experiment is that cross peaks are in-phase multiplets, by contrast with other correlation experiments (such as COSY) which yield antiphase multiplets, with the inconvenience of poor legibility and possible peak cancellation when the splitting is of the order of (or smaller than) the linewidth. Another very interesting feature is the propagation of transfers by a relay process within a given spin system for a spin-lock application much greater than $1/2J$. The response for a given spin therefore involves the total coupling network to which it belongs, hence the terminology of total correlation spectroscopy.

A final point concerns the radio-frequency field used during the spin-lock period: it can be noted that the TOCSY (or HOHAHA) sequence is basically identical to the ROESY sequence (see Section 4.1.3), which is aimed at measuring relaxation parameters in the rotating frame. Of course, the measurement of these parameters is not the goal of the TOCSY sequence; moreover, signal attenuation by relaxation should be avoided. On the other hand, the Hartmann–Hahn condition should be fulfilled for all resonances in the spectrum, implying that off-resonance effects should be kept to a minimum (or, in other words, that the effective B_1 should be approximately the same for the whole spectral width under investigation). Both objectives (reductions in relaxation and off-resonance effects) can be achieved by using windowless pulse trains of the kind employed for improving heteronuclear decoupling (see Section 1.3.5). Among these the most popular are the MLEV-17 and WALTZ-16 schemes. The latter, for instance, comprises a succession of 180° pulses, in the form of 'composite pulses', which are immune to off-resonance effects and to r.f. field inhomogeneity effects. Moreover, incessant 180° pulses have the virtue of restoring the magnetization to its equilibrium configuration, so that

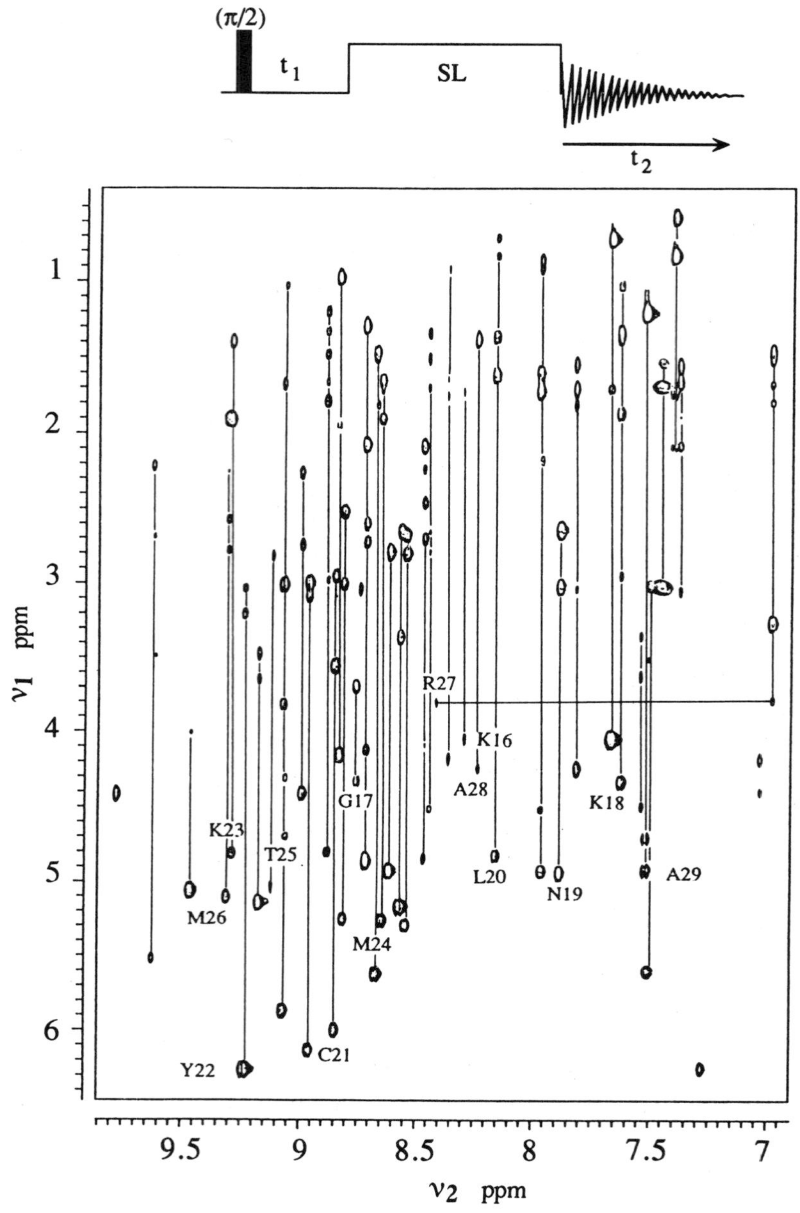

Figure 5.8 Basic scheme of the TOCSY (or HOHAHA) sequence (SL stands for the spin-lock period), illustrated by the two-dimensional map of a small protein. Each row (marked by a vertical line) corresponds to the residue associated with a given amide proton

relaxation phenomena tend to be annihilated. As a further illustration of the efficiency of the method, we show in Figure 5.9 a one-dimensional selective TOCSY spectrum displaying all multiplets belonging to the coupling network of the selectively excited spin.

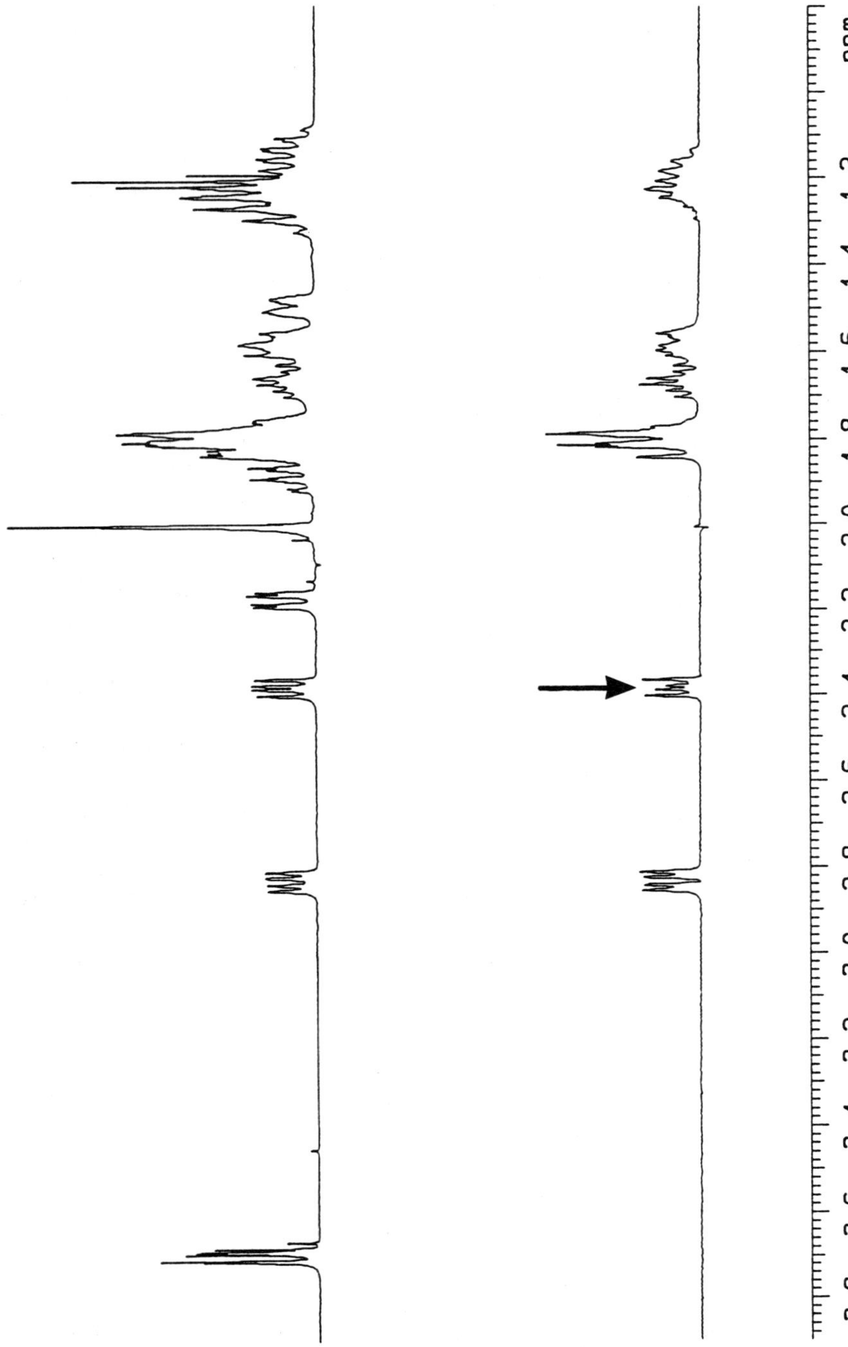

Figure 5.9 Top: conventional spectrum of an organic compound. Bottom: the TOCSY spectrum obtained by applying first a selective pulse (along the *x* axis of the rotating frame) to the multiplet marked by an arrow and then a spin-lock field (along the *y* axis of the rotating frame)

5.2.2 COHERENCE TRANSFER IN A HETERONUCLEAR SYSTEM. THE INEPT, HETCOR AND HSQC SEQUENCES

We shall consider here the creation of an antiphase A doublet via a simple $\pi/2$ pulse applied exclusively to A and followed by a precession period equal to $1/(2J_{AX})$. For simplicity (and subsequently for practical reasons of enhanced sensitivity), we shall focus on a heteronuclear AX system with, as usual, A being ^{1}H and X being ^{13}C. In a first stage, we shall further assume that A is on resonance, meaning that its resonance frequency is identical to the relevant transmitter frequency. If we identify the phase of the initial $\pi/2$ pulse (applied to A) with the x axis of the rotating frame, an antiphase A doublet exists along the x axis by the end of the precession interval of duration $1/(2J_{AX})$. The state of the spin system is thus represented by $\gamma_A(2I_x^A I_z^X) + \gamma_X I_z^X$ (spin X has been left unaffected). $\pi/2$ pulses are then applied simultaneously to A and X (Figure 5.10). The $\pi/2$ pulse applied to A must act along the y axis of the rotating frame, while the phase of the X pulse is arbitrary; without loss of generality, we can assign this phase to the x axis of the rotating frame associated with the X transmitter. The first term $\gamma_A(2I_x^A I_z^X)$ transforms into $\gamma_A(2I_z^A I_y^X)$ whereas the second term $\gamma_X I_z^X$ transforms into $\gamma_X I_y^X$. We thus arrive at an X antiphase doublet which benefits from the factor γ_A which superimposes onto the conventional X in-phase doublet (affected by γ_X). This situation is quite similar to the one depicted in Figure 5.7 but is here obtained by *coherence transfer*, since the quantity $2I_z^A\ I_y^X$ corresponds to a coherence defined as an off-diagonal element of the density matrix (see Section 2.2.3).

It can be acknowledged that these two concepts (polarization transfer and coherence transfer) are closely related, at least with regard to their consequences. The INEPT (insensitive nuclei enhanced by polarization transfer) sequence (G. A. Morris and R. Freeman, *J. Am. Chem. Soc.*, **101**, 760 (1979)) is derived from the sequence of Figure 5.10 with two π pulses (applied to A and X) in the middle of the evolution interval (Figure 5.11). According to the analysis presented earlier (Section 2.2.6) these two π pulses do not refocus J coupling effects but do refocus chemical shift effects. Consequently, A no longer needs to be on resonance. Hence the method constitutes a simple means for enhancing the X spectrum by a factor equal to γ_A/γ_X (identical to the sensitivity gain in polarization transfer experi-

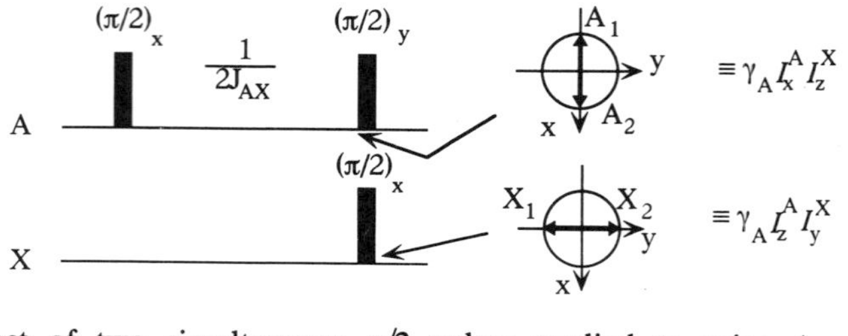

Figure 5.10 Effect of two simultaneous $\pi/2$ pulses applied to spins A and X after an evolution period of duration $1/(2J_{AX})$ which transforms the transverse A magnetization into an antiphase doublet. The first diagram (top right) depicts the A antiphase doublet which is converted into an X antiphase doublet by the two $\pi/2$ pulses (lower diagram)

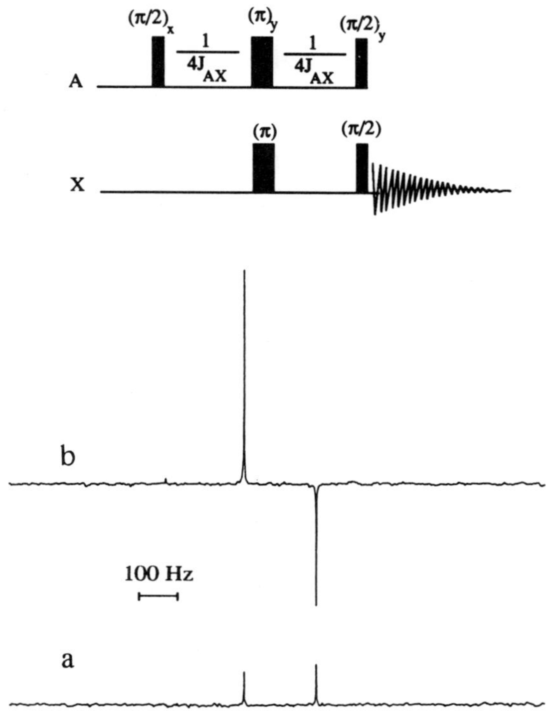

Figure 5.11 Basic scheme of the INEPT sequence and ^{13}C spectra of formic acid: (a) conventional spectrum; (b) INEPT spectrum

ments) provided that we know the order of magnitude of J_{AX} (when the time elapsed between the first and last pulses does not exactly fit $1/(2J_{AX})$, the transfer efficiency is just slightly impaired).

Nevertheless, the problem of A decoupling (proton decoupling) remains. In the sequence of Figure 5.11, if decoupling is applied during acquisition of the X free induction, the two lines in the antiphase doublet coalesce and eventually annihilate; we are left with the normal intensity of the X spectrum (Figure 5.11(a)) and we lose the benefit of the coherence transfer process. If decoupling must be employed for the purpose of spectrum legibility (and for further enhancing the sensitivity of the experiment by grouping all the lines within the considered multiplet), a refocusing period transforming the antiphase doublet into an in-phase doublet must be considered. The relevant sequence involves an evolution period of duration equal again to $1/2J_{AX}$ with X chemical shift refocusing and we end up with the so called refocused INEPT experiment shown schematically in Figure 5.12.

The two-dimensional counterpart of the INEPT sequence is deduced from the previous considerations and allows for the establishment of correlations between chemical shifts of the A and X nuclei (i.e., with our usual example, between the proton and carbon-13). Let us denote by ν_1 the dimension corresponding to proton chemical shifts, the dimension ν_2 being attributed to the physically detected signals,

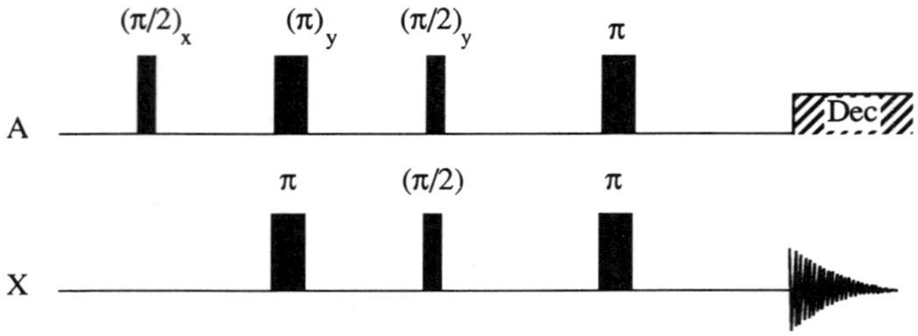

Figure 5.12 'Refocused' INEPT sequence. Each interval is set to $1/(4J_{AX})$

i.e. those of nucleus X. The relevant pulse sequence, named HETCOR (for HETeronuclear CORrelation) or H,X-COSY, is depicted in Figure 5.13. The interval t_1 is aimed at labeling transverse magnetization according to A (proton) chemical shifts; the component along the y axis of the rotating frame is modulated by $\cos(2\pi\nu_A t_1)$ just before the stage of the antiphase doublet creation. The π(X) pulse located at the middle of the t_1 interval leads to the refocusing of any AX coupling in such a way that no antiphase AX doublet can exist at the end of the evolution period (see Section 2.2.6). It can be said that A has been decoupled from X during the t_1 interval so that, in the ν_1 dimension, the A spectrum does not involve any splitting due to J couplings with X nuclei. Coherence transfer is achieved by the two $\pi/2$ pulses (applied simultaneously to A and X, or more simply cascaded without any significant interval between them) and leads to an antiphase X doublet modulated by $\cos(2\pi\nu_A t_1)$. This antiphase doublet is refocused by a second interval of duration $1/(2J_{AX})$, thus enabling A decoupling during the acquisition of the X free induction decay. It must be noted that, after the t_1 evolution period, an X chemical shift produces a dephasing (in the one-dimensional INEPT sequence, it was refocused by a π pulse). In fact, this dephasing is here of little concern because the present experiment will involve a double complex Fourier transform, implying an appropriate phase cycling for quadrature detection in t_1 (see Section 5.2.4) and a final display in the amplitude mode, and is thus insensitive to the phase alterations alluded to above.

The above discussion was concerned only with a two-spin system. In practice, we shall be dealing with A_nX spin systems (e.g. A_2X and A_3X for CH_2 and CH_3 moieties). Obviously, the so-called proton multiplicity (number of protons bound

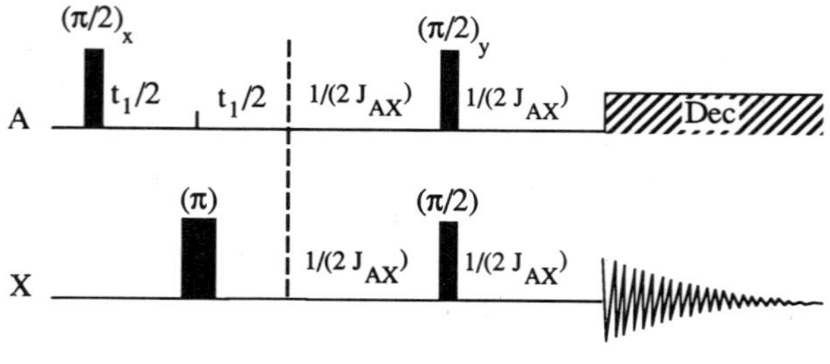

Figure 5.13 The HETCOR sequence

to a carbon) does not affect the transfer stage in an INEPT sequence because only the coupling between proton(s) and one carbon-13 is involved. However, this is no longer true for the refocusing stage in INEPT or in HETCOR experiments because, in that case, the carbon multiplet (whose structure depends on the number of bound protons) is directly involved. Consequences of proton multiplicity and their practical implications for assignment purposes will be discussed in the next section.

For the moment, we will turn to an experiment based on the INEPT concept and aimed at establishing A–X correlations with A detection (^{1}H detection), thus with improved sensitivity. This experiment employs a first INEPT sequence for transferring A coherence to X and, after an X evolution period, a second INEPT sequence for back transferring X to A (retro-INEPT). It is dubbed HSQC (heteronuclear single quantum correlation; G. Bodenhausen and D. J. Ruben, *Chem. Phys. Lett.*, **69**, 185 (1980)) by reference to the HMQC experiment (heteronuclear multiquantum correlation) to be described later. Both experiments are called 'inverse' since they enable the observation of a nucleus of low sensitivity (e.g. ^{13}C or ^{15}N), which has already benefitted from a transfer from the high gyromagnetic ratio nucleus, via detection at a higher frequency (e.g. the proton frequency), with the advantage of much improved sensitivity. The basic HSQC sequence depicted in Figure 5.14 will be briefly discussed.

The first part of the sequence (the first INEPT fragment) obviously transforms the A antiphase doublet into an X antiphase doublet (regarding the example used throughout, the antiphase doublet corresponding to ^{13}C satellites in the proton spectrum is transferred to carbon-13). During the t_1 period, the X antiphase doublet evolves according to the X chemical shift, thus yielding an X spectrum in the ν_1 dimension. However, this spectrum must be decoupled from A (i.e. from protons). This is achieved by the central π(A) pulse which refocuses all J_{AX} coupling effects. The second INEPT fragment converts the X antiphase configuration (labeled according to X chemical shifts) back to A in-phase magnetization, which is detected under X decoupling conditions in the t_2 dimension. Decoupling must cover a wide spectral range, and one generally relies upon a dedicated modulation scheme known as GARP (A. J. Shaka *et al.*, *J. Magn. Reson.*, **64**, 547

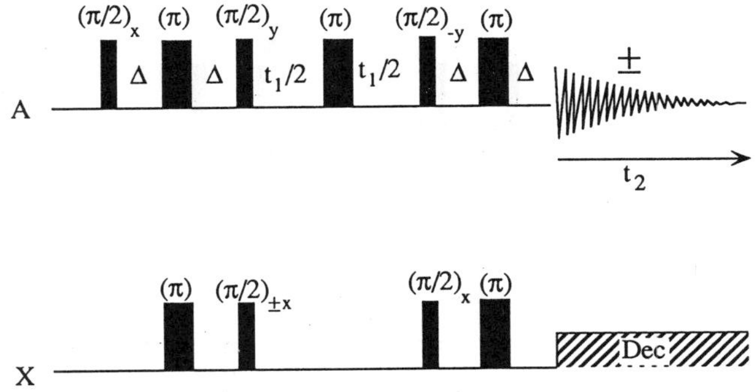

Figure 5.14 The basic HSQC sequence with its minimum phase cycling (phase alternation of the $(\pi/2)_{\pm x}$ X pulse concomitantly with acquisition sign alternation). The Δ interval is set to approximately $1/(4J_{AX})$

(1985)). Hence this A–X correlation experiment is performed throughout with the sensitivity of the high gyromagnetic ratio nucleus (with the proton sensitivity; altogether, this represents in principle a sensitivity gain by a factor of $(\gamma_A/\gamma_X)^3$ with respect to the direct observation of X): cross peaks are indicative of the existence of a J coupling between A and X and a two-dimensional correlation map similar to that of Figure 1.38 is obtained (with decoupling conditions in both dimensions). It must however be borne in mind that the experiment concerns only those molecules possessing an X active nucleus (in practice, 1% for carbon-13 at natural abundance) and that signals arising from inactive X nuclei (in practice the so-called parent signals corresponding to molecules with ^{12}C nuclei) must be removed. This is in principle achieved by the two-step phase cycle indicated in Figure 5.14: the phase of the last $\pi/2$(X) pulse in the first INEPT fragment is changed by 180°, concomitantly with sign alternation of the acquisition. This amounts to no change for the active X nuclei, since the sign of the coherence transfer is changed along with that of the acquisition. In respect of molecules without any X active nuclei, because there is no coherence transfer involved, the final result is simply subjected to a sign change leading in principle to zero. In practice, this type of difference spectroscopy suffers from instrumental imperfections, pulse misadjustments, spectrometer instabilities, etc., so that a barely acceptable result requires extensive phase cycling. The solution to the problem of parent signal elimination lies evidently in the absence of subtraction processes. Gradients acting as purges (in such a way that they selectively defocus unwanted magnetizations) can meet this goal without sensitivity loss and without additional scans for satisfying phase cycling requirements. Such a sequence (W. Wider and K. Wüthrich, *J. Magn. Reson.*, **B102**, 239 (1993)) is depicted in Figure 5.15 and commented on below.

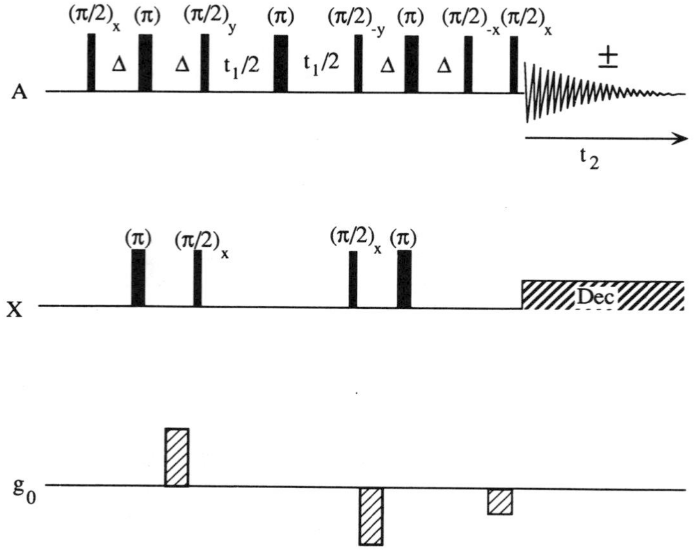

Figure 5.15 A HSQC experiment with B_0 gradients (g_0). The different intervals have the same meaning as in Figure 5.14

First, it can be noted that, prior to the application of the first gradient pulse, the state of the AX system we are dealing with is represented by $2I_z^A I_z^X$, and is thus totally immune to the B_0 gradient. By contrast, the parent magnetization is essentially transverse and will be defocussed by this gradient pulse. A similar situation prevails when the second gradient pulse occurs; it will destroy any reconstructed parent magnetization although again preserving the AX spin system state. Negative polarization of this second gradient pulse is mandatory because of the π pulse at the center of the evolution period. Finally, the third gradient pulse, whose amplitude is chosen to avoid any refocusing effect, eliminates any unwanted magnetization which may occur at this stage; the same requirements and principles are again involved for this gradient pulse (the last two $\pi/2$ pulses acting on A have the virtue of storing the quantity of interest along z while the gradient is on). Globally, the magnetization arising from protons not bound to the active X nucleus has been efficiently removed, whereas the AX spin system of interest has evolved, without sensitivity loss, and yields the desired correlation information. Finally, it should be noted that B_1 gradients can also meet the goal of removing unwanted magnetizations without the need for the sophisticated instrumentation required by B_0 gradients (see, for instance, G. Otting and K. Wüthrich, *J. Magn. Reson.*, **76**, 569 (1988)).

5.2.3 SPECTRAL EDITING AS A FUNCTION OF MULTIPLICITY. GATED DECOUPLING EXPERIMENTS. THE DEPT SEQUENCE

The major objective of the experiments to be described below is the discrimination between carbons belonging to CH_3, CH_2, CH or C groupings (we mean by 'C grouping' a carbon not directly bound to any proton). The simplest experiment does not rest on polarization or coherence transfer but merely on the specific precession behavior of multiplets associated with each of these groupings. For such an analysis the vectorial model is adequate, and we shall, in a first stage, assume that the carbon being considered is on resonance so that transverse magnetization evolves only under the effect of J_{CH} couplings. Figure 5.16 shows this evolution for a period whose duration is set to $1/J_{CH}$.

The diagrams of Figure 5.16 are easily constructed by noticing that the magnetization associated with lines positioned at $J/2$ or $3J/2$ from the multiplet center rotates by π and 3π, respectively, whereas for a line at J the rotation angle is 2π (of course, magnetization on resonance is stationary). It can be further noted that all multiplets are refocussed, so that proton decoupling can be applied during acquisition without modification of the signal intensities. However, and this is the interesting point, the sign depends on the multiplicity: signals are positive for an even multiplicity (C, CH_2) and negative for an odd multiplicity (CH, CH_3). In order to use this method for the whole set of carbons in the spectrum, chemical shift effects must be suppressed by an echo procedure. The sequence is shown in Figure 5.17 (C. Le Cocq and J. Y. Lallemand, *J. Chem. Soc. Chem. Commun.*, **150** (1981)); the application of proton decoupling during the first half of the echo prevents any evolution due to J_{CH} couplings, an evolution in agreement with the diagrams of Figure 5.16 taking place during the echo second half.

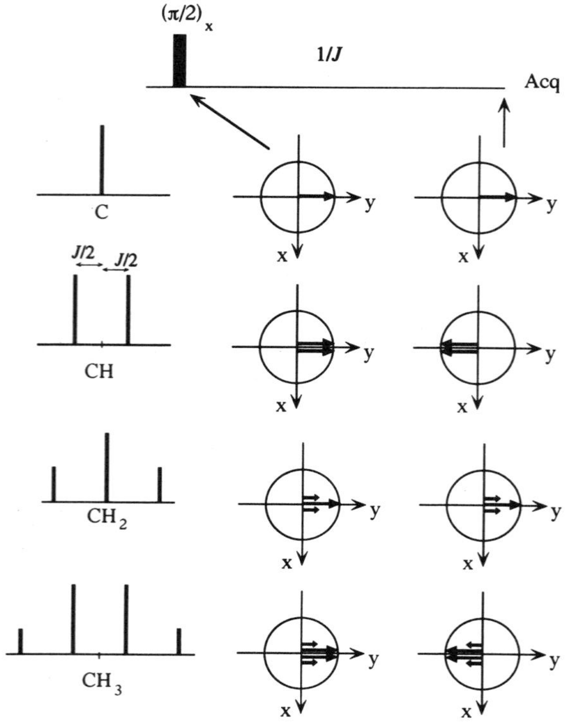

Figure 5.16 Evolution of ^{13}C multiplets corresponding to C, CH, CH_2 and CH_3 groupings for a precession time of $1/J_{CH}$. The carbon considered is assumed to be on resonance

A variant of this method has been introduced to avoid long waiting times (which are necessary for the return of magnetization to thermal equilibrium); it has been named APT (attached proton test; S. L. Patt and J. N. Shoolery, *J. Magn. Reson.*, **46**, 535 (1982)) and uses a small flip angle pulse (α) instead of the initial ($\pi/2$) pulse so as to leave a large part of magnetization along z (this is related, in standard experiments, to the choice of an optimal flip angle with regard to longitudinal relaxation time; see Section 3.1). However, magnetization is inverted by the echo central π pulse and must be restored towards $+z$ by an additional echo (Figure 5.18).

The sequences of Figures 5.17 and 5.18 have the virtue of being very simple, since they only require control of the decoupler. They do not lead, however, to a complete discrimination between the four types of carbons. Although some experiments with gated decoupling have been devised to achieve this goal, multipulse sequences relying on coherence transfer are generally preferred. The ability of such discrimination by multipulse methods can be examplified by the 'refocused INEPT' sequence of Figure 5.12. Using diagrams of the type of Figure 5.16, it can be shown that refocusing occurs for a CH grouping whereas a CH_2 grouping appears as an antiphase doublet of splitting equal to $2J$ and CH_3 grouping as an antiphase quartet (four lines of identical intensity in modulus, the two left

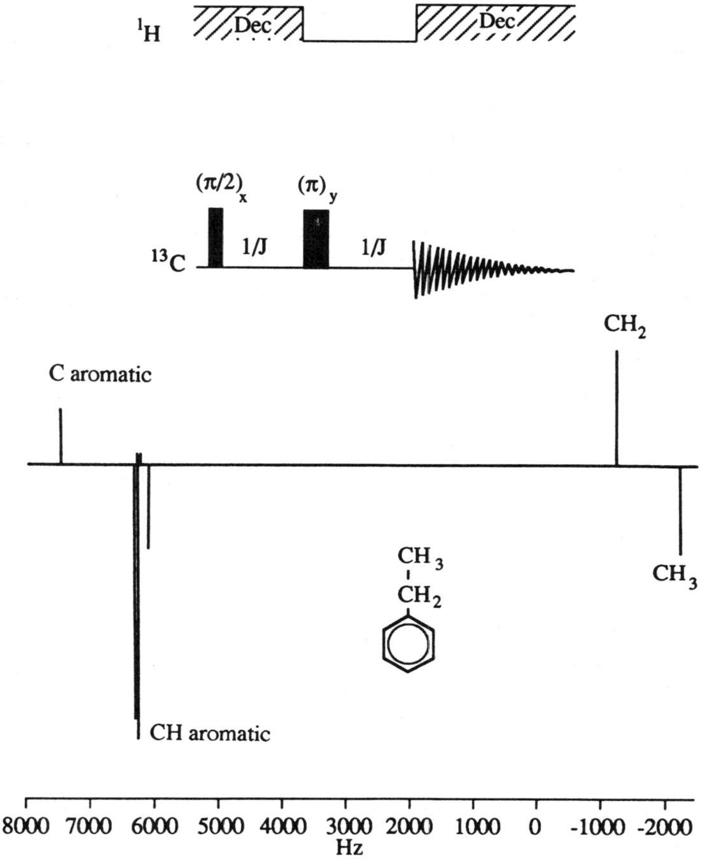

Figure 5.17 Sequence for discriminating carbons as a function of their multiplicity: positive signals for an even multiplicity (C, CH_2) and negative signals for an odd multiplicity (CH, CH_3) illustrated with the example of ethylbenzene

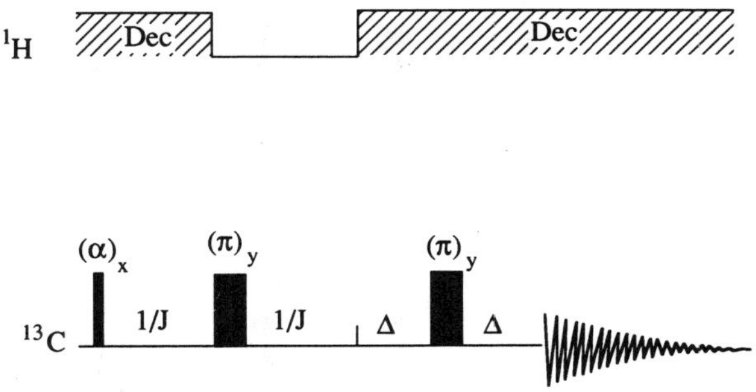

Figure 5.18 The APT pulse sequence for reducing the measuring time of the sequence shown in Figure 5.17. Δ is set to the receiver recovery time

lines and the two right lines being of opposite sign). The DEPT sequence (distorsionless enhancement polarization transfer) appears to be the method of choice for editing subspectra relating to CH, CH_2 and CH_3 groupings through three different experiments. This sequence bears some resemblance to the INEPT sequence but seems more robust regarding the setting of precession periods with respect to $1/2J_{CH}$ (it can be recalled that, as a function of hybridization, $^1J_{CH}$ ranges from 120 to 200 Hz; usually a reference value of 135 Hz is chosen, which actually corresponds to most aliphatic carbons). The sequence is depicted in Figure 5.19. The two π pulses (on the proton and carbon channels) have the sole objective of refocusing the chemical shift effects. In the following analysis, we can therefore assume that A(^{1}H) and X(^{13}C) are both on resonance.

At least three experiments must be performed for different values of the flip angle θ (generally $\pi/4$, $\pi/2$ and $3\pi/4$), yielding by appropriate linear combinations the subspectra associated with the three multiplicities CH, CH_2 and CH_3. The phase change of the θ pulse and the sign alternation of acquisition make it possible to eliminate the carbon-13 magnetization which does not originate from polarization (coherence) transfer and which could alter the presentation of the three subspectra. For a CH grouping (simple AX system), the analysis is straightforward: subsequent to the first precession interval (of duration $1/2J$), the proton doublet represented by $\gamma_A(2I_x^A I_z^X)$ is antiphase. The $\pi/2$ pulse applied to carbon-13 converts this quantity into $\gamma_A(2I_x^A I_y^X)$, which contains zero-quantum and two-quanta coherences. As stated in Chapter 2, this quantity evolves exclusively according to A and X chemical shifts whose effects are removed by the π pulses. Consequently, no evolution takes place during the second interval of duration $1/2J$. The $\theta_{\pm y}$ pulse leads to $2\gamma_A(\cos\theta I_x^A \pm \sin\theta I_z^A)I_y^X$. Only the second term corresponds to an observable quantity, an X antiphase doublet, which is refocused into an in-phase doublet (along x) at the end of the third interval of duration $1/2J$. This enables proton decoupling to be applied for observing a single peak whose intensity is proportional to $\pm\gamma_A \sin\theta$. The γ_A factor indicates that carbon-13 has benefited from a proton polarization transfer.

The case of a CH_2 grouping is more delicate, and necessitates calculations of the same type as those developed in Chapter 2. We shall denote the two protons by A and

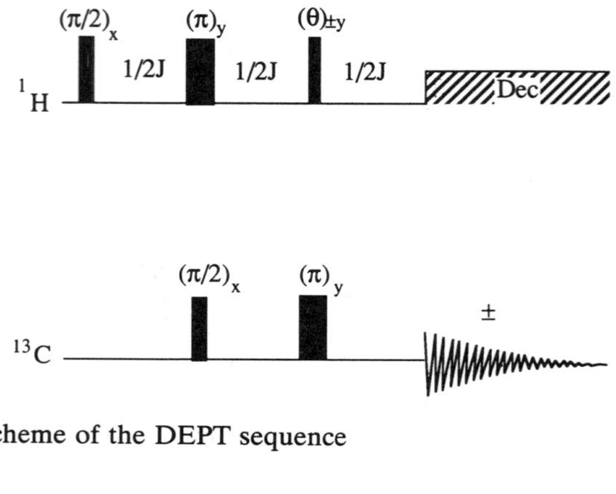

Figure 5.19 Basic scheme of the DEPT sequence

A′ in such a way that the antiphase configuration of the proton doublet, by the end of the first interval, can be represented by $2\gamma_A(I_x^A + I_x^{A'})I_z^X$; it is converted into $\gamma_A(2I_x^A I_y^X + 2I_x^{A'} I_y^X)$ by the $\pi/2$ pulse acting on X. Because of the two π pulses, the hamiltonian which governs the evolution of the spin system is reduced to the sole coupling term (between A, A′ and X):

$$\mathcal{H}_J = J(I_z^A + I_z^{A'})I_z^X$$

so that the evolution of the density operator can be evaluated by

$$\sigma(t) = \exp[-2i\pi Jt(I_z^A + I_z^{A'})I_z^X]\sigma(0)\exp[2i\pi Jt(I_z^A + I_z^{A'})I_z^X].$$

This equation can also be used for the second interval of duration $1/2J$ provided that $\sigma(0)$ is replaced by $\gamma_A[(2I_x^A I_y^X) + (2I_x^{A'} I_y^X)]$, which describes the state of the system prior to that interval. Hence the above equation can be decomposed into two terms, and we will detail below the calculations for the first of them; with $t = 1/2J$, this can be written as

$$\exp[-i\pi(I_z^A + I_z^{A'})I_z^X](2\gamma_A I_x^A I_y^X)\exp[i\pi(I_z^A + I_z^{A'})I_z^X]$$

Since and $I_z^A\, I_z^X$ and $I_z^{A'} I_z^X$ commute, the above expression can be expanded as

$$\exp(-i\pi I_z^{A'} I_z^X)\{\exp(-i\pi I_z^A I_z^X)(2\gamma_A I_x^A I_y^X)\exp(i\pi I_z^A I_z^X)\}\exp(i\pi I_z^{A'} I_z^X).$$

The term between braces is simply the one which would be involved in the evolution of an AX two spin system. We have seen before that it reduces to $2\gamma_A I_x^A I_y^X$. Therefore, it just remains to evaluate

$$\exp(-i\pi I_z^{A'} I_z^X)(2\gamma_A I_x^A I_y^X)\exp(i\pi I_z^{A'} I_z^X)$$

A′ can be considered as a passive spin whose operator $I_z^{A'}$ behaves as a simple factor in the above expression. We thus obtain

$$2\gamma_A I_x^A[-I_x^X \sin(\pi I_z^{A'}) + I_y^X \cos(\pi I_z^{A'})].$$

Making use of relationships (2.30), which provide the sine and cosine of a $1/2I_z$ spin operator, we finally obtain $-4\gamma_A I_x^A I_z^{A'} I_x^X$. By symmetry, the second term of the initial expression leads to $-4\gamma_A I_z^A I_x^{A'} I_x^X$. Subsequent to the pulse of flip angle θ, the only observable quantity which arises from these two quantities can be expressed as (according to the phase cycle):

$$\pm 4\gamma_A I_z^A I_z^{A'} I_x^X \sin 2\theta \tag{5.8}$$

This corresponds to an antiphase configuration for the X triplet with the sign of the central line opposite to the sign of the two outer lines. This antiphase configuration is refocused by the last interval of duration $1/2J$ and results in a polarization transfer from A to X, weighted by $\sin 2\theta$.

For a CH_3 grouping, a similar analysis yields $\pm 24\gamma_A I_z^A I_z^{A'} I_z^{A''} I_y^X \sin\theta\cos^2\theta$, immediately after the θ pulse. The examination of the energy diagram of an A_3X spin system indicates that this latter quantity represents an antiphase configuration with sign alternation for the successive lines in the relevant quartet. After the last interval of duration $1/2J$, refocusing occurs along the x axis (with zeroing along y) and the observed signal has as its intensity $(3\gamma_A/4)(\sin\theta + \sin 3\theta)$. It is maximum for $\theta = 35.26°$ with an intensity enhancement equal to $1.15\gamma_A/\gamma_X$.

The DEPT sequence therefore provides carbon-13 signals whose intensity is enhanced by $\gamma_H/\gamma_C \approx 4$, modulated as a function of the angle θ and of the multiplicity of the carbon considered (Table 5.2). Spectral editing is obtained by appropriate linear combinations (Figure 5.20).

Table 5.2 Relative amplitude of carbon-13 signals obtained from the sequence of Figure 5.19 for three particular values of the flip angle θ. CH, CH_2 and CH_3 subspectra can be deduced from appropriate linear combinations of these three experiments

	$\theta = \pi/4$	$\theta = \pi/2$	$\theta = 3\pi/4$
CH	$\sqrt{2}/2$	1	$-\sqrt{2}/2$
CH_2	-1	0	1
CH_3	$3\sqrt{2}/4$	0	$3\sqrt{2}/4$

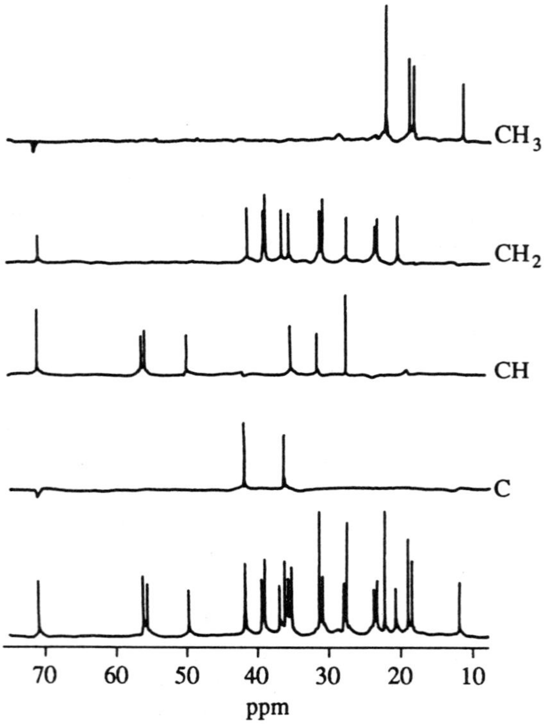

Figure 5.20 Editing as a function of carbon multiplicity C, CH, CH_2, CH_3 subspectra for the aliphatic part of the cholesterol molecule (the 'C subspectrum' corresponding to carbons not directly bound to protons, is derived from the conventional spectrum from which the three other subspectra have been subtracted)

5.2.4 HOMONUCLEAR COHERENCE TRANSFER. THE COSY SEQUENCE

This is undoubtedly the simplest and most widely used two-dimensional experiment (J. Jeener, Ampere Summer School, Basko Polje, Yugoslavia (1971); A. Bax and R. Freeman, *J. Magn. Reson.*, **44**, 542 (1981)). It leads to information mutual coupling and is therefore capable of replacing, in a single measurement, a series of spin decoupling experiments. It is shown schematically in Figure 5.21 and the sequence will first be analyzed in the case where the two $\pi/2$ pulses share the same phase (i.e. $\varphi_1 = \varphi_2 = x$).

For that purpose, we shall be looking at the evolution of spin A in the J-coupled AX spin system (spins 1/2 are assumed throughout). The evolution interval t_1 is actually a free precession period. By the end of t_1, the (A_1, A_2) doublet has evolved and its state can be represented as the superimposition of elementary in-phase and antiphase configurations which are derived through methods given in Chapter 2. This decomposition is detailed in Figure 5.21 and could be drawn in a

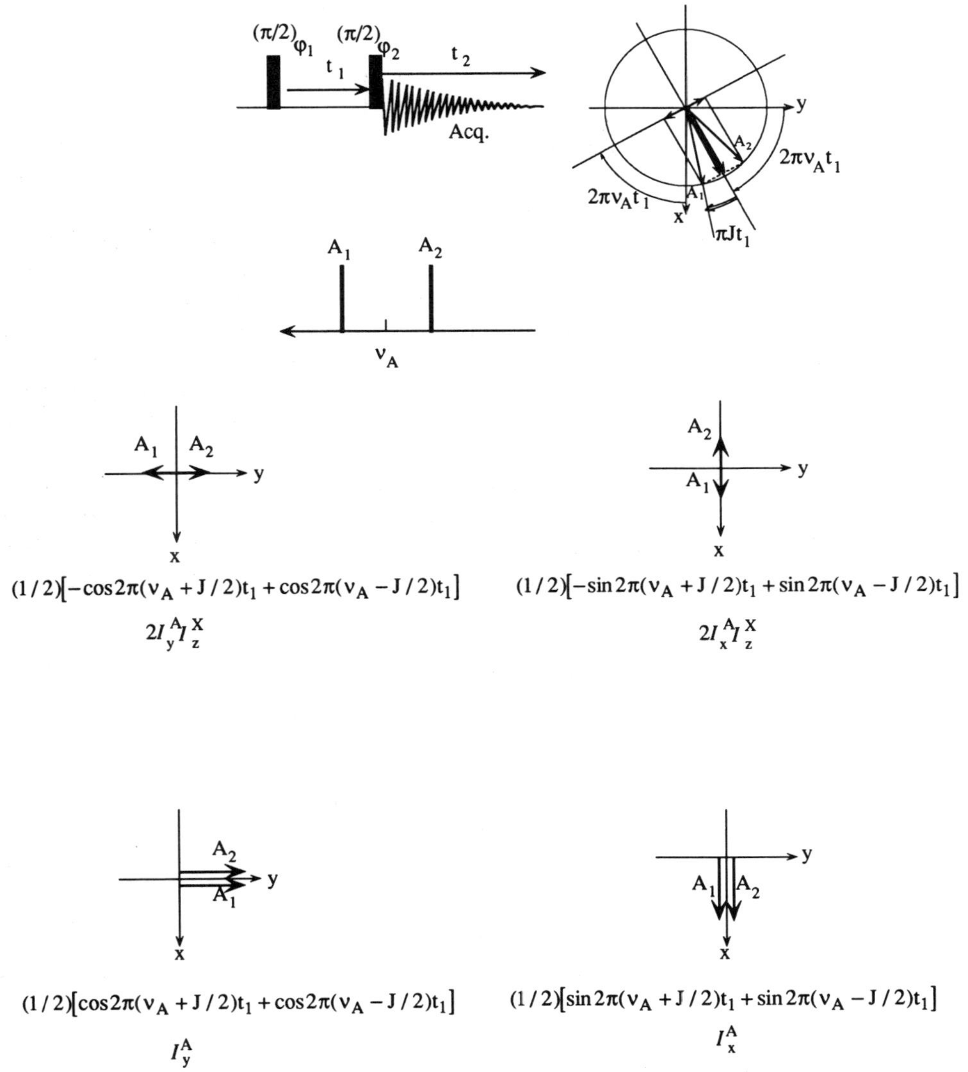

Figure 5.21 The basic COSY sequence and the decomposition into elementary configurations of the state of spin A at the end of the t_1 evolution period for $\varphi_1 = x$

similar way for the (X_1, X_2) doublet. We need to concern ourselves with the second pulse: (i) it leaves unchanged I_x^A, which evolves during t_2 as an in-phase doublet and according to ν_A and J; (ii) it changes I_y^A into I_z^A without any contribution to the detected signal; (iii) it transforms $2I_x^A I_x^X$ into $2I_x^A I_y^X$ which, because it includes one-quantum and two-quanta coherences, does not again contribute to the detected signal; (iv) it converts $2I_y^A I_z^X$ into the X antiphase coherence $-2I_z^A I_y^X$, which evolves according to ν_x and J during the detection period t_2. Hence, the detected signal arising from A magnetization can be expressed as a function of its t_1 modulation:

$$\left\{\frac{1}{2}\sin\left[2\pi\left(\nu_A + \frac{J}{2}\right)t_1\right] + \frac{1}{2}\sin\left[2\pi\left(\nu_A - \frac{J}{2}\right)t_1\right]\right\} I_x^A$$
$$\left\{\frac{1}{2}\cos\left[2\pi\left(\nu_A + \frac{J}{2}\right)t_1\right] - \frac{1}{2}\cos\left[2\pi\left(\nu_A - \frac{J}{2}\right)t_1\right]\right\}(2I_z^A I_y^X) \quad (5.9)$$

The first term corresponds to an A in-phase doublet in both the ν_1 dimension (because of the sum of the two sine functions, at frequencies $\nu_A + J/2$ and $\nu_A - J/2$, respectively) and the ν_2 dimension (because of I_x^A). If by convention (used throughout) we assign the y direction and cosine modulation to signals in absorption, we obtain for this first term a 'diagonal peak' ($\nu_1 = \nu_2 = \nu_A$) which is in-phase and dispersive. Relying on similar arguments, it can be seen that the second term generates a cross peak ($\nu_1 = \nu_A$; $\nu_2 = \nu_X$) in the form of an antiphase absorptive doublet in both dimensions. An analogous approach can be applied to magnetization originating from the X spin, which leads to an in-phase diagonal dispersive doublet, centered on ν_X ($\nu_1 = \nu_2 = \nu_X$) and a cross peak ($\nu_1 = \nu_X$; $\nu_2 = \nu_A$) symmetrically to the cross peak at $\nu_1 = \nu_A$ and $\nu_2 = \nu_X$. This is illustrated in Figure 5.22, which emphasizes that the occurrence of cross peaks indicates mutual coupling (in accordance with the objective of the COSY experiment).

We shall now discuss the different ways used to remove any ambiguity of sign in the ν_1 requency dimension. It can be recalled that the transmitter frequency

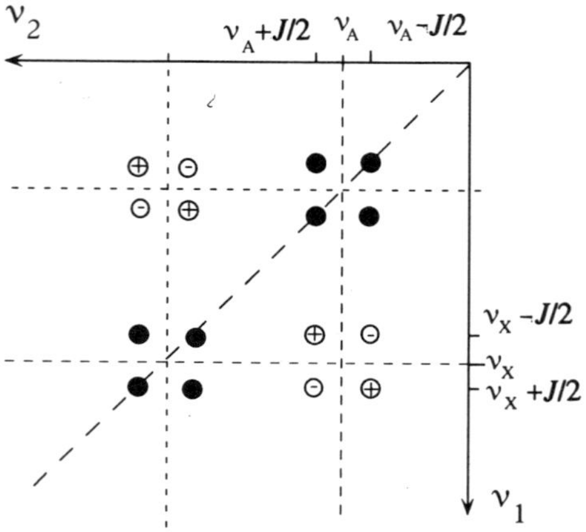

Figure 5.22 Schematic patterns resulting from a COSY experiment with pulses of identical phase applied to a weakly coupled AX spin system. Full circles represent peaks together with their signs

is usually set in the middle of the spectral window, thus permitting physical *quadrature* detection in the t_2 dimension. One method for retrieving the sign information in the ν_1 dimension is the so-called TPPI procedure (see Section 3.5.3) which consists of adding $\pi/2$ at the phase of the initial pulse for each new value of t_1 (incremented at a rate twice as fast as the normal sampling rate used in t_2). In addition, this procedure yields a final phase-sensitive spectrum, which would be of little value for the basic COSY experiment because, for absorptive cross peaks, diagonal peaks are dispersive with large wings prone to obscuring nearby cross peaks. Removal of artifacts is accomplished (i) by applying a CYCLOPS-type phase cycling, and (ii) by alternating the phase of the second pulse between φ_2 and $-\varphi_2$ without changing anything else in the sequence (this alternation can be seen to be without effect on NMR signals but helps to cancel spurious d.c. components). The relevant eight-step phase cycle is given in Table 5.3.

In fact, quadrature in the t_1 dimension can be achieved by means of a very simple phase cycling of the second pulse without increasing the number of experiments, albeit with however the necessity of performing a double complex Fourier transform. This latter feature precludes the obtention of a phase-sensitive spectrum and imposes the display of spectra in the magnitude mode, with its associated inconvenience in respect of spectral resolution (see Section 3.5.1). However, the experiment is very robust and is widely used on a routine basis. To understand properly how the φ_2 phase cycling can afford quadrature in t_1, let us re-emphasize that a coherence labeled by y (e.g. I_y^A or $2I_y^A I_z^X$) is associated with a cosine modulation, whereas a coherence labeled by x (e.g. I_x^A or $2I_x^A I_z^X$) is associated with a sine modulation (see Figure 5.21). This will be symbolized by

$$\begin{aligned} y_1 &\leftrightarrow \cos(2\pi\nu_1 t_1) \\ x_1 &\leftrightarrow \sin(2\pi\nu_1 t_1) \end{aligned} \tag{5.10a}$$

However, because in the t_2 dimension quadrature detection is physically employed, this correspondence should be written as

$$\begin{aligned} y_2 &\leftrightarrow \exp(2i\pi\nu_2 t_2) \\ x_2 &\leftrightarrow \mathrm{i}\exp(2i\pi\nu_2 t_2) \end{aligned} \tag{5.10b}$$

Now, referring to the above discussion, we can recall that, for $\varphi_1 = x$ and $\varphi_2 = x$, the only observable quantities are I_x^A (which arises from I_x^A; diagonal peaks) and

Table 5.3 Phase cycling of the phase-sensitive basic COSY sequence

φ_1	φ_2	Acq
x	x	x
x	$-x$	x
y	y	y
y	$-y$	y
$-x$	$-x$	$-x$
$-x$	x	$-x$
$-y$	$-y$	$-y$
$-y$	y	$-y$

$-2I_z^A I_y^X$ (which arises from $2I_y^A I_z^X$; cross peaks indicating coherence transfer). This can be rendered as

$$x_1 \Rightarrow x_2 \quad \text{(diagonal)}$$

$$y_1 \Rightarrow -y_2 \quad \text{(cross)}$$

Conversely, for $\varphi_1 = x$ and $\varphi_2 = y$, one has, using similar arguments:

$$y_1 \Rightarrow y_2 \quad \text{(diagonal)}$$

$$-x_1 \Rightarrow +x_2 \quad \text{(cross)}$$

If the two experiments ($\varphi_1 = x$, $\varphi_2 = x$; $\varphi_1 = x$, $\varphi_2 = y$) are coadded, we obtain for the diagonal peaks

$$x_1x_2 + y_1y_2 \rightarrow [i \sin(2\pi\nu_1 t_1) + \cos(2i\pi\nu_1 t_1)] \exp(2i\pi\nu_2 t_2)$$
$$= \exp(2i\pi\nu_1 t_1) \exp(2i\pi\nu_2 t_2) \quad (5.11)$$

since the t_2 signal has to be multiplied by the t_1 signal from which it originates. A similar result (except for a trivial sign change) holds for cross peaks. Quadrature detection is effectively achieved in t_1 since the modulation is represented by $\exp(2i\pi\nu_1\tau_1)$, thus providing the frequency sign. It can be seen that we obtain a result of type p (same signs for the arguments of the two exponentials). Generally, a type n spectrum is preferred; this can easily be obtained by applying a minus sign to the acquisition of the second experiment ($\varphi_1 = x$; $\varphi_2 = y$). Furthermore, major artifacts (corresponding to d.c. components; see Section 3.6) can be removed by noticing that $+\varphi_2$ or $-\varphi_2$ yields the same result as far as NMR signals are concerned; we are thus led to the minimal phase cycling of the basic COSY sequence given in Table 5.4. (An additional phase cycling of the CYCLOPS type, involving the initial pulse and the receiver phase, can be superimposed). An easy way of reducing artifacts is to make the final 2D map symmetrical. The amplitude mode makes this procedure possible; it consists of retaining the lowest signal for two points in symmetrical positions with respect to the diagonal. Finally, a widely used trick is to resort to a 45° flip angle for the second pulse. It can be acknowledged that this tends to reduce diagonal peaks with respect to cross peaks, although at the expense of an overall sensitivity loss. This experiment is dubbed COSY-45.

The use of pulsed (B_0) gradients enables us to suppress entirely the phase cycling given in Table 5.4 by (i) insuring quadrature in the t_1 dimension, and (ii)

Table 5.4 Minimal phase cycling of the basic COSY sequence

φ_1	φ_2	Acq
x	x	+
x	$-x$	+
x	y	−
x	$-y$	−

removing all artifacts of instrumental origin, thus speeding up the experiment, which can typically be performed in less than five minutes. The sequence, already alluded to in Section 2.2.6, is depicted in Figure 5.23. As compared with the basic COSY sequence, it essentially includes two identical gradient pulses on each side of the second $\pi/2$ pulse; it can be seen that the two r.f. pulses have identical phases. To understand how quadrature in the t_1 dimension can be achieved in a *single* scan per increment, let us consider quantities present by the end of the evolution period (limited to those pertaining to the A nucleus): I_x^A, I_y^A, $2I_x^A I_z^X$, $2I_y^A I_z^X$ (see Figure 5.21). Referring to the rules given in Section 2.2.6, we first examine how I_x, I_y and I_z are transformed

$$I_x \xrightarrow{g_0} I_x \cos\theta - I_y \sin\theta \xrightarrow{(\pi/2)_x} I_x \cos\theta + I_z \sin\theta \xrightarrow{g_0} I_x \cos^2\theta - I_y \sin\theta\cos\theta + I_z \sin\theta$$
$$I_y \xrightarrow{g_0} I_x \sin\theta + I_y \cos\theta \xrightarrow{(\pi/2)_x} I_x \sin\theta - I_z \cos\theta \xrightarrow{g_0} I_x \sin\theta\cos\theta - I_y \sin^2\theta - I_z \cos\theta$$
$$I_z \xrightarrow{g_0} I_z \xrightarrow{(\pi/2)_x} I_y \xrightarrow{g_0} I_x \sin\theta + I_y \cos\theta$$

where θ represents the precession angle at a given location within the sample produced by the application of the gradient pulse.

Assuming that the gradient is strong enough and that it is applied for a sufficiently long period, we have for the averages over the whole sample:

$$\langle \sin\theta \rangle = \langle \cos\theta \rangle = \langle \sin 2\theta \rangle = \langle \cos 2\theta \rangle = 0$$

and

$$\langle \sin^2\theta \rangle = \langle \cos^2\theta \rangle = 1/2$$

(see also Table 2.1), so that the quantities of interest are actually transformed into:

$$I_x^A \rightarrow I_x^A/2$$
$$I_y^A \rightarrow -I_y^A/2$$
$$2I_x^A I_z^X \rightarrow (2I_z^A I_x^X)/2$$
$$2I_y^A I_z^X \rightarrow -(2I_z^A I_y^X)/2$$

Obviously, the first two quantities will lead to diagonal peaks and the last two quantities are associated with coherence transfers and will lead to cross peaks. In any case, it can be seen that $x_1 \rightarrow x_2$ and $y_1 \rightarrow -y_2$. Retaining the conventions adopted for the standard COSY sequence (i.e. $y_1 \leftrightarrow \cos(2\pi\nu_1 t_1)$; $x_1 \leftrightarrow \sin(2\pi\nu_1 t_1)$; $y_2 \leftrightarrow \exp(2i\pi\nu_2 t_2)$; $x_2 \leftrightarrow i\exp(2i\pi\nu_2 t_2)$), and recognizing that both x_1 and y_1

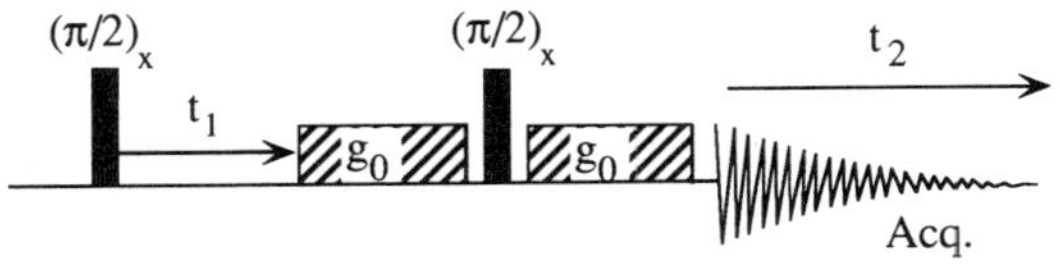

Figure 5.23 Basic COSY sequence with two identical B_0 gradient pulses (denoted by g_0) aimed at removing phase cycling

components (with their specific modulation) are present at the end of the evolution period, we obtain for the final signal

$$x_1x_2 - y_1y_2 \rightarrow i\sin(2\pi\nu_1t_1)\exp(2i\pi\nu_2t_2) - \cos(2\pi\nu_1t_1)\exp(2i\pi\nu_2t_2)$$
$$= \exp(-2i\pi\nu_1t_1)\exp(2i\pi\nu_2t_2) \quad (5.12)$$

which, except for the sign of the t_1 exponential argument (leading to a type n spectrum), is analogous to (5.11). Quadrature is therefore achieved in the t_1 dimension with a single scan per t_1 increment. It can however be seen that such an experiment, which is of the phase modulation type, can result only in magnitude mode spectra. Also, gradients entail a sensitivity loss by a factor of 2. It can finally be mentioned that a homologous sequence with B_1 gradients can be designed by resorting to the equivalences given in Figure 2.25.

Numerous variants of the COSY sequence exist. The most widely employed is the DQF COSY experiment (DQF for double quantum filtered; U. Piantini *et al.*, *J. Am. Chem. Soc.*, **104**, 6800 (1982)), whose advantages are twofold: (i) diagonal and cross peaks have the same phase (e.g. both are absorptive), and (ii) non-coupled diagonal peaks are in principle suppressed (in practice attenuated, owing to fast repetition rates used for shortening the measuring time and not allowing a complete return to thermal equilibrium, and owing also to the inevitable imperfections of the subtraction process). The method involves a third $\pi/2$ pulse which immediately follows the second pulse of the conventional COSY sequence and on a four step phase cycle which leads to the above mentioned properties (Figure 5.24).

The reference to double-quantum coherences is somewhat formal, since the latter do not evolve in this experiment. Rather, they are involved as an intermediate ($2I_x^AI_y^X$) which will be selected and further reconverted into observable coherences by the last $\pi/2$ pulse. In order to understand how the double-quantum filter works, let us notice that the last $\pi/2$ pulse is phase cycled according to the usual four steps, i.e. x, y, $-x$ and $-y$, whereas the acquisition which immediately follows is *in quadrature* with respect to this last pulse. This means that single spin operators systematically cancel, as demonstrated by the fate of I_x^A at the end of the evolution period: I_x^A is evidently unchanged by the second pulse of the sequence and also by the third pulse with phases $+x$ and $-x$; however, because of the detection which is in quadrature, I_x^A is 'seen' as $-y_2$ and $+y_2$, respectively, and

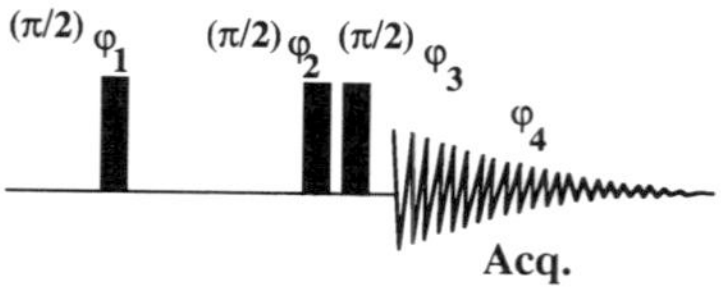

φ1	φ2	φ3	φ4
x	x	x	y
x	x	y	x
x	x	-x	-y
x	x	-y	-x

Figure 5.24 The DQF COSY sequence and its basic phase cycle

thus cancels. On the other hand, for the phases y and $-y$ of the third pulse, I_x^A is converted into the unobservable states I_z^A and $-I_z^A$, respectively. In a similar way, I_y^A at the end of the evolution period is first transformed into $-I_z^A$ which, owing to the phase cycling of the third $\pi/2$ pulse and of the acquisition, is seen as $-x_2$, x_2, $-x_2$, x_2. Likewise the antiphase coherence $2I_y^A I_z^X$, converted into $-2I_z^A I_y^X$ by the second $(\pi/2)_x$ pulse, leads to zero as the net result of the four-step phase cycle. Conversely, the antiphase coherence $2I_x^A I_z^X$, which exists at the end of the evolution period and which is converted into $2I_x^A I_y^X$ by the second $(\pi/2)_x$ pulse, provides diagonal peaks in addition to cross peaks (resulting from transfer from A to X) with identical phases owing to the acquisition phases. All these results can be better understood by examination of Table 5.5, which emphasizes the role of the so-called double-quantum filter. This terminology is in fact excessive for two reasons: (i) if it is true that quantities of the type $2I_x^A I_y^X$ are effectively retained, it can be recognized that they in fact encompass zero-quantum coherences (arising from product operators of the type $I_+^A I_-^X$) and two-quanta coherences (arising from product operators of the type $I_+^A I_+^X$ or $I_-^A I_-^X$, and (ii) the filter is in fact low-pass in nature, since coherences of higher order (greater than or equal to two quanta) will be present in the final spectrum. Finally, an interesting feature arises from the identical phase of diagonal and cross peaks which allows for a pure absorption display, with its inherent advantage of minimum overlap between diagonal peaks and nearby cross peaks. Phase-sensitive spectra can be achieved experimentally by the TPPI procedure (see Section 3.5.3).

A complete phase cycling scheme, along the lines developed above, aimed at reducing artifacts is given in Table 5.6. This optimum phase cycling scheme involves eight steps, with the inconvenience of a lengthy measuring time; this can be considerably reduced by the use of B_0 gradients, although at the expense of poorer sensitivity. The double-quantum filter $(\pi/2)_x$ $(\pi/2)_{\varphi 3}$ $(\text{Acq})_{\varphi 4}$ can be replaced by the sequence represented in Figure 5.25. This requires only a single

Table 5.5 The fate of quantities present at the end of the evolution period in the DQF COSY experiment as a function of the phase of the third $\pi/2$ pulse (φ_3). The final phase of the actual signal (taking into account the receiver phase) is indicated by the subscript 2 attached to x or y. For convenience, transformations of these quantities under the second $(\pi/2)$ pulse are also given

	φ_3			
	x	y	$-x$	$-y$
$I_x^A \xrightarrow{(\pi/2)_x} I_x^A$	I_x^A $-y_2$	I_z^A	I_x^A y_2	$-I_z^A$
$I_y^A \xrightarrow{(\pi/2)_x} -I_z^A$	$-I_y^A$ $-x_2$	I_x^A x_2	I_y^A $-x_2$	$-I_x^A$ x_2
$2I_x^A I_z^X \xrightarrow{(\pi/2)_x} 2I_x^A I_y^X$	$-2I_x^A I_z^X$ y_2	$2I_z^A I_y^X$ y_2	$2I_x^A I_z^X$ y_2	$-2I_z^A I_y^X$ y_2
$2I_y^A I_z^X \xrightarrow{(\pi/2)_x} -2I_z^A I_y^X$	$2I_y^A I_z^X$ x_2	$2I_x^A I_y^X$	$2I_y^A I_z^X$ $-x_2$	$-2I_x^A I_y^X$

Table 5.6 The complete phase cycling of the DQF COSY sequence with the notations of Figure 5.24

φ_1	φ_2	φ_3	φ_4
y	x	y	x
y	x	$-x$	$-y$
y	x	$-y$	$-x$
y	x	x	y
x	y	$-x$	$-y$
x	y	$-y$	$-x$
x	y	x	y
x	y	y	x

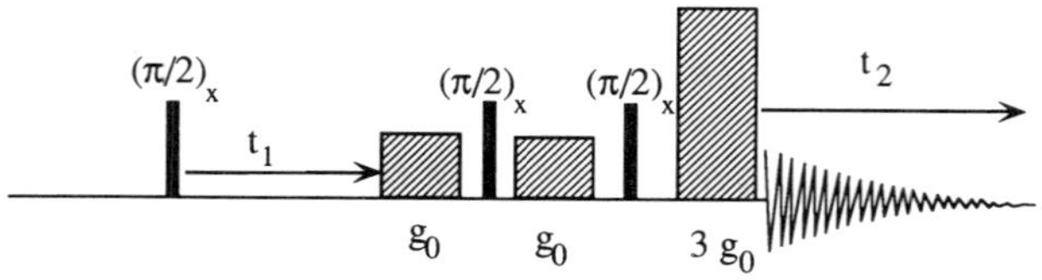

Figure 5.25 The DQF COSY sequence with B_0 gradient pulses. No phase cycling is necessary

scan per t_1 increment and has the further advantage that gradient pulses clean up some effects arising from instrumental artifacts.

Without entering into details of the sequence, we can recognize how double-quantum coherences (with their usual 'excessive' meaning; that is, quantities of the type $2I_x^AI_y^XI_x^AI_y^X$) are retained by gradient pulses and converted back to observable quantities, whereas other quantities are simply cancelled by the successive application of gradient pulses. At the end of the t_1 evolution period the antiphase coherence $2I_x^AI_z^X$ is affected by $\cos\theta$ because of the first gradient pulse of amplitude g_0 (as before, θ is the precession angle produced by the gradient for a given location within the sample); this quantity is transformed into $2I_x^AI_y^X$ by the second $(\pi/2)_x$ pulse which is modulated according to $\cos^2\theta$ by the second g_0 gradient pulse. With the previous $\cos\theta$ modulation, this leads to a $\cos^3\theta$ modulation which is retained when $2I_x^AI_y^X$ is converted back into the observable quantity $2I_x^AI_z^X$ by the third $(\pi/2)_x$ pulse. Clearly, such a modulation in $\cos^3\theta$ can be refocused only by means of a gradient pulse of amplitude $3g_0$ (see Table 2.1). It can be seen that all single-quantum coherences, and also coherences of the type $2I_y^AI_z^X$, are eliminated by the gradient pulses. The coherence $2I_x^AI_z^X$, whose fate has been followed above, leads to a diagonal peak. Conversely, starting from the same antiphase coherence $2I_x^AI_z^X$ and considering its sine modulation, i.e. its transformation into $-2I_y^AI_z^X\sin\theta$ under the first gradient pulse, we would arrive at a refocused coherence which corresponds to a transfer from A to X (cross peak).

An equivalent of the DQF COSY sequence exists with B_1 gradients (J. Brondeau *et al.*, *J. Magn. Reson.*, **100**, 611 (1992)); it is, however, based on different principles. The pulse sequence (shown in Figure 5.26) is fairly simple and

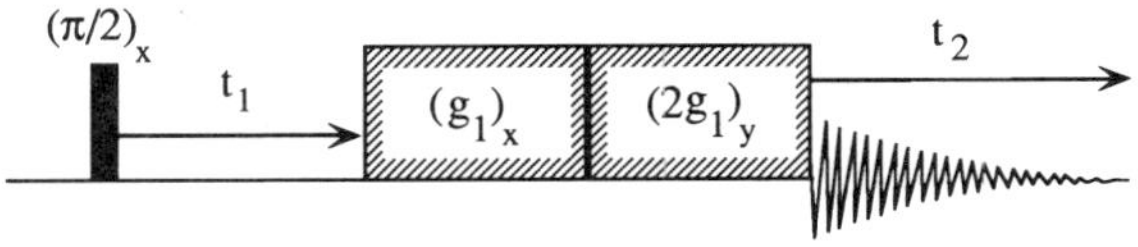

Figure 5.26 The equivalent of the DQF COSY experiment with B_1 gradient pulses: $(g_1)_x$ stands for a gradient pulse with the radio-frequency field acting along the x axis of the rotating frame and $(2g_1)_y$ stands for a gradient pulse of duration twice as large with the radio-frequency field acting along the y axis of the rotating frame

takes advantage of the filtering properties of the concatenation of two B_1 gradient pulses of phase x and y respectively. It is a simple matter to determine the fate of the coherences present at the end of the evolution period by performing averages on the nutation angle θ (with the usual relationships of the type $\langle \sin\theta \rangle = \langle \cos\theta \rangle = 0$, $\langle \sin^2\theta \rangle = \langle \cos^2\theta \rangle = 1/2$, which can be found more extensively in Table 2.1). We obtain:

$$I_x \xrightarrow{(g_1)_x(2g_1)_y} 0$$
$$I_y \xrightarrow{(g_1)_x(2g_1)_y} 0$$
$$2I_x^A I_z^X \xrightarrow{(g_1)_x(2g_1)_y} 0$$
$$2I_y^A I_z^X \xrightarrow{(g_1)_x(2g_1)_y} (2I_y^A I_z^X)/4 + (2I_z^A I_y^X)/4$$

In this latter transformation, the first term leads to a diagonal peak, whereas coherence transfer (from A to X) is represented by the second term and evidently leads to a cross peak. Some comments can be made about these results: (i) a single transient per t_1 value is necessary, thus eliminating the need for phase cycling; (ii) cross peaks and diagonal peaks have the same phase, thus allowing the display of phase-sensitive spectra (the TPPI procedure can be applied to the initial $\pi/2$ pulse so that the experiment is sensitive to the frequency sign in the ν_1 dimension); and (iii) only antiphase coherences are involved in the process without the need to create intermediate double-quantum states. High resolution COSY spectra are obtained by this method, as exemplified by Figure 5.27.

5.3 Multiquanta Spectroscopy. The Inadequate and HMQC Experiments

5.3.1 THE INADEQUATE EXPERIMENT

As we have demonstrated in the DQF COSY sequence, the state of a two spin 1/2 system represented by $(2I_x^A I_y^X)$ or $(2I_y^A I_x^X)$ is unobservable, as it involves coherences of order different from ± 1 (it can be recalled that only single-quantum coherences are physically detectable). Also, in the latter sequence, such coherences were used only as intermediate states occurring for filtering purposes. In fact, if we refer to the definition of raising and lowering operators given in Chapter 1 (relationships (1.5)), we can recognize that $(2I_x^A I_y^X) + (2I_y^A I_x^X)$ can be written as

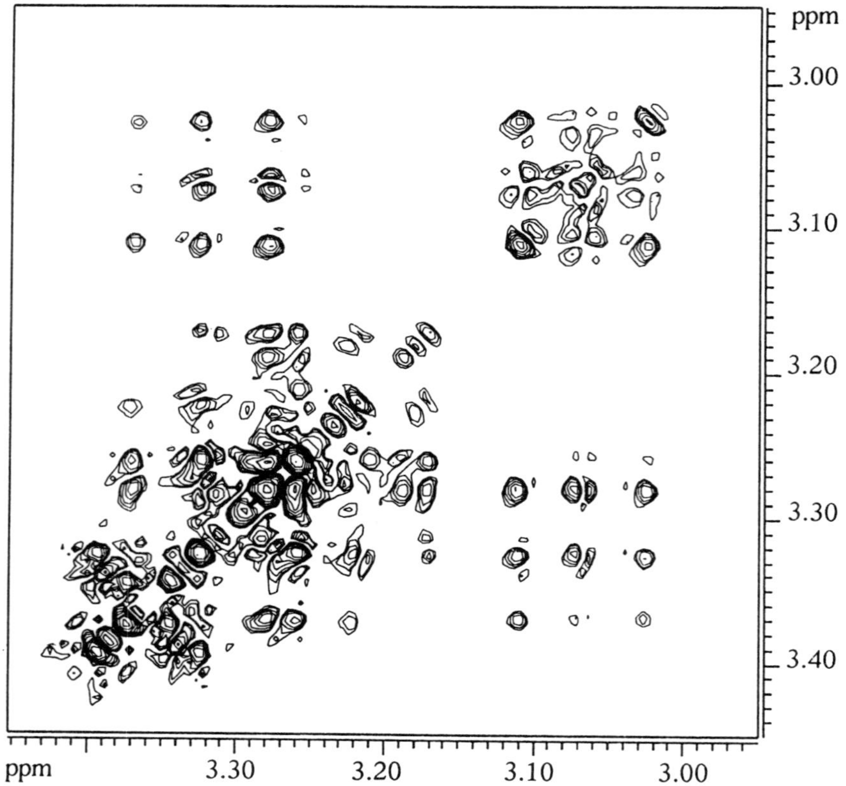

Figure 5.27 A detail of the 2D map of glucose as obtained by the sequence of Figure 5.26

$(I_+^A I_+^X - I_-^A I_-^X)/i$. It is obvious that the scalar product $\langle \varphi | I_+^A I_+^X | \psi \rangle$ is non-zero only for $\varphi = \alpha_A \alpha_X$ and $\psi = \beta_A \beta_X$, with a similar result for $I_-^A I_-^X$ by interchanging φ and ψ. Thus, for switching from φ to ψ, one has simultaneously to invert the A and X spins, which basically would involve two quanta, hence the terminology of 'two-quanta coherences'.

Now, we shall show that the pulse sequence of Figure 5.28 enables one to generate two-quanta coherences for two coupled spins A and X and to remove (or to filter out) one-quantum coherences (corresponding to a spin Y which is not coupled to any other nucleus of the same isotopic nature). First, we can recall that the central π pulse eliminates all chemical shift effects so that, for the diagrams shown in Figure 5.28 (and established according to the practical rules given in Chapter 2), A can be considered as being on resonance. Similar diagrams could be devised for describing the behavior of nucleus X.

Each experiment in the phase cycling applied to $\pi/2$ pulses yields for A: $-2I_x^A I_y^X$, $-2I_y^A I_x^X$, $-2I_x^A I_y^X$, $-2I_y^A I_x^X$ and for X: $-2I_y^A I_x^X$, $-2I_x^A I_y^X$, $-2I_y^A I_x^X$, $-2I_x^A I_y^X$. Each step therefore produces the spin state represented by $-2I_x^A I_y^X - 2I_y^A I_x^X$, whereas the uncoupled spin alternates along $-z$ or $+z$. The first two steps would then be sufficient to eliminate contributions from the uncoupled spin Y. The two subsequent phase steps ($-x$, $-y$) help to improve the elimination process and are still active as far as the creation of double-quantum coherences is concerned.

Returning to the system of two spin 1/2 nuclei, we can recognize that a read-pulse [for instance $(\pi/2)_x$] transforms the coherence obtained at each step of the preceding scheme (i.e. $-2I_x^A I_y^X - 2I_y^A I_x^X$) into $-2I_x^A I_z^X - 2I_z^A I_x^X$, and thus into two antiphase doublets centered on ν_A and ν_X respectively, whereas the uncoupled spin simply disappears. Hence, as in the DQF COSY experiment, only coupled spins lead to observable signals. Elimination of resonances arising from uncoupled spins is excellent; a major application of this experiment to natural abundance carbon-13 spectra is to retain those antiphase doublets corresponding to two consecutive carbon-13 nuclei (the experiment is of course carried out under broadband proton decoupling). The latter are associated with one molecule in 10 000, whereas the usual singlets (corresponding to one molecule 100) are removed by the phase cycling. The sequence (A. Bax *et al.*, *J. Magn. Reson.*, **41**, 349 (1980)) has been named INADEQUATE (Incredible Natural Abundance Double QUantum Transfer Experiment), and can be extended to a two-dimensional version (Figure 5.29) which depends on the evolution of double-quantum coherence at frequency $(\nu_A + \nu_X)$ (A. Bax *et al.*, *J. Magn. Reson.*, **43**, 478 (1981)).

More precisely, as we have seen in Section 2.2.6, the evolution of the quantity $2I_x^A I_y^X$ can be seen as the rotations of a vector **A**, initially along the x axis of the rotating frame, and of a vector **X**, initially along y, at frequencies ν_A and ν_X, respectively (Figure 5.30). By the end of the interval t_1, the new configuration can be represented by

$$2[I_x^A \cos(2\pi\nu_A t_1) - I_y^A \sin(2\pi\nu_A t_1)][I_y^X \cos(2\pi\nu_X t_1) + I_x^X \sin(2\pi\nu_X t_1)]$$

Because only $2I_x^A I_y^X$ and $2I_y^A I_x^X$ can be converted into observable quantities by the last $(\pi/2)_x$ pulse, we shall retain from the previous expression

$$(2I_x^A I_y^X)\cos(2\pi\nu_A t_1)\cos(2\pi\nu_X t_1) - (2I_y^A I_x^X)\sin(2\pi\nu_A t_1)\sin(2\pi\nu_X t_1).$$

Conversely, as far as spin X is concerned, the quantity $2I_y^A I_x^X$ yields

$$-(2I_x^A I_y^X)\sin(2\pi\nu_A t_1)\sin(2\pi\nu_X t_1) + (2I_y^A I_x^X)\cos(2\pi\nu_A t_1)\cos(2\pi\nu_X t_1).$$

Accounting for contributions from both A and X spins, we obtain

$$(2I_x^A I_y^X + 2I_y^A I_x^X)\cos[2\pi(\nu_A + \nu_X)t_1]$$

which after the read pulse is converted into

$$(2I_x^A I_z^X + 2I_z^A I_x^X)\cos[2\pi(\nu_A + \nu_X)t_1]$$

where the sign affecting the coherences prior to the t_1 interval has been reinserted. The A antiphase coherence (i.e. $2I_x^A I_z^X$) evolves in t_2 according to ν_A and the X antiphase coherence (i.e. $2I_z^A I_x^X$) evolves according to ν_X. In the two-dimensional spectrum, two signals appear at frequencies ($\nu_1 = \nu_A + \nu_X$; $\nu_2 = \nu_A$) and ($\nu_1 = \nu_A + \nu_X$; $\nu_2 = \nu_X$), respectively. Each of these signals is composed of anti-

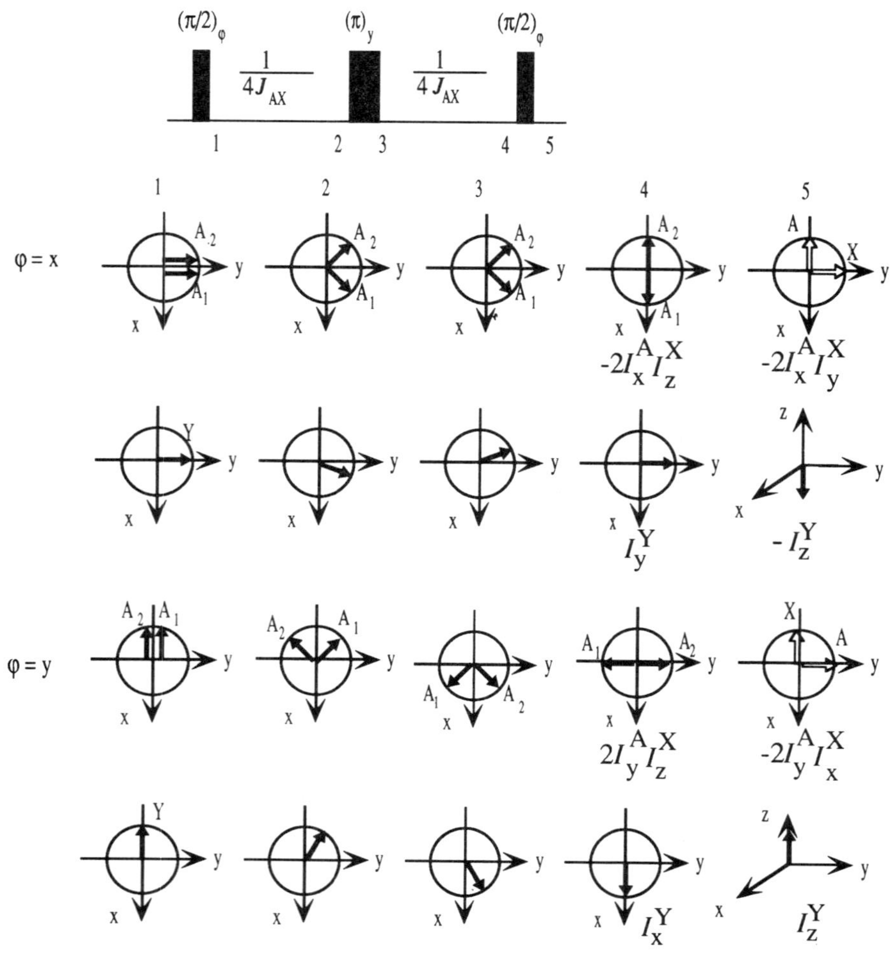

Figure 5.28 Double quantum filter. For the four phases ($\varphi = x, y, -x, -y$), the behavior of a spin A coupled to a spin X is shown diagrammatically (upper row) whereas the behavior of an uncoupled spin Y is displayed in the row immediately below

phase doublets in both dimensions. As the one-bond J_{CC} coupling constant is often of the order of 35 Hz, the INADEQUATE method is the proper approach for identifying carbon–carbon bonds and thus obtaining the carbon backbone of any molecule; however, it has the major drawback of low sensitivity. A variant of the method (D. L. Turner, *J. Magn. Reson.*, **49**, 175 (1982)), which involves delaying the acquisition by another interval of duration t_1, affords a more legible display in the manner of a COSY two-dimensional map, that is, with cross peaks (antiphase doublets) at frequencies ($\nu_1 = \nu_A$; $\nu_2 = \nu_X$) and ($\nu_1 = \nu_X$; $\nu_2 = \nu_A$) but in principle without diagonal peaks (Figure 5.31). Phase cycling to achieve quadrature in the t_1 dimension is essentially as in the COSY experiment and will not be discussed further here.

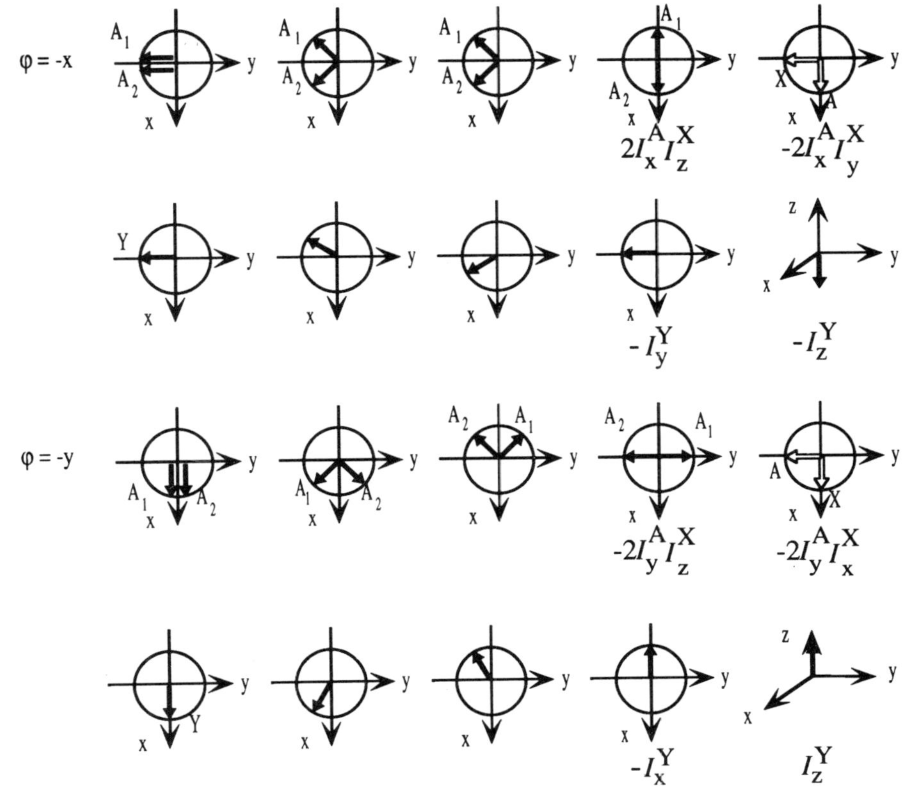

5.3.2 THE HMQC EXPERIMENT

We turn now to a heteronuclear version of the evolution of two-quantum coherences with an objective similar to that of the HSQC experiment (see Section 5.2.2), i.e. correlation spectroscopy in the 'inverse' mode, which takes full advantage of the higher sensitivity of proton detection. Again, J coupling mediates correlations between an X nucleus (e.g. carbon-13 or nitrogen-15) and a proton. The pulse sequence is relatively simple but with the mandatory condition of eliminating those resonances which arise from protons not coupled to the X nucleus (sometimes called 'parent signals'), and are overwhelming when carbon-13 or nitrogen-15 present at the natural abundance level; in turn, this implies the preservation of 'satellites' which correspond to the scarce protons coupled with the X nucleus (Section 1.3.5); hence the terminology sometimes employed of 'X filter'. The basic sequence (R. Freeman *et al.*, *J. Magn. Reson.*, **42**, 341 (1981)) depicted in Figure 5.32 in principle meets these requirements. After the first interval of duration equal to $1/2J_{XH}$, the proton doublet is antiphase and can be represented by $-2I_x^A I_z^X$ (A is ^{1}H). This antiphase doublet actually appears along the x axis thanks to the refocusing $(\pi)_y$ pulse, which makes it possible to ignore chemical shift effects. If no pulse is applied to X (i.e. for the phase step $(\pi/2)_x(\pi/2)_{-x}$), this

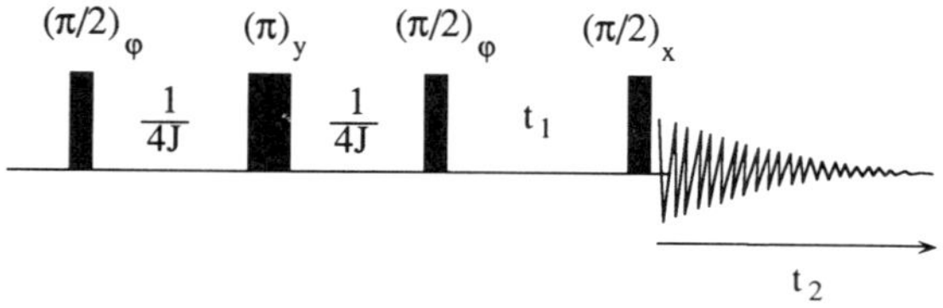

Figure 5.29 The two-dimensional version of the INADEQUATE sequence. The phase cycle (φ) is identical to that given in Figure 5.28

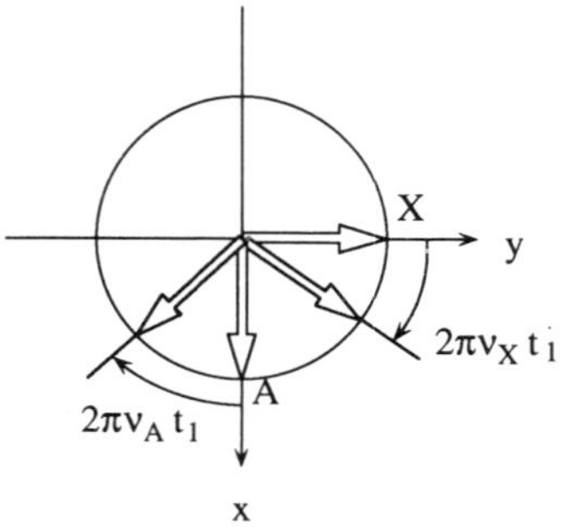

Figure 5.30 Vectorail representation of the fate of the quantity $2I_x^{\mathrm{A}}I_y^{\mathrm{X}}$ during an evolution interval t_1

latter quantity (after the central $(\pi)_y$ pulse) becomes $2I_x^{\mathrm{A}}I_z^{\mathrm{X}}$ and is refocused into an in-phase doublet along $+y$ at the end of the second interval of duration $1/2J_{\mathrm{XH}}$. Conversely, for the other phase step $(\pi/2)_x(\pi/2)_x$, a π pulse is effectively applied to X with the consequence that $-2I_x^{\mathrm{A}}I_z^{\mathrm{X}}$ is unchanged, leading to an in-phase doublet along $-y$. Since the acquisition sign is alternated concomitantly with the phase cycling $(\pi/2)_x(\pi/2)_{\pm x}$, the satellite signals are added at each step, whereas other proton signals cancel by subtraction (Figure 5.33). This experiment offers optimal sensitivity and, because it involves in-phase doublets, X decoupling can be used during acquisition so as to bring together both satellites into a single signal.

Extension to the two-dimensional version of the sequence of Figure 5.32 is straightforward. This is the well known HMQC experiment (A. Bax *et al.*, *J. Magn. Reson.*, **55**, 301 (1983)) depicted in Figure 5.34 which can be analyzed as follows: at the beginning of the t_1 interval, the state of the spin system is represented by $-2I_x^{\mathrm{A}}I_y^{\mathrm{X}}$ which is again a double-quantum coherence, just as in the INADEQUATE experiment, hence the terminology adopted for this sequence (Heteronuclear MultiQuanta Correlation). Because of the central $(\pi)_y$ pulse applied to A (A being ^{1}H), only I_y^{X} evolves in such a way that, by the end of the t_1 interval, the state of the system is described by

$$-2I_x^{\mathrm{A}}[I_y^{\mathrm{X}}\cos(2\pi\nu_{\mathrm{X}}t_1) + I_x^{\mathrm{X}}\sin(2\pi\nu_{\mathrm{X}}t_1)] \qquad (5.13)$$

The last pulse, $(\pi/2)_{\mp x}$, yields a proton doublet antiphase, along +y or $-y$, modulated according to $\cos(2\pi\nu_{\mathrm{X}}t_1)$. Thus, a double Fourier transform of the final result (after the interval $1/2J_{\mathrm{XH}}$, which allows for refocusing of the proton doublet) leads to a correlation peak at frequencies $\nu_1 = \nu_{\mathrm{X}}$ and $\nu_2 = \nu_{\mathrm{A}}$. The frequency sign in the ν_1 dimension can be obtained through the TPPI procedure, which in addition

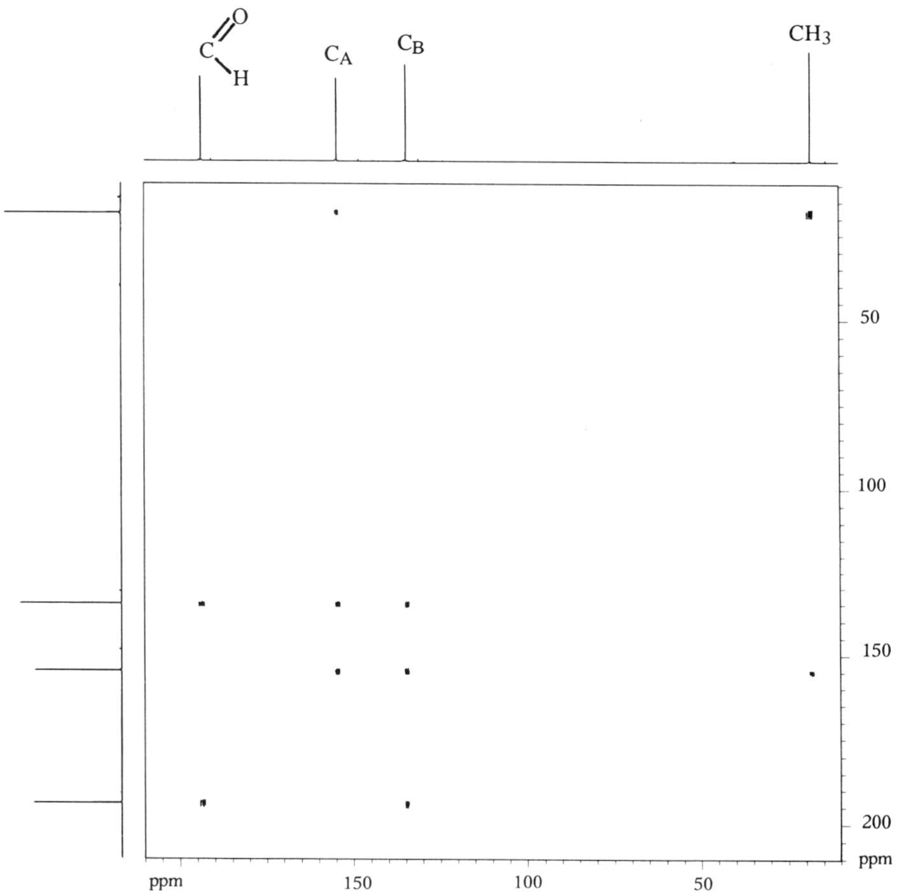

Figure 5.31 The two-dimensional INADEQUATE experiment applied to *trans*-crotonaldehyde. Diagonal peaks arise from residual signals of molecules containing a single carbon-13. Cross peaks appear in the form of doublets whose splitting is equal to J_{CC} (hardly visible due to an insufficient digital reduction)

yields phase-sensitive spectra. An illustrative example of this heteronuclear correlation spectroscopy was given in Chapter 1 (Figure 1.38). The complete elimination of signals arising from protons not coupled to X (the parent signals) is not at all trivial. The subtraction process of the basic sequence of Figure 5.34 usually entails an incomplete suppression due to instrumental instabilities, which result in more or less important ridges in the ν_1 dimension. A first remedy is to invert the parent signals selectively (and thus to leave unchanged the magnetization of interest); a time lapse of $T_1 \ln 2$ is then allowed so that the parent magnetization goes through zero as in an inversion-recovery experiment. Therefore the normal HMQC is started with the advantage of minimized parent signals. This elegant procedure may, however, suffer from a spread in T_1 values. The selective inverting procedure is known under the acronym of BIRD (J. R. Garbow *et al.*, *Chem. Phys. Lett.*, **93**,

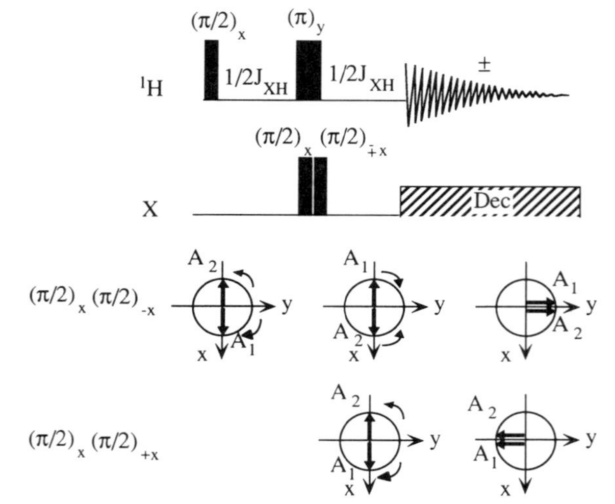

Figure 5.32 The basic sequence for removing all proton signals not coupled to X. A_1 and A_2 stand for the two lines in the proton doublet and their evolution is indicated for the two steps of the phase cycle

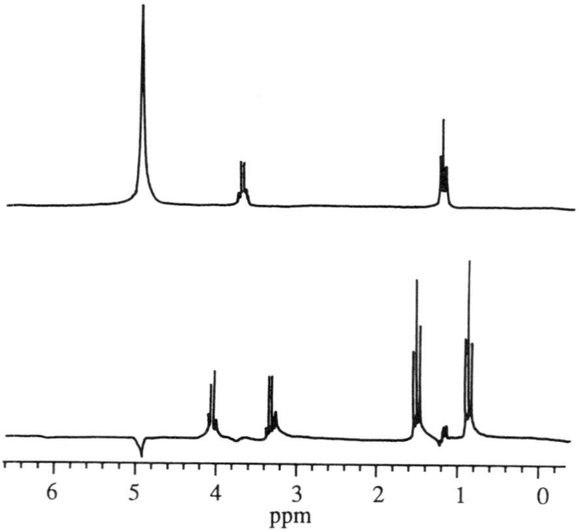

Figure 5.33 Top, the normal proton spectrum of ethanol. Bottom, the carbon-13 satellites with, in principle, the elimination of all other signals by means of the sequence depicted in Figure 5.32

504 (1982)) and can be easily understood from the rules given previously in this chapter (Figure 5.35).

A very efficient and simple way to remove the parent signals stems from the use of gradient pulses. As an illustration, we shall present the simplest implementation (Figure 5.36), acknowledging that there exist numerous variants (for a complete discussion see J. Ruiz-Cabello *et al.*, *J. Magn. Reson.*, **100**, 282 (1992)). The idea

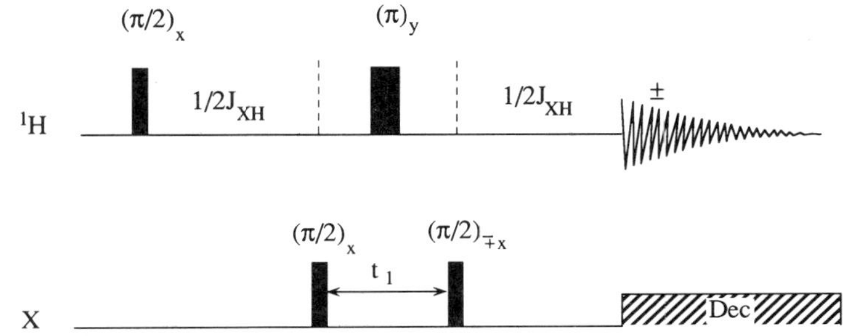

Figure 5.34 The basic scheme of the HMQC experiment

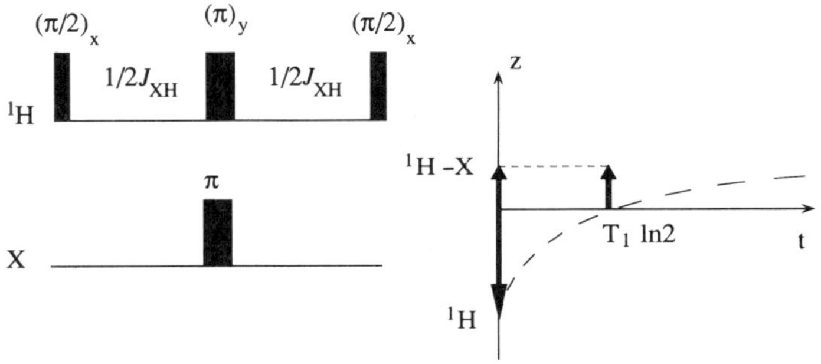

Figure 5.35 The BIRD sequence and its application in view of minimizing signals from protons not coupled to X

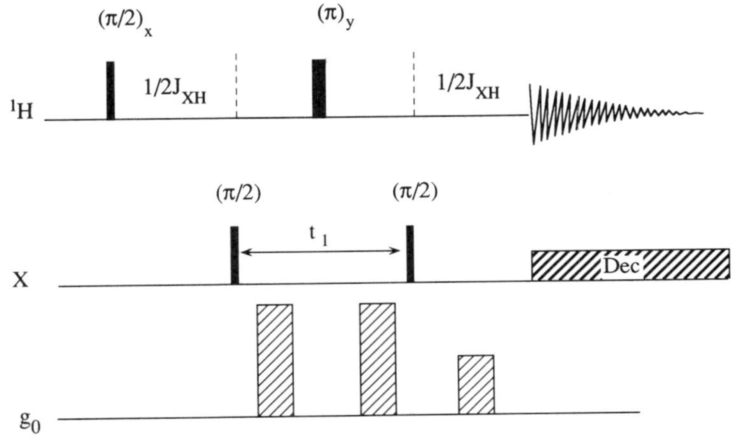

Figure 5.36 A possible version of the HMQC experiment with B_0 gradient pulses (g_0) for avoiding phase cycling and improving the elimination of unwanted resonances. The amplitudes of the gradient pulses are in the ratio $\gamma_H/2\gamma_X : \gamma_H/2\gamma_X : 1$

is to defocus I_y^X and I_x^X involved in the coherences which evolve during the incremented time t_1 (i.e. in $-2I_x^A I_y^X$ and $-2I_x^A I_x^X$; see above). After the last $\pi/2$ pulse applied to X, these coherences are converted into proton antiphase coherences, which must be refocused by the third gradient pulse. The purpose of this gradient pulse is also to defocus all one-quantum coherences such as those arising from the parent signals or from the solvent resonance. It can be noted that the two gradient pulses which bracket the central π pulse do not affect I_x^A (or I_y^A) precisely because of the refocusing π pulse which acts only on the proton, so that the quantities of interest, $2I_x^A I_y^X$ and $2I_x^A I_x^X$, become respectively

$$2I_x^A(I_y^X \cos 2\theta + I_x^X \sin 2\theta)$$

and

$$2I_x^A(I_x^X \cos 2\theta - I_y^X \sin 2\theta)$$

where θ is the X *precession angle* (at a given location in the sample) due to *one* gradient pulse. When one of the above quantities has been converted into a proton antiphase coherence, it can be refocused only if a gradient pulse produces a precession angle of 2θ in such a way that we can rely upon the average values: $\langle \cos^2 2\theta \rangle$ or $\langle \sin^2 2\theta \rangle$ (which are equal to 1/2). Because, at this stage of the sequence, we are dealing with quantities of the type $2I_x^A I_z^X$ (or $2I_y^A I_z^X$), only proton precession is involved, and the gradient amplitude must be reduced by the factor $2\gamma_X/\gamma_H$. Finally, it can be noted that the experiment is as sensitive as the conventional HMQC experiment and that evidently no phase cycle is needed.

5.3.3 THE HMBC EXPERIMENT

The acronym means Heteronuclear Multiple Bond Correlation; the basic experiment (A. Bax and M. F. Summers, *J. Am. Chem. Soc.*, **108**, 2093 (1986)), depicted in Figure 5.37, is aimed at establishing correlations between protons and an X nucleus via long range couplings (e.g. $^2J_{CH}$ or $^3J_{CH}$, denoted in the following by $^nJ_{XH}$), implying that correlations due to one-bond coupling (^1J, much larger than any long range coupling) must be cancelled. This is accomplished by the first stage of the sequence, up to the first $(\pi/2)_X$ pulse (applied to X), which, thanks to its phase alternation, constitutes a sort of ^1J filter.

In fact, at this point, proton magnetization can be decomposed (i) in part, into an antiphase configuration, according to $^1J_{XH}$, which is converted into a double-quantum coherence by the first $(\pi/2)^X$ pulse and further filtered out as indicated

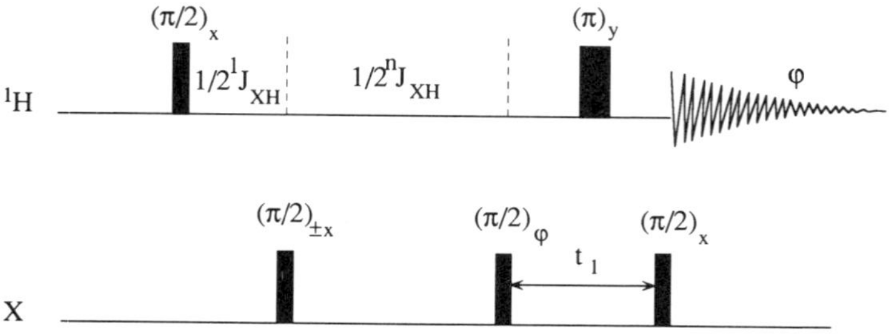

Figure 5.37 The basic HMBC experiment. φ stands for a CYCLOPS-type, phase cycle (see Section 3.6) which is superimposed onto the phase cycling of the first $(\pi/2)_{\pm x}$ pulse

above, (ii) in part, into an in-phase configuration due to $^nJ_{XH}$ couplings (as $1/2^1J \ll 1/2^nJ$). This in-phase configuration will evolve toward an antiphase configuration during the interval $1/(2^nJ_{XH})$. It is then converted into double-quantum coherences which evolve during t_1 just as in the HMQC experiment, with the difference that they originate here from long range couplings. They are finally transformed, by the third $(\pi/2)^X$ pulse, back into an antiphase proton configuration and detected. The CYCLOPS phase cycle which is applied to the second $(\pi/2)^X$ pulse and concomitantly to the receiver has two objectives: cancellation of unwanted one-quantum signals (parent or solvent peak) and sign sensitivity in the ν_1 dimension. Two further points can be mentioned: (i) because of the long duration of the interval $1/(2^nJ_{XH})$, homonuclear proton–proton couplings will also develop and tend to decrease the overall sensitivity, and (ii) in the same way, it appears difficult to let the final antiphase doublet refocus (which would necessitate another $1/(2^nJ_{XH})$ interval). This precludes the use of X decoupling during acquisition and the obtention of phase-sensitive spectra. The result of a HMBC experiment is therefore displayed in the magnitude mode.

5.4 Basic Sequences of NMR Imaging

This is undoubtedly a very broad topic, still subject to important advances and developments in the near future, and it will be considered here within the framework of multidimensional NMR. In fact, obtaining an image by NMR proceeds fundamentally from the approach employed for producing purely spectroscopic two- (or three-)dimensional diagrams as those described in the previous sections.

5.4.1 ONE-DIMENSIONAL DETERMINATION OF THE SPIN DENSITY

In this section, we will describe from mathematical and instrumental viewpoints the basic principles of MRI (magnetic resonance imaging) which were suggested in Section 1.3.1. For that purpose, we shall rely upon a hypothetical experiment which aims at providing the nuclear spin distribution along a given spatial direction (denoted by X), and we shall assume that the static induction B_0 varies linearly along that direction:

$$B_0(X) = B_{00} + g_0X$$

as shown in Figure 5.38, where it is further assumed that $B_{00} = B_0(0)$ corresponds to a location within the object under investigation.

In the previous expression, we have retained the usual notation of g_0, which denotes a *uniform* field gradient; it is delivered by coils of appropriate geometry and is superimposed on the induction B_{00}, which is assumed to be perfectly homogeneous within the volume of the sample. In order to understand properly how a field gradient can lead to the representation of the spin density, we shall rely on the fundamental Larmor equation $\nu_0 = \gamma B_0/2\pi$, from which we have purposely removed the shielding effect (which can be thought as being included in γ), and suppose, in accordance, that the sample under investigation involves only one

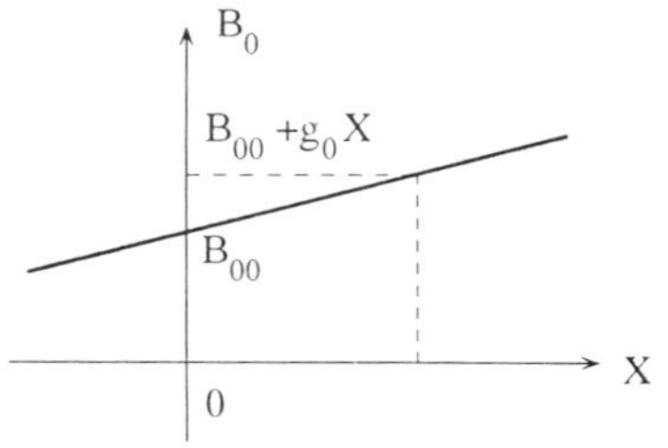

Figure 5.38 Linear variation of the static magnetic field along the X spatial direction

chemical species (e.g. water, as is the case in biological systems). At the outset a radio-frequency pulse is applied to the system so as to take the nuclear magnetization into the (x, y) plane (where small letters denote the rotating frame associated with the radio-frequency field whereas capital letters X, Y, Z denote spatial directions; in other words, the laboratory frame). Immediately after the r.f. pulse, a static field gradient g_0 effective along the X direction is switched on (Figure 5.39(a) in such a way that the *precession frequency*, in the rotating frame, becomes *dependent upon the spatial direction X*. If we denote by $\rho(X)$ the spin density along

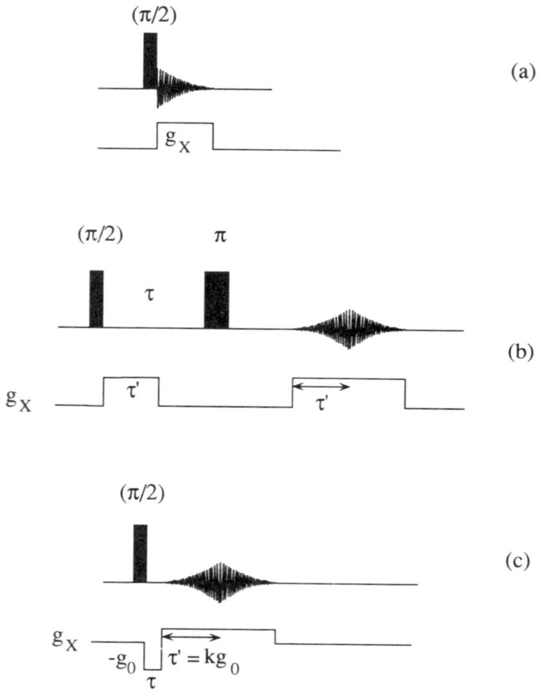

Figure 5.39 Imaging sequences for obtaining the spin density profile along the X spatial direction by means of a uniform static field gradient g_X acting along that direction. (a) Direct observation of the free induction decay; (b) by spin echo; (c) by gradient echo

X (i.e. integrated over the other two dimensions Y and Z), the detected signal can be expressed as:

$$S(t) = \int_{\text{object}} \rho(X) \exp[2i\pi\nu'(X)t] \exp(-t/T_2^*)\,dX \tag{5.14}$$

with $\nu'(X) = \gamma g_0 X/(2\pi)$ if we assume that, in the absence of gradient, the carrier frequency ν_{ref} is set on resonance for the chemical species employed for producing the image (or the spin density profile), i.e. $\nu_{ref} = \gamma B_{00}/(2\pi)$. It can be noted that ν' depends linearly upon X; hence frequency has been spatially encoded. More formally, we can introduce the so-called k-space by defining the variable k_X as:

$$k_X = (2\pi)^{-1}\gamma g_X t \tag{5.15}$$

(where g_X is the gradient acting along X) and ignore the damping factor $\exp(-t/T_2^*)$. Equation (5.14) can then be expressed as:

$$S(k_X) = \int \rho(X) \exp(2i\pi k_X X)\,dX \tag{5.16}$$

which demonstrates the Fourier relationship between the NMR signal $S(k_X)$ and the spin density $\rho(X)$. Indeed, the inverse Fourier transform of (5.16) yields

$$\rho(X) = \int S(k_X) \exp(-2i\pi k_X X)\,dk_X \tag{5.17}$$

To make this property less formal, let us discretize the expression (5.14); thus, let us denote by ρ_l the spin density pertaining to the lth slice. We assume that the slices are thin enough so that the overall envelope appears quasi-continuous (Figure 5.40). This approach is reminiscent of that used for converting the Fourier integral into a discrete summation. The signal $S(t)$ can then be written:

$$S(t) = \sum_{l=1}^{N} \rho_l \exp[i\gamma g_0 X_l t] \exp[-t/T_2^*] \tag{5.18}$$

This is simply a classical interferogram made up of N signals of amplitude ρ_l at the frequency $\gamma g_0 X_l/2\pi$. The Fourier transform of $S(t)$ therefore yields a spectrum of lines of intensity proportional to ρ_l and at frequency X_l provided that the frequency scale is expressed in $\gamma g_0/2\pi$ units. Superimposition of these elementary Lorentzian lines in fact provides an overall (continuous) broad line which reflects the spin distribution along X, and thus the object profile pertaining to that direction. This is simply an illustration of the space-frequency duality from which is

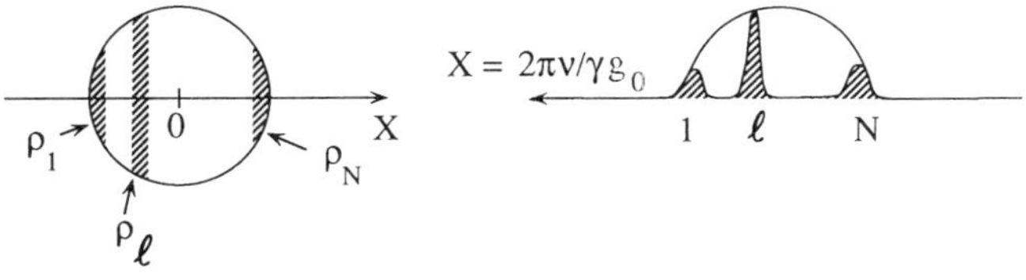

Figure 5.40 Cylindrical object with slices defined perpendicularly to the X direction (left). The Fourier transform of free induction decay obtained under the application of a uniform gradient g_0 along the X direction (right)

deduced the condition for examining the whole object. The sampling rate (or dwell time Dw) must fulfill the condition $Dw \leq 1/2\nu_{max}$ with, here, $\nu_{max} = \gamma g_0 X_{max}/(2\pi)$. This condition determines what is usually called the *field of view*. Moreover, it can be recognized that the spectral width is directly proportional to g_0. Since the spectral resolution is related to T_2^* and is expressed in frequency units, it is obvious that the gradient strength determines the *spatial resolution*. Of course, the digital resolution (i.e. the number of points in the spectral or spatial dimension) should not be lower than the theoretical spatial resolution (if necessary the technique of zero-filling can be employed; see Section 3.7.1).

In practice, because of the particular structure of r.f. coils which surround the object under investigation, relatively weak r.f. fields are produced which are unable to match the large spectral window resulting from the application of gradients (of the order of several gauss or tens of gauss per centimeter). Gradients are thus generally applied in the form of pulses during intervals which do not involve any r.f. pulses. It must be borne in mind that, unlike r.f. coils, which are inserted in tuned and impedance-matched circuits, the rise time of a gradient is not negligible, since the coil delivering the gradient is fed directly by a current generator. Moreover, gradient switching induces *eddy currents* into materials constituting the probe and the magnet; those in turn perturb the detection of the NMR signals. These are among the reasons for which one usually resorts to a spin echo technique (Figure 5.39(b)) whose virtue is to postpone acquisition of the NMR signal. This prevents artifacts arising from eddy currents at the expense of signal attenuation by transverse relaxation and translational diffusion phenomena. Generally, almost the entire echo is acquired (this means that acquisition starts well before the echo maximum, sometimes after the π pulse of the $(\pi/2-\tau-\pi-\tau)$ sequence), as it can be shown (Chapter 3) that the amplitude spectrum obtained after Fourier transformation of the entire echo is free from phase correction and does not suffer from extra broadening. An alternative to the delayed acquisition method is the so-called gradient echo technique, which avoids the π refocusing r.f. pulse (Figure 5.39(c)). It first should be noted that a gradient echo cannot 'refocus' the residual inhomogeneities arising from the static magnetic field, owing to the lack of a radio-frequency refocusing pulse. Apart from this unfortunate feature, the way in which a echo gradient works is very similar to that of a classical spin echo: the gradient $g_X = -g_0$ applied for a time τ produces a precession by an angle of $-\gamma g_0 X\tau$ (for magnetization located at the abscissa X). At the end of this interval τ, the gradient is switched from $-g_0$ to kg_0 in a time as short as possible (in practice, k may be set to 1/2 so as to reduce instrumental instabilities and to ensure a proper signal acquisition). Precession is therefore inverted and, after a time $\tau' = \tau/k$, every elementary magnetization is back to its initial position. This manifests itself by echo formation whose maximum amplitude occurs at time τ'. Again, processing of the whole echo can be considered as in the classical spin echo method.

5.4.2 TWO-DIMENSIONAL IMAGING

Methods developed in the previous section provide the spin density profile along a given spatial direction, that of the applied gradient. We shall assume here (for simplicity) that the spatial dimension Z can be ignored, meaning that the object is

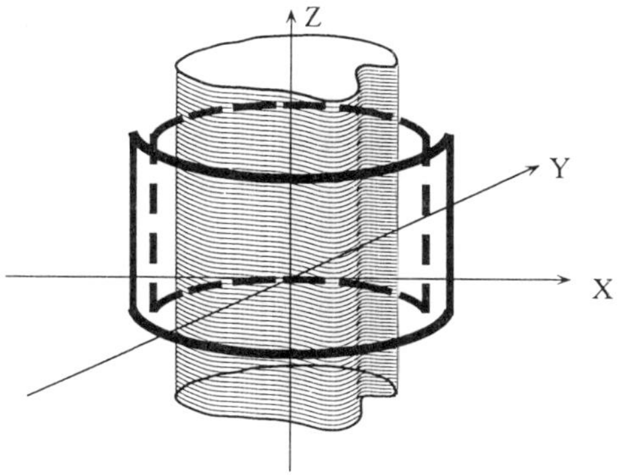

Figure 5.41 Object supposed to be homogeneous along the Z direction but showing up a differential spin distribution with respect to X and Y. The detection coil surrounding the object is assumed to limit the volume under investigation

approximately homogeneous along the direction Z within the volume sensed by the NMR detection coil (Figure 5.41), the aim being to determine the spin distribution in the XY plane.

A procedure must thus be devised for examining both spatial dimensions X and Y. A first approach consists of 'rotating' the gradient in order to explore the XY plane. As a matter of fact, the first imaging experiment designed by P. C. Lauterbur (*Nature*, **242**, 190 (1973)) relied on this principle in a fashion similar to the X scanner method widely used for clinical observations. Schematically, one may resort to the simplest of the procedures used for producing a density profile (Figure 5.39(a)) with the help of two gradients acting along the X and Y directions, respectively. In the course of an appropriate series of experiments, the amplitudes of these two gradients are modified so that the resulting gradient effectively rotates step-by-step in the XY plane and altogether operates a 2π rotation (Figure 5.42). The free induction decay obtained for each experiment is Fourier transformed, yielding a profile (or a *projection*) corresponding to the effective gradient direction. An algorithm of 'projection-reconstruction' can therefore be employed so as to end up with a two-dimensional image. It may be noted that this method proceeds from *polar encoding*.

Conversely, one may consider *cartesian encoding*. The first approach of that kind

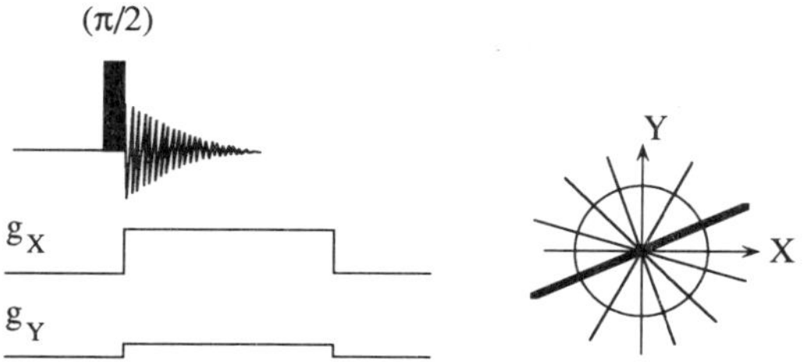

Figure 5.42 Schematic view of the examination of successive polar directions in the XY plane by varying the amplitudes of the gradients g_X and g_Y which operate in the X and the Y directions, respectively. The direction printed in bold face (right) corresponds to the effective gradient axis obtained by the actual amplitudes of the two X and Y gradients (left)

is due to Ernst and coworkers (A. Kumar, D. Welti and R. R. Ernst, *J. Magn. Reson.*, **18**, 69 (1975)) and is an extension of purely spectroscopic two-dimensional methods (Figure 5.43). It consists of applying two gradients g_Y and g_X of identical amplitude, one during an incrementable interval t_1, the other during signal acquisition (in the time dimension t_2). A double Fourier transform, with respect to t_1 and t_2, yields a (ν_1, ν_2) diagram, where ν_1 and ν_2 are linearly related to the spatial dimensions Y and X, respectively. The intensity of 'cross peaks' is here simply substituted by the spin density at the location (X,Y). A variant of the method now in widespread use is the so-called '*spin-warp*' technique (W. A. Edelstein *et al.*, *Med. Biol.*, **25**, 748 (1980)), which involves keeping t_1 at some fixed value and incrementing the amplitude of the gradient applied along the Y direction (Figure 5.44).

It can be seen that in both cases (constant amplitude gradient with incrementation of its time of application or constant time application with incrementation of its amplitude) one is dealing with phase encoding. In order to clarify this statement, let us consider the first approach (t_1 incrementation, Figure 5.43): the magnetization associated with the Y location precesses through an angle $\gamma g_Y Y t_1$ (relaxation is ignored); as a consequence, just before signal acquisition, there exist two magnetization components along the x and y axes of the rotating frame of amplitude $\cos(\gamma g_Y Y t_1)$ and $\sin(\gamma g_Y Y t_1)$, respectively. Hence this amounts to a phase modulation amenable to a complex Fourier transform. The phase angle can be expressed as $\gamma g_Y Y k \mathrm{Dw}_1$, where Dw_1 governs the t_1 increment: $t_1 = k\mathrm{Dw}_1$. This is equivalent to keeping t_1 constant (the duration of the gradient application being denoted by τ) and incrementing the gradient amplitude according to $g'_Y \mathrm{Dw}_1$, with $g'_Y = g_Y/\tau$, as expressed more formally by the following relationships:

$$\gamma g_Y Y t_1 = \gamma g_Y Y k \mathrm{Dw}_1 = \gamma \tau Y (g_Y/\tau) \mathrm{Dw}_1 = \gamma \tau Y g'_Y \mathrm{Dw}_1.$$

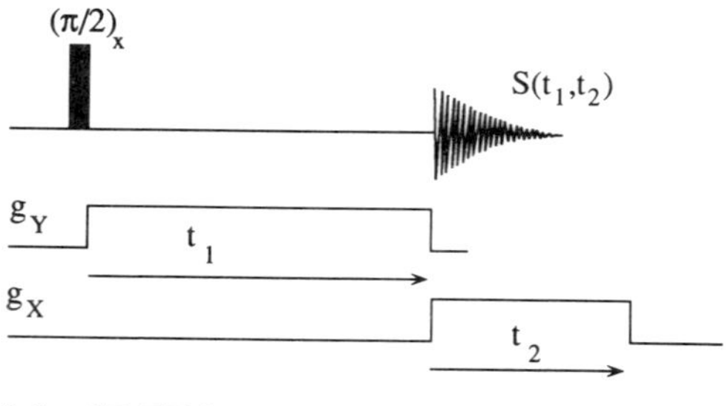

Figure 5.43 Principle of the 2D FT imaging method

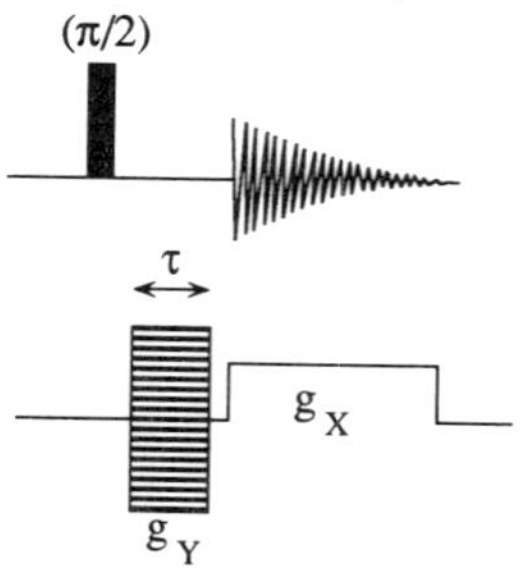

Figure 5.44 Principle of the spin-warp method: instead of incrementing the gradient g_Y duration (t_1), it is applied for a constant (τ) and its amplitude is varied from one experiment to the other; this is schematized by the series of horizontal lines

The advantage of gradient amplitude incrementation is twofold:

- it is possible to span negative and positive values of the gradient, and thus to explore negative and positive values of the Y spatial variable without sign ambiguity;
- since the gradient is applied for a constant time, relaxation acts in a constant fashion as far as the first dimension is concerned, preventing extra line broadening and thus contributing to improved spatial resolution. The phase encoding employed in the 'spin warp' method can, of course, be combined with the spin echo or gradient echo detection mode, as shown in the two sequences depicted in Figure 5.45.

5.4.3 THE THIRD SPATIAL DIMENSION. FULL IMAGING SEQUENCES

In order to examine the third spatial dimension (Z in our current notation), one may rely upon an incremented g_Z gradient (used as a second phase gradient) and proceed by a triple Fourier transformation. Such an experimental procedure approach actually exists, but entails considerable measuring times which may become unacceptable in biomedical applications. In practice, the method favored consists of selecting a slice perpendicular to the Z direction; this *slice selection* procedure is then followed by one of the two-dimensional sequences depicted in Figure 5.45. This can be repeated for a limited number of slices, possibly by interleaving the experiments relating to different slices, i.e. measuring one slice

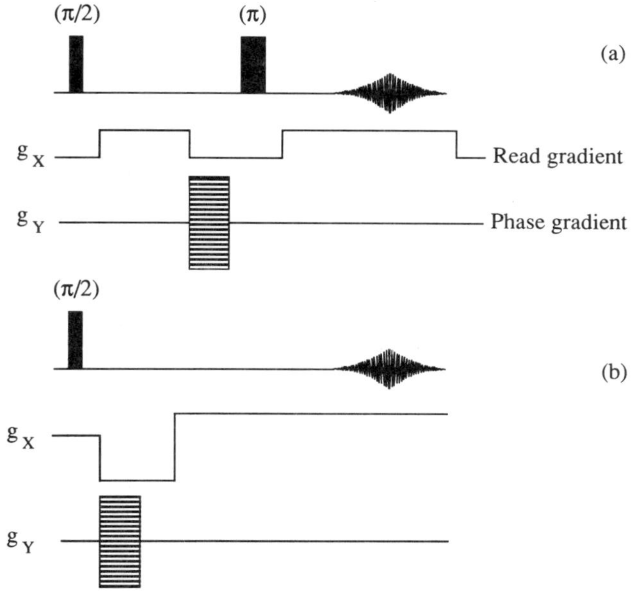

Figure 5.45 Combination of a phase gradient (encoding along the Y spatial direction) with, for the X spatial direction (read gradient), a spin echo sequence (a) or a gradient echo sequence (b). The double Fourier transform of the echo leads to a spin density image

while the magnetization of another recovers to equilibrium. Such experimental protocols bring the measuring time down to a few minutes, or even to a few tens of seconds. Slice selection procedures depend on the application of a selective radio-frequency pulse (Section 5.1.1) while a gradient in the Z direction is on. In that way, magnetization at ordinate Z is tipped to the *xy* plane of the rotating frame, provided that $\gamma g_Z Z/2\pi$ is equal to the frequency of the selective pulse (with respect to the resonance frequency in the absence of a gradient). The selective pulse is necessarily of weak amplitude and is applied for a time required to produce a 90° flip angle. Selectivity is improved by choosing an appropriate pulse shape (e.g. a sinc function; see the discussion in Section 5.1.1). The selected magnetization is necessarily defocused to the length of the selective r.f. pulse applied simultaneously with the g_Z gradient. For sensitivity reasons, its refocusing is mandatory, and is usually achieved by inverting the gradient polarity for an appropriate duration (Figure 5.46).

The full sequence which produces a two-dimensional image of the selected slice can be designed by combining sequences as those of Figure 5.46 and 5.45(a)); this is depicted in Figure 5.47. However, the method requires a relatively long measuring time (of the order of several minutes) since, in principle, full recovery to thermal equilibrium should be achieved prior to each experiment. This inconvenience can be circumvented by the FLASH sequence (Fast Low Angle SHot; J. Frahm, *Naturwissenschaften*, **74**, 415 (1987)), which can be considered as a variant of the method of Figure 5.47 which makes use of a gradient echo together with a single radio-frequency pulse whose flip angle is small, allowing much faster repetition and the production of an image in a few tens of seconds (Figure 5.48).

Another way of obtaining NMR images in a very short time involves the so-called 'echo planar' imaging concept due to P. Mansfield and I. L. Pykett, (*J. Magn. Reson.*, **29**, 355 (1978)). Although a full description is beyond the scope of this book, it should be mentioned that the method is becoming more and more popular, as evidenced by its implementation in whole-body imagers, its major feature being that image can be produced within one second. The essential characteristics of the experiment involve fast sweeping of the two-dimensional *k*-space (k_X and k_Y) by a series of g_Y gradient pulses (phase encoding) simultaneously with recurrent g_X gradient echoes (read gradient), the NMR signal being acquired in a one-shot fashion at each point where an echo is formed.

Finally, without further developing this point, we can mention that images arising from all these methods can be contrasted according to various parameters (T_1, T_2, chemical shifts, self-diffusion coefficients). For instance, in the classical method (Figure 5.47), T_2 contrast can be achieved through the experimental

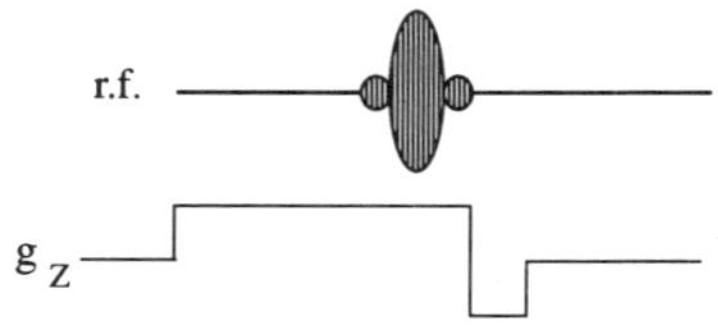

Figure 5.46 A slice selection procedure which includes a sinc-shaped radio-frequency pulse applied in the presence of a positive gradient along the Z direction. The subsequent negative gradient is aimed at magnetization refocusing through a gradient echo process

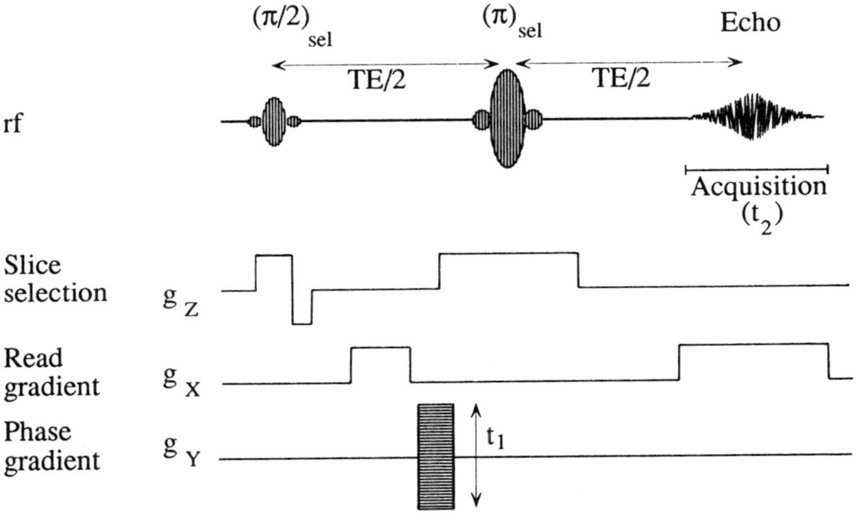

Figure 5.47 The standard 2D FT imaging pulse sequence. TE: echo time; both r.f. pulses are selective (soft pulses shaped according to a sinc or Gaussian functions); the totality of the spin echo is sampled; the incrementation of the g_Y amplitude is equivalent to the t_1 dimension of a usual 2D experiment (spin-warp). A double Fourier transformation with respect to t_1 and t_2 yields an image showing the variation of spin density in the XY plane

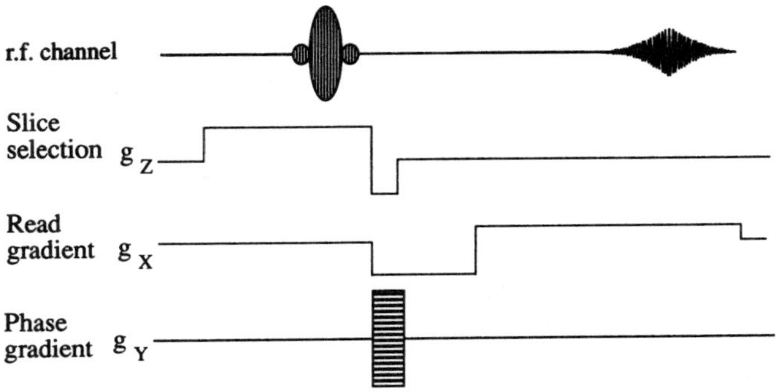

Figure 5.48 The FLASH sequence

parameter T_E and T_1 contrast would arise from the adjustment of the waiting time between two consecutive experiments.

5.4.4 IMAGING BY B_1 GRADIENTS

NMR imaging can be performed by resorting only to B_1 gradients (r.f. gradients). The method is a sort of transposition of procedures using B_0 gradients. Advantages and pitfalls of B_1 gradients support the conclusion that, at the present time, B_1 gradients are better used for NMR microscopy, with applications essentially in the field of materials science. This basic scheme of 2D imaging by B_1 gradients (P. Maffei *et al.*, *J. Magn. Reson.*, **A107**, 40 (1994)) involves at the outset a slice selection procedure. This is carried out by the r.f. gradient (along the Z direction) of a specially designed saddle coil, such as that used in a conventional NMR probe for transmit–receive operations. The r.f. read gradient (acting along a transverse direction) is generated by a single-turn coil which is orthogonal to the saddle coil. The XY plane is examined by means of projections along the gradient axis, collected for different orientations of the object with respect to the Z axis (the NMR signal is acquired by means of the saddle coil in short windows during which the read gradient is switched off). To this end, the experiment, including the initial slice selection, is repeated for as many sample rotations as necessary, performed in a stepped manner around Z. The NMR signal pertaining to one projection (defined by φ, which specifies the angle by which the object has been rotated with respect to its initial position) relies on the nutation angle produced by the r.f. read gradient and can be expressed as:

$$S(k, \varphi) = \iint \rho(\mathrm{X}, \mathrm{Y}) \cos\left[2\pi k(\mathrm{X} \cos\varphi + \mathrm{Y} \sin\varphi + D)\right] \mathrm{dX}\,\mathrm{dY}$$

where X and Y are the coordinates of the elementary magnetization being considered, D is the distance between the rotating axis and the virtual point where B_1 is zero, and k is related to the time t during which the gradient has been applied $[k = (2\pi)^{-1}\gamma g_1 t]$. The whole set of data (projections) can be treated with the help of projection-reconstruction methods.

Bibliography

R. R. Ernst, G. Bodenhausen and A. Wokaun: *Principles of Nuclear Magnetic Resonance in One and Two Dimensions*, Clarendon, Oxford, 1987, Chapter 7–10.

A. E. Derome: *Modern NMR Techniques for Chemistry Research*, Pergamon, Oxford, 1987

N. Chandrakumar and S. Subramanian: *Modern Techniques in High Resolution FT-NMR*, Springer, New York, 1987

J. K. M. Sanders and B. K. Hunter: *Modern NMR Spectroscopy*, University Press, Oxford, 1987

H. Kessler, M. Gehrke and C. Griesinger: Two-dimensional NMR spectroscopy: background and overview of the experiments, *Angew. Chem. Int. Ed. Engl.*, **27**, 490 (1988)

P. G. Morris: *Nuclear Magnetic Resonance Imaging in Medicine and Biology*, Clarendon, Oxford, 1987

W. S. Brey (Ed.): *Pulse Methods in 1D and 2D Liquid Phase NMR*, Academic Press, San Diego, 1988

M. Goldman: *Quantum Description of High Resolution NMR in Liquids*, Clarendon, Oxford, 1988

R. Freeman: *A Handbook of Nulcear Magnetic Resonance*, Longman, Harlow, 1988

M. Guéron, P. Plateau and M. Décorps: Solvent suppression in NMR, *Prog. NMR Spectrosc.*, **23**, 135 (1991)

R. Freeman: Selective excitation in high resolution NMR, *Chem. Rev.*, **91**, 1397 (1991)

H. Kessler, S. Mronga and G. Gemmeker: Multidimensional NMR experiments using selective pulses, *Magn. Reson. Chem.* **29**, 527 (1991)

P. T. Callaghan: *Principles of Nuclear Magnetic Resonance Microscopy*, Clarendon, Oxford, 1991

H. Friebolin: *Basic One- and Two-dimensional NMR Spectroscopy*, VCH, Weinheim, 2nd ed., 1993

J. W. Hennel and J. Klinowski: *Fundamentals of Nuclear Magnetic Resonance*, Longman Scientific and Technical, New York, 1993

INDEX